# Advances in Computer Vision and Pattern Recognition

More information about this series at http://www.springer.com/series/4205

Luis Enrique Sucar

# Probabilistic Graphical Models

## Principles and Applications

Second Edition

Luis Enrique Sucar
Instituto Nacional de Astrofísica,
Óptica y Electrónica (INAOE)
San Andrés Cholula, Puebla, Mexico

ISSN 2191-6586 ISSN 2191-6594 (electronic)
Advances in Computer Vision and Pattern Recognition
ISBN 978-3-030-61942-8 ISBN 978-3-030-61943-5 (eBook)
https://doi.org/10.1007/978-3-030-61943-5



This Springer imprint is published by the registered company Springer Nature Switzerland AG
The registered company address is: Gewerbestrasse 11, 6330 Cham, Switzerland

*To my family, Doris, Edgar and Diana,*
*for their unconditional love and support.*

# Foreword

Probabilistic graphical models (PGMs), and their use for reasoning intelligently under uncertainty, emerged in the 1980s within the statistical and artificial intelligence reasoning communities. The Uncertainty in Artificial Intelligence (UAI) conference became the premier forum for this blossoming research field. It was at UAI-92 in San Jose that I first met Enrique Sucar—both of us graduate students—where he presented his work on relational and temporal models for high-level vision reasoning. Enrique's impressive research contributions to our field over the past 25 years have ranged from the foundational work on objective probabilities, to developing advanced forms of PGMS such as temporal and event Bayesian networks, to the learning of PGMs, for example, his more recent work on Bayesian chain classifiers for multi-dimensional classification.

Probabilistic graphical models are now widely accepted as a powerful and mature technology for reasoning under uncertainty. Unlike some of the ad hoc approaches taken in early experts systems, PGMs are based on the strong mathematical foundations of graph and probability theory. They can be used for a wide range of reasoning tasks including prediction, monitoring, diagnosis, risk assessment and decision making. There are many efficient algorithms for both inference and learning available in open-source and commercial software. Moreover, their power and efficacy have been proven through their successful application to an enormous range of real-world problem domains. Enrique Sucar has been a leading contributor in this establishment of PGMs as practical and useful technology, with his work across a wide range of application areas. These include medicine, rehabilitation and care, robotics and vision, education, reliability analysis and industrial applications ranging from oil production to power plants.

The first authors to drawn upon the early research on Bayesian networks and craft it into compelling narratives in the book form were Judea Pearl in *Probabilistic Reasoning in Intelligent Systems* and Rich Neapolitan in *Probabilistic Reasoning in Expert Systems*. This monograph from Enrique Sucar is a timely addition to the body of literature following Pearl and Neapolitan, with up-to-date coverage of a broader range of PGMs than other recent texts in this area: various classifiers, hidden Markov models, Markov random fields, Bayesian networks and its dynamic, temporal and causal variants, relational PGMs, decision graphs and Markov decision process. It presents these PGMs, and the associated methods for reasoning (or inference) and learning, in a clear and accessible manner, making it suitable for advanced students as well as researchers or practitioners from other disciplines interested in using probabilistic models. The text is greatly enriched by the way Enrique has drawn upon his extensive practical experience in modelling with PGMs, illustrating their use across a diverse range of real-world applications from bioinformatics to air pollution to object recognition. I heartily congratulate Enrique on this book and commend it to potential readers.

Melbourne, Australia
May 2015

Ann E. Nicholson

# Preface

## Highlights of the Second Edition

- A new chapter on Partially Observable Markov Decision Process has been incorporated, which includes a detailed introduction to these models, approximate solution techniques and application examples.
- The chapter on causal models has been extended and divided into two chapters, one on causal graphical models, including causal inference, and other on causal discovery. Several causal discovery techniques are presented, and additional application examples.
- It includes a new chapter that gives an introduction to deep neural networks and their relation with probabilistic graphical models; presenting an analysis of different schemes for integrating deep neural networks and probabilistic graphical models, and some examples of the application of these hybrid models in different domains.
- Additional types of classifiers are described, including Gaussian Naive Bayes, Circular Chain Classifiers, and Hierarchical Classifiers with Bayesian Networks.
- The chapter on Hidden Markov Models incorporates Gaussian Hidden Markov Models.
- A knowledge transfer scheme for learning Bayesian networks is described.
- Sampling techniques for Dynamic Bayesian Networks including Particle Filters have been added.
- It incorporates an additional method to solve Influence Diagrams based on the transformation to a Decision Tree.
- Several additional application examples have been incorporated.
- The number of problems in each chapter has been increased by 50%.
- A Python Library for inference and learning for Probabilistic Graphical Models has been developed, implementing several of the algorithms described in the book.

## Overview

Probabilistic graphical models have become a powerful set of techniques used in several domains. This book provides a general introduction to probabilistic graphical models (PGMs) from an engineering perspective. It covers the fundamentals of the main classes of PGMs: Bayesian classifiers, hidden Markov models, Bayesian networks, dynamic and temporal Bayesian networks, Markov random fields, influence diagrams, Markov decision processes and partially observable Markov decision processes; including representation, inference and learning principles for each one. It describes several extensions of PGMs: relational probabilistic models, causal models, and hybrid models. Realistic applications for each type of model are covered in the book.

Some key features are the following:

- The main classes of PGMs are presented in a single monograph under a unified framework.
- The book covers the fundamental aspects: representation, inference and learning for all the techniques.
- It illustrates the application of the different techniques in practical problems, an important feature for students and practitioners.
- It includes some of the latest developments in the field, such as multidimensional and hierarchical Bayesian classifiers, relational graphical models, causal models and causal discovery, and hybrid deep neural networks-graphical models.
- Each chapter has a set of exercises, including suggestions for research and programming projects.

Motivating the application of probabilistic graphical models to real-world problems is one of the goals of this book. This requires not only knowledge of the different models and techniques, but also some practical experience and domain knowledge. To help the professionals in different fields gain some insight into the use of PGMs for solving practical problems, the book includes many examples of the application of the different types of models in a wide range of domains, including:

- Computer vision.
- Biomedical applications.
- Industrial applications.
- Information retrieval.
- Intelligent tutoring systems.
- Bioinformatics.
- Environmental applications.
- Robotics.

- Human-computer interaction.
- Information validation.
- Caregiving.

## Audience

This book can be used as a textbook for an advanced undergraduate or a graduate course in probabilistic graphical models for students of computer science, engineering, physics, etc. It could also serve as a reference book for professionals who want to apply probabilistic graphical models to different areas, or anyone who is interested in knowing the basics of these techniques.

It does not have specific prerequisites, although some background in probability and statistics is recommended. It is assumed that the reader has a basic knowledge of mathematics at the high school level, as well as a certain background in computing and programming. The programming exercises require some knowledge and experience with any programming language such as C, C++, JAVA, Python, MATLAB.

## Exercises

Each chapter (except the introduction) includes a set of exercises. Some of these exercises are questions and problems designed to reinforce the understanding of the concepts and techniques presented in the chapter. There are also a few suggestions for research or programming projects (marked with "***") in each chapter, which could be used as projects for a course

## Software

Several of the algorithms for learning and inference of the different classes of probabilistic graphical models have been implemented in Python and are publicly available. A description of the Python Library and how to access it is given in an Appendix.

## Organization

The book is divided into four parts. The first part provides a general introduction and motivation for PGMs, and reviews the required background in probability and graph theory. The second part describes the models which do not consider decisions or utilities: Bayesian classifiers, hidden Markov models, Markov random fields, Bayesian networks, and dynamic and temporal Bayesian networks. The third part starts with a brief introduction to decision theory, and then describes the models which support decision-making, including decision trees, influence diagrams, Markov decision processes, and partially observable Markov decision processes. Finally, the fourth part presents several extensions to the *standard* PGMs, including relational probabilistic graphical models and causal models (causal inference and causal discovery); and a chapter that gives an introduction to deep learning and its relation with PGMs.

The *dependency relations* between the chapters are shown in the Figure below. An arc from chapter $X$ to chapter $Y$, $X \rightarrow Y$, indicates that chapter $X$ is required (or at least recommended) for understanding chapter $Y$. This graphical representation

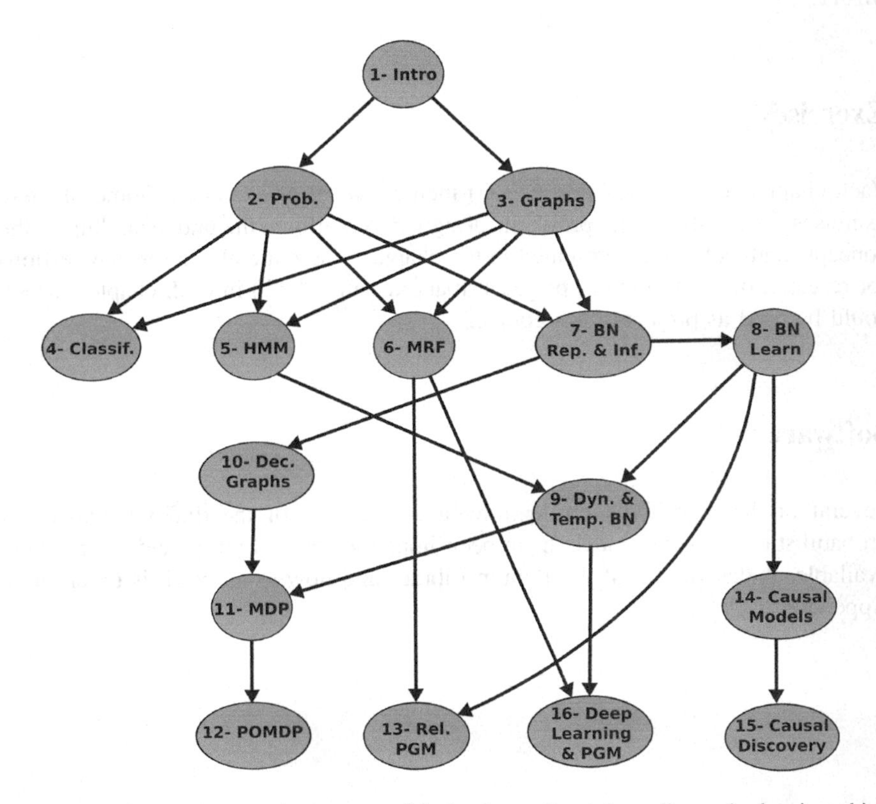


**Fig. 1** This figure represents the structure of the book as a directed acyclic graph, showing which chapters are prerequisites for other chapters

of the book gives a lot of information, in an analogous way to the graphical models that we will cover later.

From the Figure, we can deduce different ways of reading this book. First, it is recommended that you read the introduction and the fundamental Chaps. 2 and 3. Then you can study relatively independently the different models in Part II: classification (Chap. 4), hidden Markov models (Chap. 5), Markov random fields (Chap. 6), and Bayesian networks (Chaps. 7–9). Before reading about learning Bayesian networks, it is necessary to read Chap. 7—representation and learning; and both chapters are required before going into dynamic and temporal Bayesian networks.

The topics in part III and IV require some of the chapters in part II. For Chap. 10, which covers decision trees and influence diagrams, you should at least read the first chapter on Bayesian networks. For Chaps. 11 and 12, which cover sequential decision-making (MDPs and POMDPs), it is recommended that you have covered decision graphs (Chap. 10), hidden Markov models and dynamic and temporal Bayesian networks. Relational PGMs (Chap. 13), are based on Markov random fields and Bayesian networks, so both, Chaps. 6 and 8 are required. Causal models, included in Chap. 14, are based on Bayesian networks including the learning techniques, and causal discovery, Chap. 15 requires previous knowledge of causal models. Finally, Chap. 16 that contrasts deep leaning and PGMs requires dynamic and temporal Bayesian networks and Markov random fields.

If there is not enough time in a course to cover all the book, there are several alternatives. One is to focus on probabilistic models without considering decisions or the more advanced extensions, covering Parts I and II. Another alternative is to focus on decision models, including Part I, the necessary prerequisites from Part II and Part III. Or you can design your course *à la carte*, only respecting the dependencies in the graph. However, if you have the time and desire, I suggest you read all the book in order. Enjoy!

San Andrés Cholula, Puebla, Mexico
September 2020

Luis Enrique Sucar

# Acknowledgements

This book grew out of a course that I have been teaching for several years to graduate students. It initiated as a course in Uncertain Reasoning at Tec de Monterrey in Cuernavaca, and became a course on Probabilistic Graphical Models when I moved to INAOE, Puebla in 2006. During these years, my students have been the main motivation and source of inspiration for writing this book. I thank them all for their interest, questions, and frequent corrections to my notes. This book is dedicated to all my students, past, present and future.

I would like to acknowledge those students with whom I have collaborated for their bachelor's, master's or Ph.D. thesis. Some of the novel aspects of this book and most of the application examples originated from their work. I thank them all, and will mention just a few whose work was more influential for this manuscript: Gustavo Arroyo, Shender Ávila, Héctor Hugo Avilés, Rodrigo Barrita, Sebastián Bejos, Leonardo Chang, Ricardo Omar Chávez, Elva Corona, Francisco Elizalde, Hugo Jair Escalante, Iván Feliciano, Lindsey Fiedler, Giovani Gómez, Carlos Hernández, Pablo Hernández, Yasmín Hernández, Pablo Ibargüengoytia, Roger Luis-Velásquez, Miriam Martínez, Jose Antonio Montero, Samuel Montero, Julieta Noguez, Arquímides Méndez, Annette Morales, Miguel Palacios, Mallinali Ramírez, Alberto Reyes, Joel Rivas, Verónica Rodríguez, Elías Ruiz, Jonathan Serrano-Pérez, Sergio Serrano, Gerardo Torres-Toledano and Julio Zaragoza. Special thanks to Lindsey Fiedler who helped me with all the figures and revised the English of the first edition; and Jonathan Serrano-Pérez who has helped me with the additional figures of the second edition, revised the text and implemented the Python library. Also thanks to my son Edgar who revised earlier drafts of some chapters of the second edition.

I would also like to thank my collaborators, the common research projects and technical discussions have enriched my knowledge of many topics, and have helped me write this manuscript. In particular, I would like to mention my colleagues and friends: Juan Manuel Ahuactzin, Olivier Aycard, Nadia Berthouze, Concha Bielza, Roberto Ley Borrás, Cristina Conati, Douglas Crockett, Javier Díez, Hugo Jair Escalante, Duncan Gillies, Jesús González, Miguel González, Edel García, Jesse Hoey, Pablo Ibargüengoytia, Leonel Lara-Estrada, Pedro Larrañaga, Ron Leder,

Jim Little, José Martínez-Carranza, José Luis Marroquín, Oscar Mayora, Manuel Montes, Eduardo Morales, Enrique Muñoz de Cote, Rafael Murrieta, Felipe Orihuela, Luis Pineda, David Poole, Alberto Reyes, Andrés Rodríguez, Carlos Ruiz, Sunil Vadera and Luis Villaseñor. I appreciate the comments that Edel, Felipe and Pablo have made of some earlier versions of this book.

I thank my friend and colleague Anne Nicholson, for taking the time to read the book and write the Foreword.

I would like to acknowledge the support from my institution, Instituto Nacional de Astrofísica, Óptica y Electrónica (INAOE), which provides an excellent environment for doing research and teaching, and has given me all the facilities to dedicate part of my time to write this book.

Last, but not least, I would like to thank my family. My parents, Fuhed† and Aida, who promoted the desire to learn and to work hard, and supported my studies. In particular my Father, who wrote several wonderful books, and gave me the inspiration (and probably the genes) for writing. My brother Ricardo†, and my sisters, Shafía and Beatriz, who have always supported and motivated my dreams. And specially to my wife Doris and my children, Edgar and Diana, who have suffered the long hours I have dedicated to the book instead of them, and whose love and support is what keeps me going.

# Contents

## Part IV Relational, Causal and Deep Models

# Acronyms

ADD Algebraic Decision Diagram
AI Artificial Intelligence
BAN Bayesian network Augmented Naive Bayes classifier
BCC Bayesian Chain Classifier
BCCD Bayesian Constraint-based Causal Discovery
BN Bayesian Network
CBN Causal Bayesian Network
CMI Conditional Mutual Information
CPT Conditional Probability Table
CRF Conditional Random Field
DAG Directed Acyclic Graph
DBN Dynamic Bayesian Network
DBNC Dynamic Bayesian Network Classifier
DD Decision Diagram
DDN Dynamic Decision Network
DT Decision Tree
EC Expected Cost
EM Expectation-Maximization
FN False Negative
FP False Positive
GHMM Gaussian Hidden Markov Model
GRM Gibbs Random Field
HMM Hidden Markov Model
ICM Iterative Conditional Modes
ID Influence Diagram
ILP Inductive Logic Programming
KB Knowledge Base
LIMID Limited Memory Influence Diagram
MAG Maximal Ancestral Graph
MAP Maximim a Posteriori Probability

| | |
|---|---|
| MB | Markov Blanket |
| MBC | Multidimensional Bayesian network Classifier |
| MC | Markov Chain |
| MDL | Minimum Description Length |
| MDP | Markov Decision Process |
| MLN | Markov Logic Network |
| MN | Markov Network |
| MPE | Most Probable Explanation |
| MPM | Maximum Posterior Marginals |
| MRF | Markov Randorm Field |
| NBC | Naïve Bayes Classifier |
| PAG | Parental Ancestral Graph |
| PBVI | Point-based Value Iteration |
| PGM | Probabilistic Graphical Model |
| PL | Pseudo-Likelihood |
| POMDP | Partially Observable Markov Decision Process |
| PRM | Probabilistic Relational Model |
| RPGM | Relational Probabilistic Graphical Model |
| SA | Simulated Annealing |
| SNBC | Semi-Naïve Bayesian Classifier |
| TAN | Tree Augmented Naive Bayes classifier |
| TEN | Temporal Event Network |
| TN | Temporal Node |
| TNBN | Temporal Nodes Bayesian Network |
| WFF | Well Formed Formula |

# Notation

| | |
|---|---|
| $T$ | True |
| $F$ | False |
| $A$, $B$, $C$, ... | Propositions (binary variables) |
| $\neg A$ | Not A (negation) |
| $A \wedge B$ | A and B (conjunction) |
| $A \vee B$ | A or B (disjunction) |
| $A \rightarrow B$ | B if A (implication) |
| $A \leftrightarrow B$ | A if B and B if A (double implication) |
| $\forall(X)$ | Universal quantifier: for all $X$ |
| $\exists(X)$ | Existential quantifier exists an $X$ |
| $C \cup D$ | Union of two sets |
| $C \cap D$ | Intersection of two sets |
| $X$ | A random variable |
| $x$ | A particular value of a random variable, $X = x$ |
| $\mathbf{X}$ | A vector of random variables, $\mathbf{X} = X_1, X_2, ..., X_N$ |
| $\mathbf{x}$ | A particular realization of vector $\mathbf{X}$, $\mathbf{x} = x_1, x_2, ..., x_N$ |
| $X_{1:T}$ | Vector of variable $X$ from $t = 1$ to $t = T$, $X_{1:T} = X_1, X_2, ..., X_T$ |
| $P(X = x)$ | Probability of $X$ being in state $x$; for short $P(x)$ |
| $P(\mathbf{X} = \mathbf{x})$ | Probability of $\mathbf{X}$ being in state $\mathbf{x}$; for short $P(\mathbf{x})$ |
| $P(x, y)$ | Probability of $x$ and $y$ |
| $P(x \vee y)$ | Probability of $x$ or $y$ |
| $P(x \mid y)$ | Conditional probability of $x$ given $y$ |
| $P(x) \sim y$ | The probability of $x$ is *proportional* to y, that is $P(x) = k \times y$ |
| $\mathbf{P}(X)$ | Cumulative distribution function of a discrete variable $X$ |
| $P(X)$ | Probability function of a discrete variable $X$ |
| $F(X)$ | Cumulative distribution function of a continuous variable $X$ |
| $f(X)$ | Probability density function of a continuous variable $X$ |
| $I(X, Y, Z)$ | $X$ independent of $Z$ given $Y$ |
| $G(V, E)$ | Graph $G$ with set of vertices $V$ and set of edges $E$ |
| $E(V_j, V_k)$ | Edge $E$ between vertices $V_j$ and $V_k$ in a graph |

| | |
|---|---|
| $Adj(V)$ | Vertices adjacent to vertex $V$ in a graph |
| $Pa(X)$ | Parents of node $X$ in a directed graph |
| $Nei(X)$ | Neighbors of node $X$ in a graph |
| $n!$ | Factorial of $n$, $n! = n \times (n-1) \times (n-2) \times ...1$ |
| $\binom{n}{r}$ | Combinations of $r$ from $n$, $\binom{n}{r} = \frac{n!}{r!(n-r)!}$ |
| $exp(x)$ | Exponential of $x$, $exp(x) = e^x$ |
| $\|X\|$ | The dimension or number of states of a discrete variable $X$ |
| $\sum_i X_i$ | Summation over $X$ |
| $\prod_i X_i$ | Product over $X$ |
| $\Omega$ | Sample espace |
| $\mu$ | Mean |
| $\sigma^2$ | Variance |
| $\sigma$ | Standard deviation |
| $N(\mu, \sigma^2)$ | Normal distribution with mean $\mu$ and standard deviation $\sigma$ |
| $I(m)$ | Information |
| $H(M)$ | Entropy |
| $E(X)$ | Expected value of a random variable $X$ |
| $ArgMax_x F(X)$ | The value of $X$ for which the function $F$ is maximum |
| $\lambda = \{A, \Pi\}$ | A Markov chain with vector of prior probabilities $\Pi$ and matrix of transition probabilities $A$ |
| $\lambda = \{A, B, \Pi\}$ | A HMM with vector of prior probabilities $\Pi$, matrix of transition probabilities $A$ and matrix of observation probabilities $B$ |
| $A \succ B$ | State $A$ is preferred over $B$ |
| $A \sim B$ | States $A$ and $B$ have the same preference (indifference) |
| $U(A)$ | Utility of state $A$ |
| $\pi$ | A policy for an MDP or POMDP: mapping from states to actions |
| $V^\pi$ | Value function following policy $\pi$ |
| $\gamma$ | Discount factor |

# Part I
# Fundamentals

The first chapters of the book include a general introduction to probabilistic graphical models, and provide the theoretical foundations required for the rest of the book: probability theory and graph theory.

# Chapter 1
# Introduction

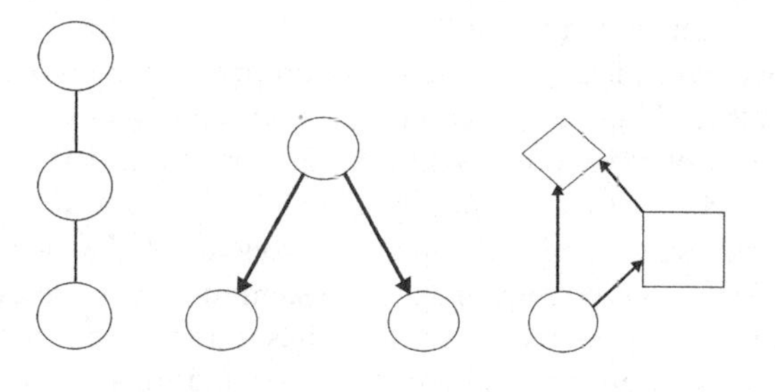

## 1.1 Uncertainty

For achieving their goals, intelligent agents, natural or artificial, have to select a course of actions among many possibilities. That is, they have to make decisions based on the information that they can obtain from their environment, their previous knowledge and their objectives. In many cases, the information and knowledge is incomplete or unreliable, and the results of their decisions are not certain, that is they have to make *decisions under uncertainty*. For instance, in an emergency, a medical doctor must act promptly even if she has limited information on the patient's state; an autonomous vehicle that detects what might be an obstacle in its way, must decide if it should turn or stop without being certain about the obstacle's distance, size and velocity; or a financial agent that needs to select the best investment according to its vague predictions on the expected return of the different alternatives and its clients' requirements.

One of the goals of artificial intelligence is to develop systems that can reason and make decisions under uncertainty. Reasoning under uncertainty presented a challenge to early intelligent systems, as traditional paradigms were not well suited for managing uncertainty.



L. E. Sucar, *Probabilistic Graphical Models*, Advances in Computer Vision and Pattern Recognition, https://doi.org/10.1007/978-3-030-61943-5_1

### 1.1.1 Effects of Uncertainty

Early artificial intelligence systems were based on classical logic, in which knowledge can be represented as a set of logic clauses or rules. These systems have two important properties, *modularity* and *monotonicity*, which help to simplify knowledge acquisition and inference.

A system is modular if each *piece* of knowledge can be used independently to arrive to conclusions. That is, if the premises of any logical clause or rule are true, then we can assert its conclusion without needing to consider other elements in the knowledge base. For example, if we have the rule, $\forall X, stroke(X) \rightarrow impaired\text{-}arm(X)$, then if we know that $Mary$ had a stroke we know that she has an impaired-arm.

A system is monotonic if its knowledge always increases monotonically: that is, any deduced fact or conclusion is maintained even if new facts are known by the system. For example, if there is a rule such as $\forall X, bird(X) \rightarrow flies(X)$, then if *Tweety* is a bird, we can assert that it flies.

However, if we have uncertainty these two properties are not true in general. In medical systems there is usually uncertainty about the diagnosis of a patient, so if a person suffers a stroke, her arm might not be impaired; it depends on the part of the brain affected by the stroke. Similarly, not all birds fly, so if we later learn that *Tweety* is a penguin, we will need to *retract* the conclusion that she flies.

The loss of these two properties makes a system that has to reason under uncertainty more complex. In principle, the system has to take into account all available knowledge and facts when deriving a conclusion, and must be able to change its conclusions when acquiring new data.

## 1.2 A Brief History

From an artificial intelligence perspective, we can consider the following stages in the development of uncertainty management techniques:

*Beginnings* (1950s and 60s)—artificial intelligence (AI) researchers focused on solving problems such as theorem proving, games like chess, and the "blocks world" planning domain, which do not involve uncertainty, making it unnecessary to develop techniques for managing uncertainty. The symbolic paradigm dominated AI in the beginnings.

*Ad hoc techniques* (1970s)—the development of expert systems for realistic applications such as medicine and mining, required the development of uncertainty management approaches. Novel ad hoc techniques were developed for specific expert systems, such as MYCIN's certainty factors [16] and Prospector's pseudo-probabilities [4]. It was later shown that these techniques had a set of implicit assumptions which limited their applicability [6]. Also in this period, alternative theories were proposed to manage uncertainty in expert systems, including fuzzy logic [18] and the Dempster-Shafer theory [15].

*Resurgence of probability* (1980s)—probability theory was used in some initial expert systems, however it was later discarded because its application in naive ways implies a high computational complexity (see Sect. 1.3). New developments, in particular Bayesian networks [10], make it possible to build complex probabilistic systems in an efficient manner, starting a new era for uncertainty management in AI.

*Diverse formalisms* (1990s)—Bayesian networks continued and were consolidated with the development of efficient inference and learning algorithms. Meanwhile, other techniques such as fuzzy and non-monotonic logics were considered as alternatives for reasoning under uncertainty.

*Probabilistic graphical models* (2000s)—several techniques based on probability and graphical representations were consolidated as powerful methods for representing, reasoning and making decisions under uncertainty, including Bayesian networks, Markov networks, influence diagrams and Markov decision processes, among others.

## 1.3 Basic Probabilistic Models

Probability theory provides a well established foundation for managing uncertainty, therefore it is natural to use it for reasoning under uncertainty. However, if we apply probability in a naive way to complex problems, we are soon deterred by computational complexity

In this section we will show how we can model a problem using a naive probabilistic approach based on a *flat* representation; and then how we can use this representation to answer some probabilistic queries. This will help to understand the limitations of the basic approach, motivating the development of probabilistic graphical models.[1]

Many problems can be formulated as a set of variables, $X_1, X_2, ...X_n$ such that we know the values for some of these variables and the others are unknown. For instance, in medical diagnosis, the variables might represent certain symptoms and the associated diseases; usually we know the symptoms and we want to find the most probable disease(s). Another example could be a financial institution developing a system to help decide the amount of credit given to a certain customer. In this case the relevant variables are the attributes of the customer, i.e. age, income, previous credits, etc.; and a variable that represents the amount of credit to be given. Based on the customer attributes we want to determine, for instance, the maximum amount of credit that is safe to give to the customer. In general there are several types of problems that can be modeled in this way, such as diagnosis, classification, and perception problems; among others.

[1] This and the following sections assume that the reader is familiar with some basic concepts of probability theory; a review of these and other concepts is given in Chap. 2.

Under a probabilistic framework we can consider that each attribute of a problem is a random variable, such that it can take a certain value from a set of values.[2] Let us consider that the set of possible values is finite; for example, $X = \{x_1, x_2, ..., x_m\}$ might represent the $m$ possible diseases in a medical domain. Each value of a random variable will have a certain probability associated in a context; in the case of $X$ it could be the probability of each disease within certain population (which is called the *prevalence* of the disease), i.e.; $P(X = x_1)$, $P(x = x_2)$, ..., for short $P(x_1)$, $P(x_2)$, ....

If we consider two random variables, $X, Y$, then we can calculate the probability of $X$ taking a certain value and $Y$ taking a certain value, that is $P(X = x_i \wedge Y = y_j)$ or just $P(x_i, y_j)$; this is called the joint probability of $X$ and $Y$. The idea can be generalized to $n$ random variables, where the joint probability is denoted as $P(X_1, X_2, ...X_n)$. We can think of $P(X_1, X_2, ...X_n)$ as a function that assigns a probability value to all possible combinations of values of the variables $X_1, X_2, ...X_n$.

Therefore, we can represent a domain as:

1. A set of random variables, $X_1, X_2, ...X_n$.
2. A joint probability distribution associated to these variables, $P(X_1, X_2, ...X_n)$.

Given this representation, we can answer some queries with respect to certain variables in the domain, such as:

*Marginal probabilities:* the probability of one of the variables taking a certain value. This can be obtained by summing over all the other variables of the joint probability distribution. In other words, $P(X_i) = \sum_{\forall X \neq X_i} P(X_1, X_2, ...X_n)$. This is known as *marginalization*. Marginalization can be generalized to obtain the marginal joint probability of a subset of variables by summing over the rest.

*Conditional probabilities:* by definition the conditional probability of $X_i$ given that we know $X_j$ is $P(X_i \mid X_j) = P(X_i, X_j)/P(X_j)$, $P(X_j) \neq 0$. $P(X_i, X_j)$ and $P(X_j)$ can be obtained via marginalization, and from them we can obtain conditional probabilities.

*Total Abduction or MPE:* given that a subset (**E**) of variables is known, abduction consists in finding the values of the rest of variables (**J**) that maximize their conditional probability given the evidence, $max\, P(\mathbf{J} \mid \mathbf{E})$. That is $ArgMax_J[max\, P(X_1, X_2, ...X_n)/P(\mathbf{E})]$.

*Partial abduction or MAP:* in this case there are 3 subsets of variables: the evidence, **E**, the query variables that we want to maximize, **J**, and the rest of the variables, **K**, such that we want to maximize $P(\mathbf{J} \mid \mathbf{E})$. This is obtained by marginalizing over **K** and maximizing over **J**, that is $ArgMax_J[\sum_{X \in K} P(X_1, X_2, ...X_n)/P(\mathbf{E})]$.

Additionally, if we have data from the domain of interest, we might obtain a *model* from this data, that is, estimate the joint probability distribution of the relevant variables.

Next we illustrate the basic approach with a simple example.

[2] Random variables are formally defined later on.

### 1.3.1 An Example

We will use the traditional *golf* example to illustrate the basic approach. In this problem we have 5 variables: *outlook, temperature, humidity, windy, play*. Table 1.1 shows some data for the golf example; all variables are discrete so they can take a value from a finite set of values, for instance Outlook could be sunny, overcast or rain. We will now illustrate how we can calculate the different probabilistic queries mentioned before for this example.

First, we will simplify the example using only two variables, Outlook and Temperature. From the data in Table 1.1 we can obtain the joint probability of Outlook and Temperature as depicted in Table 1.2. Each entry in the table corresponds to the joint probability $P(Outlook, Temperature)$, for example, $P(Outlook = S, Temp. = H) = 0.143$.

Let us first obtain the marginal probabilities for the two variables. If we sum per row (marginalizing Temperature) then we obtain the marginal probabilities for Out-

**Table 1.1** A sample data set for the golf example

| Outlook | Temperature | Humidity | Windy | Play |
|---|---|---|---|---|
| Sunny | High | High | False | No |
| Sunny | High | High | True | No |
| Overcast | High | High | False | Yes |
| Rain | Medium | High | False | Yes |
| Rain | low | Normal | False | Yes |
| Rain | low | Normal | True | No |
| Overcast | low | Normal | True | Yes |
| Sunny | Medium | High | False | No |
| Sunny | low | Normal | False | Yes |
| Rain | Medium | Normal | False | Yes |
| Sunny | Medium | Normal | True | Yes |
| Overcast | Medium | High | True | Yes |
| Overcast | High | Normal | False | Yes |
| Rain | Medium | High | True | No |

**Table 1.2** Joint probability distribution for Outlook and Temperature. Notation: H-high, M-medium, L-low; S-sunny, O-overcast, R-rain

| | Temp. | | |
|---|---|---|---|
| Outlook | H | M | L |
| S | 0.143 | 0.143 | 0.071 |
| O | 0.143 | 0.0.71 | 0.071 |
| R | 0 | 0.214 | 0.143 |

look, $P(Outlook) = [0.357, 0.286, 0.357]$; and if we sum per column we obtain the marginal probabilities for Temperature, $P(Temperature) = [0.286, 0.428, 0.286]$. From these distributions, we obtain that the most probable Temperature is $M$ and the most probable values for Outlook are $S$ and $R$.

Now we can calculate the conditional probabilities of Outlook given Temperature and vice-versa. For instance:

$$P(Temp. \mid Outlook = R) = P(Temp. \wedge Outlook = R)/P(Outlook = R) = [0, 0.6, 0.4]$$

$$P(Outlook \mid Temp. = L) = P(Outlook \wedge Temp. = L)/P(Temp. = L) = [0.25, 0.25, 0.5]$$

Given these distributions, the most probable Temperature given that the Outlook is Rain is Medium, and the most probable Outlook given that the Temperature is Low is Rain.

Finally, the most probable combination of Outlook and Temperature is $\{Rain, Medium\}$, which in this case can be obtained directly from the joint probability table.

Although it is possible to compute the different probabilistic queries for this small example, this approach becomes impractical for complex problems with many variables, as the size of the table and the direct computation of marginal and conditional probabilities grow exponentially with the number of variables in the model.

Another disadvantage of this naive approach is that to have good estimates for the joint probabilities from data, we will require a very large database if there are many variables in the model. A rule of thumb is that the number of instances (records) is at least 10 times the number of possible combination values for the variables, so if we consider 50 binary variables, it will require at least $10 \times 2^{50}$ instances!

Finally, the joint probability table does not say much about the problem to a human; so this approach also has cognitive limitations.

The problems seen with the basic approach are some of the motivations for the development of probabilistic graphical models.

## 1.4 Probabilistic Graphical Models

Probabilistic Graphical Models (PGMs) provide a framework for managing uncertainty based on probability theory in a computationally efficient way. The basic idea is to consider only those independence relations that are valid for a certain problem, and include these in the probabilistic model to reduce complexity in terms of memory requirements and computational time. A natural way to represent the dependence and independence relations between a set of variables is using graphs, such that variables that are directly dependent are connected, and the independence relations are implicit in this dependency graph.

A Probabilistic Graphical Model is a compact representation of a joint probability distribution, from which we can obtain marginal and conditional probabilities. It has several advantages over a flat representation:

- It is generally much more compact (space).
- It is generally much more efficient (time).
- It is easier to understand and communicate.
- It is easier to learn from data or to construct based on expert knowledge.

A probabilistic graphical model is specified by two aspects: (i) a graph, $G(V, E)$, that defines the structure of the model; and (ii) a set of local functions, $f(Y_i)$, that define the parameters, where $\mathbf{Y_i}$ is a subset of $\mathbf{X}$. The joint probability is obtained by the product of the local functions:

$$P(X_1, X_2, \ldots, X_N) = K \prod_{i=1}^{M} f(Y_i) \tag{1.1}$$

Where $K$ is a normalization constant (it makes the probabilities sum to one).

This representation in terms of a graph and a set of local functions (called *potentials*) is the basis for inference and learning in PGMs:

Inference: obtain the marginal or conditional probabilities of any subset of variables $\mathbf{Z}$ given any other subset $\mathbf{Y}$.

Learning: given a set of data values for $\mathbf{X}$ (that can be incomplete) estimate the structure (graph) and parameters (local functions) of the model.

We can classify probabilistic graphical models according to three dimensions:

1. Directed or undirected
2. Static or dynamic
3. Probabilistic or decisional

The first dimension has to do with the type of graph used to represent the dependence relations. Undirected graphs represent symmetric relations, while directed graphs represent relations in which the direction is important. Given a set of random variables with the corresponding conditional independence relations, it is not possible to represent all the relations with one type of graph [10]; thus, both types of models are useful.

The second dimension defines if the model represents a set of variables at a certain point in time (static) or across different times (dynamic). The third dimension distinguishes between probabilistic models that only include random variables, and decisional models that also include decision and utility variables.

The most common classes of PGMs and their type according to the previous dimensions are summarized in Table 1.3.

All these types of models will be covered in detail in the following chapters, and also some extensions that consider more expressive models (relational probabilistic graphical models) or represent causal relations (causal Bayesian networks).

**Table 1.3** Main types of probabilistic graphical models

| Type | Directed/undirected | Static/dynamic | Prob./decisional |
|---|---|---|---|
| Bayesian classifiers | D/U | S | P |
| Markov chains | D | D | P |
| Hidden Markov models | D | D | P |
| Markov random fields | U | S | P |
| Bayesian networks | D | S | P |
| Dynamic Bayesian networks | D | D | P |
| Influence diagrams | D | S | D |
| Markov decision processes (MDPs) | D | D | D |
| Partially observable MDPs | D | D | D |

## 1.5 Representation, Inference and Learning

There are three main aspects for each class of probabilistic graphical model, representation, inference and learning.

The representation is the basic property of each model, and it defines which entities constitute it and how these are related. For instance, all PGMs can be represented as graphs that define the structure of the model and by local functions that describe its parameters. However, the type of graph and the local functions vary for the different types of models.

Inference consists in answering different probabilistic queries based on the model and some evidence. For instance, obtaining the posterior probability distribution of a variable or set of variables given that other variables in the model are known. The challenge is how to do this efficiently.

To construct these models there are basically two alternatives: to build it "by hand" with the aid of domain experts or to induce the model from data. The emphasis in recent years has been to induce the models based on machine learning techniques, because it is difficult and costly to do it with the aid of experts. In particular obtaining the parameters for the models is usually done based on data, as humans tend to be *bad* estimators of probabilities.

An important property of these techniques from an application point of view is that they tend to separate the inference and learning techniques from the model. That is, as in other artificial intelligence representations such as logic and production rules, the reasoning mechanisms are general and can be applied to different models. As a result, the techniques developed for probabilistic inference and learning in each class of PGM, can be applied directly for different models in a variety of applications.

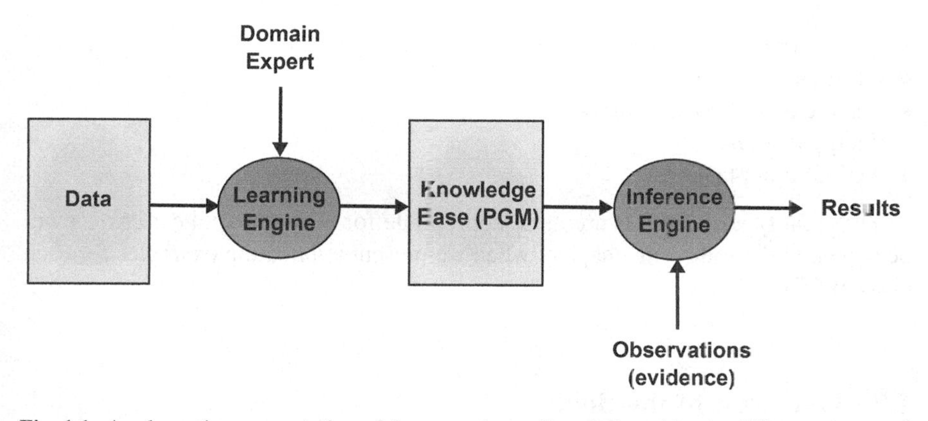


**Fig. 1.1** A schematic representation of the general paradigm followed by the different classes of PGMs, in which there is a clear separation between the learning and inference engines, which are generic, and the knowledge base, which depends on the particular application

This idea is illustrated schematically in Fig. 1.1. Based on data or expert knowledge or a combination of both, the knowledge base, in this case a probabilistic graphical model, is built using the learning engine. Once we have the model, we can use it to do probabilistic reasoning through the inference engine; based on the observations and the model, the inference engine derives the results. The learning and inference engines are generic for a class of PGMs, so they can be applied for modeling and reasoning in different domains.

For each type of PGM presented in the book, we will first describe its representation and then present some of the most common inference and learning techniques.

## 1.6 Applications

Most real-world problems imply dealing with uncertainty and usually involve a large number of factors or variables to be considered when solving them. Probabilistic graphical models constitute an ideal framework to solve complex problems with uncertainty, so they are applied in a wide range of domains such as:

- Medical diagnosis and decision making.
- Mobile robot localization, navigation and planning.
- Diagnosis for complex industrial equipment such as turbines and power plants.
- User modeling for adaptive interfaces and intelligent tutors.
- Speech recognition and natural language processing.
- Pollution modeling and prediction.
- Reliability analysis of complex processes.
- Modeling the evolution of viruses.
- Object recognition in computer vision.

- Error correction in communications.
- Information retrieval.
- Gesture and activity recognition.
- Energy markets.
- Agricultural planning.

Different types of PGMs are more appropriate for different applications, as will be shown in the following chapters when we present application examples for each class of PGM.

## 1.7 Overview of the Book

The book is divided in four parts.

Part I provides the mathematical foundations for understanding the models and techniques presented in the following chapters. Chapter 2 presents a review of some basic concepts in probability and information theory which are important for understanding probabilistic graphical models. Chapter 3 gives an overview of graph theory, with emphasis on certain aspects that are important for representation and inference in probabilistic graphical models; including, among others, cliques, triangulated graphs and perfect orderings.

Part II covers the different types of probabilistic models that only have random variables, and do not consider decisions or utilities in the model. This is the largest part and it includes the following types of PGMs:

- Bayesian classifiers
- Markov chains and hidden Markov models
- Markov random fields
- Bayesian networks
- Dynamic Bayesian networks and temporal networks

A chapter is dedicated to each type of model (except for Bayesian networks which is divided into two chapters), including representation, inference and learning; also practical application examples are shown.

Part III presents those models that consider decisions and utilities, and as such are focused on aiding the decision maker to take the optimal actions under uncertainty. This part includes three chapters. The first chapter covers modeling techniques for when there are one or few decisions, including decision trees and influence diagrams. The second chapter covers sequential decision problems considering full observability, in particular Markov decision processes. The third chapter is dedicated to sequential decision problems with partial observability, that is partially observable Markov decision processes.

Part IV considers alternative paradigms that can be thought as extensions to the traditional probabilistic graphical models, as well as their relations with deep learning. It includes four chapters. The first one is dedicated to relational probabilistic models, which increase the representational power of *standard* PGMs, by combining the expressive power of first–order logic with the uncertain reasoning capabilities of probabilistic models. The second one introduces causal graphical models that go beyond representing probabilistic dependencies, to express cause and effect relations. The third chapter focuses on causal discovery, that is learning causal models from data. Finally, the fourth chapter presents a brief introduction to deep learning models, focusing in their relation and potential combination with probabilistic graphical models.

## 1.8 Additional Reading

In this book we present a broad perspective on probabilistic graphical models. There are few other books that have a similar coverage. One is by Koller and Friedman [8], which presents the models under a different structure with less emphasis on applications. Another is by Lauritzen [9], which has a more statistical focus. Bayesian programming [2] provides an alternative approach to implement graphical models based on a programming paradigm. Barber [1] presents several classes of PGMs with emphasis on machine learning.

There are several books that cover one or few types of models in more depth, such as: Bayesian networks [3, 11, 12], decision graphs [7], Markov random fields [10], Markov decision processes [14], relational probabilistic models [5] and causal models [13, 17].

## References


1. Barber, D.: Bayesian Reasoning and Machine Learning. Cambridge University Press, Cambridge (2012)
2. Bessiere, P., Mazer, E., Ahuactzin, J.M., Mekhnacha, K.: Bayesian Programming. CRC Press, Boca Raton (2014)
3. Darwiche, A.: Modeling and Reasoning with Bayesian Networks. Cambridge University Press, New York (2009)
4. Duda, R.O., Hart, P.A., Nilsson, N.L.: Subjective Bayesian methods for rule-based inference systems. In: Proceeding of the National Computer Conference, vol. 45, pp. 1075–1082 (1976)
5. Getoor, L., Taskar, B.: Introduction to Statistical Relational Learning. MIT Press, Cambridge (2007)
6. Heckerman, D.: Probabilistic interpretations for MYCIN's certainty factors. In: Proceedings of the First Conference on Uncertainty in Artificial Intelligence (UAI), pp. 9–20 (1985)
7. Jensen, F.V.: Bayesian Networks and Decision Graphs. Springer, New York (2001)
8. Daphne Koller, D., Friedman. N.: Probabilistic Graphical Models: Principles and Techniques. MIT Press. Cambridge (2009)
9. Lauritzen, S.L.: Graphical Models. Oxford University Press, Oxford (1996)
10. Li, S.Z.: Markov Random Field Modeling in Image Analysis. Springer, London (2009)
11. Neapolitan, R.E.: Probabilistic Reasoning in Expert Systems. Wiley, New York (1990)
12. Pearl, J.: Probabilistic Reasoning in Intelligent Systems: Networks of Plausible Inference. Morgan Kaufmann, San Francisco (1988)

13. Pearl, J.: Causality: Models, Reasoning and Inference. Cambridge University Press, New York (2009)
14. Puterman, M.L.: Markov Decision Processes: Discrete Stochastic Dynamic Programming. Wiley, New York (1994)
15. Shafer, G.: A Mathematical Theory of Evidence. Princeton University Press, Princeton (1976)
16. Shortliffe, E.H., Buchanan, B.G.: A model of inexact reasoning in medicine. Math. Biosci. **23**, 351–379 (1975)
17. Spirtes, P., Glymour, C., Scheines, R.: Causation, Prediction, and Search. MIT Press, Cambridge (2000)
18. Zadeh, L.A.: Knowledge representation in fuzzy logic. IEEE Trans. Knowl. Data Eng. **1**(1), 89–100 (1989)

# Chapter 2
# Probability Theory

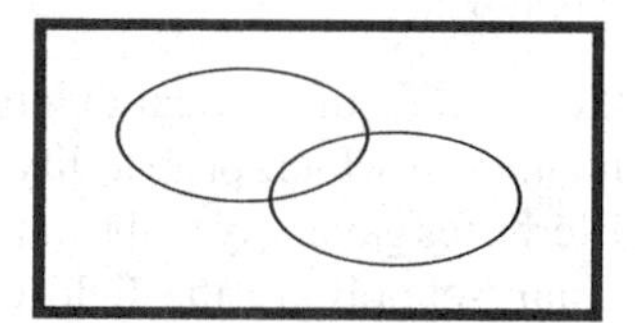

## 2.1 Introduction

Probability theory was originated in games of chance and has a long and interesting history; it has developed into a mathematical language for quantifying uncertainty.

Consider a certain *experiment*, such as throwing a die; this experiment can have different results, we call each result an *outcome* or element. In the die example, the possible outcomes or elements are the following: $\{1, 2, 3, 4, 5, 6\}$. The set of all possible outcomes of an experiment is called the sample space, $\Omega$. An *event* is a set of elements or subset of $\Omega$. Continuing with the die example, one event could be that the die shows an even number, that is $\{2, 4, 6\}$.

Before we mathematically define probability, it is worth discussing what the meaning or interpretation of probability is. Several definitions or interpretations of probability have been proposed, starting from the *classical* definition by Laplace, including the *limiting frequency*, the *subjective*, the *logical* and the *propensity* interpretations [2]:

Classical: probability has to do with equiprobable events; if a certain experiment has $N$ possible outcomes, the probability of each outcome is $1/N$.

Logical: probability is a measure of rational belief; that is, according to the available evidence, a rational person will have a certain belief regarding an event, which will define its probability.



L. E. Sucar, *Probabilistic Graphical Models*, Advances in Computer Vision and Pattern Recognition, https://doi.org/10.1007/978-3-030-61943-5_2

Subjective: probability is a measure of the personal degree of belief in a certain event; this could be measured in terms of a betting factor –the probability of a certain event for an individual is related to how much that person is willing to bet on that event.

Frequency: probability is a measure of the number of occurrences of an event given a certain experiment, when the number of repetitions of the experiment tends to infinity.

Propensity: probability is a measure of the number of occurrences of an event under repeatable conditions; even if the experiment only occurs once.

These interpretations can be grouped in what are the two main approaches in probability and statistics:

- Objective (classical, frequency, propensity): probabilities exist in the *real* world and can be measured.
- Epistemological (logical, subjective): probabilities have to do with human knowledge, they are measures of belief.

Both approaches follow the same mathematical axioms defined below; however there are differences in the manner in which probability is applied, in particular in statistical inference. These differences gave way to the main two schools for statistics: the frequentist and the Bayesian schools. In the field of artificial intelligence, in particular in expert systems, the preferred approach tends to be the epistemological or subjective one; however, the objective approach is also used [5].

We will consider the logical or normative approach and define probabilities in terms of the degree of plausibility of a certain proposition given the available evidence [3]. Based on Cox's work, Jaynes establishes some basic desiderata that this degree of plausibility must follow [3]:

- Representation by real numbers.
- Qualitative correspondence with common sense.
- Consistency.

Based on these intuitive principles, we can derive the three axioms of probability:

1. $P(A)$ is a continuous monotonic function in [0, 1].
2. $P(A, B \mid C) = P(A \mid C)P(B \mid A, C)$ (product rule).
3. $P(A \mid B) + P(\neg A \mid B) = 1$ (sum rule).

Where $A, B, C$ are propositions (binary variables) and $P(A)$ is the probability of proposition $A$. $P(A \mid C)$ is the probability of $A$ given that $C$ is known, which is called *conditional probability*. $P(A, B \mid C)$ is the probability of $A$ AND $B$ given $C$ (logical conjunction) and $P(\neg A \mid C)$ is the probability of NOT $A$ (logical negation) given $C$. These rules are equivalent to the most commonly used Kolmogorov axioms. From these axioms, all conventional probability theory can be derived.

## 2.2 Basic Rules

The probability of the disjunction (logical sum) of two propositions is given by the *sum rule*: $P(A + B \mid C) = P(A \mid C) + P(B \mid C) - P(A, B \mid C)$; if propositions $A$ and $B$ are mutually exclusive given $C$, we can simplify it to: $P(A + B \mid C) = P(A \mid C) + P(B \mid C)$. This can be generalized for $N$ mutually exclusive propositions to:

$$P(A_1 + A_2 + \cdots A_N \mid C) = P(A_1 \mid C) + P(A_2 \mid C) + \cdots + P(A_N \mid C) \quad (2.1)$$

In the case that there are $N$ mutually exclusive and exhaustive hypothesis, $H_1, H_2, \ldots, H_N$, and if the evidence $B$ does not favor any of them, then according to the principle of indifference: $P(H_i \mid B) = 1/N$.

According to the logical interpretation there are no *absolute* probabilities, all are conditional on some background information[1]. $P(H \mid B)$ conditioned only on the background $B$ is called a *prior* probability; once we incorporate some additional information $D$ we call it a *posterior* probability, $P(H \mid D, B)$. From the product rule we obtain:

$$P(D, H \mid B) = P(D \mid H, B)P(H \mid B) = P(H \mid D, B)P(D \mid B) \quad (2.2)$$

From which we obtain:

$$P(H \mid D, B) = \frac{P(H \mid B)P(D \mid H, B)}{P(D \mid B)} \quad (2.3)$$

This last equation is known as the *Bayes rule* and the term $P(D \mid H, B)$ as the *likelihood*, $L(D)$.

In some cases the probability of $H$ is not influenced by the knowledge of $D$, so it is said that $H$ and $D$ are *independent* given some background $B$, therefore $P(H, D \mid B) = P(H \mid B)$. In the case in which $A$ and $B$ are independent, the product rule can be simplified to: $P(A, B \mid C) = P(A \mid C)P(B \mid C)$, and this can be generalized to $N$ mutually independent propositions:

$$P(A_1, A_2, \ldots, A_N \mid B) = P(A_1 \mid B)P(A_2 \mid B) \cdots P(A_N \mid B) \quad (2.4)$$

If two propositions are independent given only the background information they are *marginally* independent; however if they are independent given some additional evidence, $E$, then they are *conditionally* independent: $P(H, D \mid B, E) = P(H \mid B, E)$. For example, consider that $A$ represents the proposition *watering the garden*, $B$ the *weather forecast* and $C$ *raining*. Initially, watering the garden is not independent

[1]It is commonly written $P(H)$ without explicit mention of the conditioning information. In this case we assume that there is still some context under which probabilities are considered even if it is not written explicitly.

of the weather forecast; however, once we observe rain, they become independent. That is, $P(A, B \mid C) = P(A \mid C)$.

Probabilistic graphical models are based on these conditions of marginal and conditional independence.

The probability of a conjunction of $N$ propositions, that is $P(A_1, A_2, \ldots, A_N \mid B)$, is usually called the *joint* probability. If we generalize the product rule to $N$ propositions we obtain what is known as the *chain* rule:

$$P(A_1, A_2, \ldots, A_N \mid B) = P(A_1 \mid A_2, A_3, \ldots, A_N, B)P(A_2 \mid A_3, A_4, \ldots, A_N, B) \cdots P(A_N \mid B) \tag{2.5}$$

Thus the joint probability of $N$ propositions can be obtained by this rule. Conditional independence relations between the propositions can be used to simplify this product; that is, for instance if $A_1$ and $A_2$ are independent given $A_3, \ldots, A_N, B$, then the first term in Eq. 2.5 can be simplified to $P(A_1 \mid A_3, \ldots, A_N, B)$.

Another important relation is the rule of *total probability*. Consider a partition, $B = \{B_1, B_2, \ldots B_n\}$, on the sample space $\Omega$, such that $\Omega = B_1 \cup B_2 \cup \ldots \cup B_n$ and $B_i \cap B_j = \emptyset$. That is, $B$ is a set of mutually exclusive events that cover the entire sample space. Consider another event $A$; $A$ is equal to the union of its intersections with each event $A = (B_1 \cap A) \cup (B_2 \cap A) \cup \ldots \cup (B_n \cap A)$. Then, based on the axioms of probability and the definition of conditional probability we can derive the rule of total probability:

$$P(A) = \sum_i P(A \mid B_i)P(B_i) \tag{2.6}$$

Given the total probability rule, we can rewrite Bayes rule as (omitting the background term):

$$P(B \mid A) = \frac{P(B)P(A \mid B)}{\sum_i P(A \mid B_i)P(B_i)} \tag{2.7}$$

This last expression is commonly known as Bayes Theorem.

## 2.3 Random Variables

If we consider a finite set of exhaustive and mutually exclusive propositions[2], then a discrete variable $X$ can represent this set of propositions, such that each value $x_i$ of $X$ corresponds to one proposition. If we assign a numerical value to each proposition $x_i$, then $X$ is a *discrete random variable*. For example, the outcome of the toss of a die is a discrete random variable with 6 possible values $1, 2, \ldots, 6$. The probabilities for all possible values of $X$, $P(X)$ is the probability distribution of $X$. Considering the die example, for a fair die the probability distribution will be:

[2]This means that one and only one of the propositions has a value of TRUE.

| $x$ | 1 | 2 | 3 | 4 | 5 | 6 |
|---|---|---|---|---|---|---|
| $P(x)$ | 1/6 | 1/6 | 1/6 | 1/6 | 1/6 | 1/6 |

This is an example of a *uniform* probability distribution. There are several probability distributions which have been defined. Another common distribution is the *binomial* distribution. Assume we have an urn with $N$ colored balls, red and black, of which $M$ are red, so the fraction of red balls is $\pi = M/N$. We draw a ball at random, record its color, and return it to the urn, mixing the balls again (so that, in principle, each draw is independent from the previous one). The probability of getting $r$ red balls in $n$ draws is:

$$P(r \mid n, \pi) = \binom{n}{r} \pi^r (1-\pi)^{n-r}, \tag{2.8}$$

where $\binom{n}{r} = \frac{n!}{r!(n-r)!}$.

This is an example of a binomial distribution which is applied when there are $n$ independent trials, each with two possible outcomes (success or failure), and the probability of success is constant over all trials. There are many other distributions, we refer the interested reader to the additional reading section at the end of the chapter.

There are two important quantities that in general help to characterize a probability distribution. The expected value or *expectation* of a discrete random variable is the average of the possible values, weighted according to their probabilities:

$$E(X \mid B) = \sum_{i=1}^{N} P(x_i \mid B) x_i \tag{2.9}$$

The *variance* is defined as the expected value of the square of the variable minus its expectation:

$$\sigma^2(X \mid B) = \sum_{i=1}^{N} P(x_i \mid B)(x_i - E(X \mid B))^2 \tag{2.10}$$

Intuitively, the variance gives a measure of how *wide* or *narrow* the probabilities are distributed for a certain random variable. The square root of the variance is known as the standard deviation, which is usually more intuitive as its units are the same as those of the variable.

So far we have considered discrete variables, however the rules of probability can be extended to continuous variables. If we have a continuous variable $X$, we can divide it into a set of mutually exclusive and exhaustive intervals, such that $P = (a < X \leq b)$ is a proposition, thus the rules derived so far apply to it. A *continuous random variable* can be defined in terms of a *probability density function*, $f(X \mid B)$, such that:

$$P(a < X \leq b \mid B) = \int_a^b f(X \mid B) dx \tag{2.11}$$

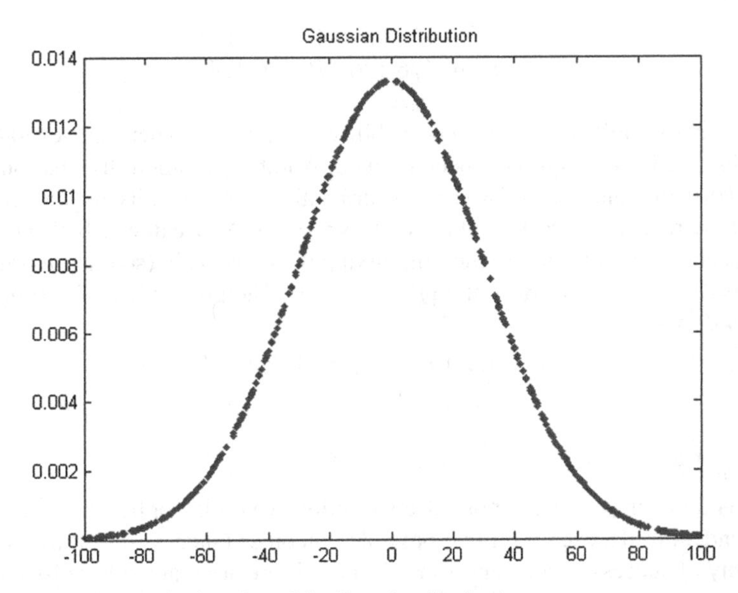


**Fig. 2.1** Probability density function of the Gaussian distribution

The probability density function must satisfy $\int_{-\infty}^{\infty} f(X \mid B)dx = 1$.

An example of a continuous probability distribution is the *Normal* or Gaussian distribution. This distribution plays an important role in many applications of probability and statistics, as many phenomena in nature have an approximately Normal distribution; it is also prevalent in probabilistic graphical models due to its mathematical properties.

A Normal distribution is denoted as $N(\mu, \sigma^2)$, where $\mu$ is the *mean* (center) and $\sigma$ is the *standard deviation* (spread); and it is defined as:

$$f(X \mid B) = \frac{1}{\sigma\sqrt{2\pi}} exp\left\{-\frac{1}{2\sigma^2}(x-\mu)^2\right\} \tag{2.12}$$

The density function of the Gaussian distribution is depicted in Fig. 2.1.

Another important continuous distribution is the Exponential distribution; as an example, the time that it takes for a certain piece of equipment to fail is usually modeled by an exponential distribution. The exponential distribution is denoted as $Exp(\beta)$; it has a single parameter $\beta > 0$, and it is defined as:

$$f(X \mid B) = \frac{1}{\beta} e^{-x/\beta}, x > 0 \tag{2.13}$$

An example of an exponential density function is shown in Fig. 2.2.

It is common to represent probability distributions, in particular for continuous variables, using the cumulative distribution function, $F$. The cumulative distribution

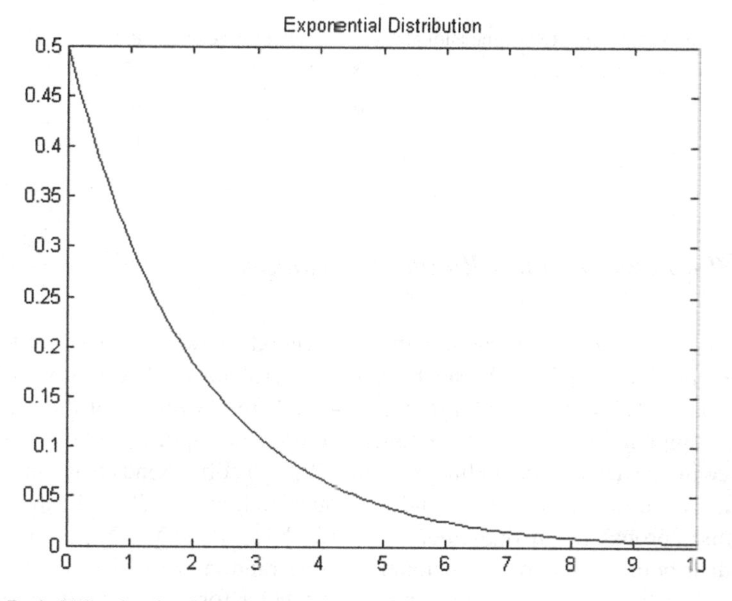


**Fig. 2.2** Probability density function of the exponential distribution

function of a random variable, $X$, is the probability that $X \leq x$. For a continuous variable, it is defined in terms of the density function as:

$$F(X) = \int_{-\infty}^{x} f(X) \tag{2.14}$$

The following are some properties of cumulative distribution functions:

- In the interval [0, 1]: $0 \leq F(X) \leq 1$
- Non-decreasing: $F(x_1) \leq F(x_2)$ if $x_1 < x_2$
- Limits: $lim_{x \to -\infty} = 0$ and $lim_{x \to \infty} = 1$

In the case of discrete variables, the cumulative probability, $P(X \leq x)$ is defined as:

$$\mathbf{P}(x) = \sum_{x=-\infty}^{X=x} P(X) \tag{2.15}$$

It has similar properties as the cumulative distribution function.

**Table 2.1** An example of a two-dimensional discrete probability distribution

| | X = 1 | X = 2 | X = 3 |
|---|---|---|---|
| Y = 1 | 0.1 | 0.3 | 0.3 |
| Y = 2 | 0.2 | 0.1 | 0 |

### 2.3.1 Two Dimensional Random Variables

The concept of a random variable can be extended to two, or more dimensions. Given two random variables, $X$ and $Y$, their joint probability distribution is defined as $P(x, y) = P(X = x \wedge Y = y)$. For example, $X$ might represent the number of products completed in one day in product line one, and $Y$ the number of products completed in one day in product line two, thus $P(x, y)$ corresponds to the probability of producing $x$ products in line one and $y$ products in line two. $P(X, Y)$ must follow the axioms of probability, in particular: $0 \leq P(X, Y) \leq 1$ and $\sum_X \sum_Y P(X, Y) = 1$.

The distribution for two-dimensional discrete random variables (known as the *bivariate* distribution) can be represented in a tabular form. For instance, consider the example of the two product lines, and assume that line one ($X$) may produce 1, 2 or 3 products per day, and line two ($Y$), 1 or 2 products. Then a possible joint distribution, $P(X, Y)$ is shown in Table 2.1.

Given the joint probability distribution, $P(X, Y)$, we can obtain the distribution for each individual random variable, what is known as the marginal probability:

$$P(x) = \sum_y P(X, Y); \; P(y) = \sum_x P(X, Y) \tag{2.16}$$

For instance, if we consider the joint distribution of Table 2.1, we can obtain the marginal probabilities for $X$ and $Y$. For example, $P(X = 2) = 0.3 + 0.1 = 0.4$ and $P(Y = 1) = 0.1 + 0.3 + 0.3 = 0.7$.

We can also calculate the conditional probabilities of $X$ given $Y$ and vice-versa:

$$P(X \mid Y) = P(X, Y)/P(Y); \; P(Y \mid X) = P(X, Y)/P(X) \tag{2.17}$$

Following the example in Table 2.1:
$P(X = 3 \mid Y = 1) = P(X = 3, Y = 1)/P(Y = 1) = 0.3/0.7 = 0.4286$

The concept of independence can be applied to two-dimensional random variables. Two random variables, $X, Y$, are independent if their joint probability distribution is equal to the product of their marginal distributions (for all values of $X$ and $Y$):

$$P(X, Y) = P(X)P(Y) \rightarrow Independent(X, Y) \tag{2.18}$$

Another useful measure is called *correlation*; it is a measure of the degree of linear relation between two random variables, $X, Y$, and is defined as:

$$\rho(X, Y) = E\{[X - E(X)][Y - E(Y)]\}/(\sigma_X \sigma_Y) \tag{2.19}$$

where $E(X)$ is the expected value of $X$ and $\sigma_x$ its standard deviation. The correlation is in the interval $[-1, 1]$; a positive correlation indicates that as $X$ increases, $Y$ tends to increase; and a negative correlation that as $X$ increases, $Y$ tends to decrease.

Note that a correlation of zero does not necessarily imply independence, as the correlation only measures a linear relationship. So it could be that $X$ and $Y$ have a zero correlation but are related through a higher order function, and thus they are not independent.

## 2.4 Information Theory

Information theory was originated in the area of communications, although it is relevant for many different fields. In the case of probabilistic graphical models, it is mainly applied in learning. In this section we will cover the basic concepts of information theory.

Assume that we are communicating the occurrence of a certain event. Intuitively we can think that the amount of *information* from communicating an event is inverse to the probability of the event. For example, consider that a message is sent informing about one of the following events:

1. It is raining in New York.
2. There was an earthquake in New York.
3. A meteorite fell over New York City.

The probability of the first event is higher than the second, and that of the second is higher than the third. Thus, the message for event 1 has the lowest amount of information and the message for event 3 gives the highest amount of information.

Lets now see how we can formalize the concept of information. Assume we have a source of information that can send $q$ possible messages, $m_1, m_2, ...m_q$; where each message corresponds to an event with probabilities $P_1, P_2, ...P_q$. We want to find a function $I(m)$ based on the probability of $m$.The function must satisfy the following properties:

- The information ranges from zero to infinity: $I(m) \geq 0$.
- The information increases as the probability decreases: $I(m_i) > I(m_j)$ if $P(m_i) < P(m_j)$.
- The information tends to infinity as the probability tends to zero: $I(m) \rightarrow \infty$ if $P(m) \rightarrow 0$.
- The information of two messages is equal to the sum of the of the information of the individual messages if these are independent: $I(m_i + m_j) = I(m_i) + I(m_j)$ if $m_i$ is independent of $m_j$.

A function that satisfies the previous properties is the logarithm of the inverse of the probability, that is:

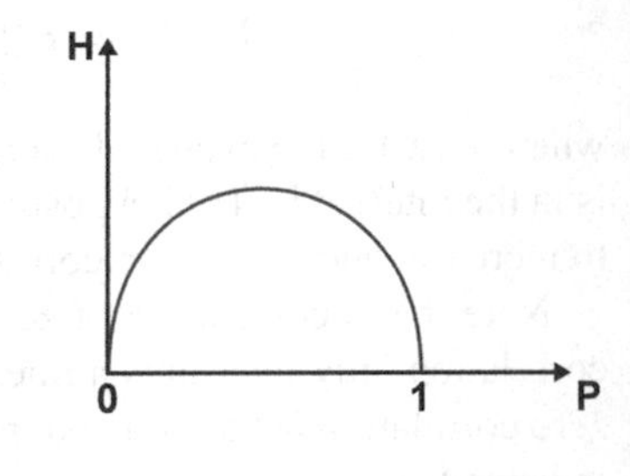


**Fig. 2.3** Entropy vs. probability for a binary source. The entropy is at its maximum when the probability is 0.5, and at its minimum when the probability is zero and one

$$I(m_k) = log(1/P(m_k)) \tag{2.20}$$

It is common to use base two logarithms, so the information is measured in “bits”:

$$I(m_k) = log_2(1/P(m_k)) \tag{2.21}$$

For example, if we assume that the probability of the message $m_r$ “raining in New York” is $P(m_r) = 0.25$, the corresponding information is $I(m_r) = log_2(1/0.25)=2$.

Once we have defined information for a particular message, another important concept is the average information for the $q$ messages; that is, the expected value of the information also known as *entropy*. Given the definition of expected value, the average information of $q$ messages or entropy is defined as:

$$H(m) = E(I(m)) = \sum_{i=1}^{q} P(m_i)log_2(1/P(m_i)) \tag{2.22}$$

This can be interpreted as on average $H$ bits of information will be sent.

An interesting question is: When will $H$ have its maximum and minimum values? Consider a binary source such that there are only two messages, $m_1$ and $m_2$; with $P(m_1) = p_1$ and $P(m_2) = p_2$. Given that there are only two possible messages, $p_2 = 1 - p_1$, so $H$ only depends on one parameter, $p_1$ (or just $p$). Figure 2.3 shows a graph of $H$ with respect to $p$. Observe that $H$ is at its maximum when $p = 0.5$ and at its minimum (zero) when $p = 0$ and $p = 1$. In general, the entropy is at its maximum when there is a uniform probability distribution for the events; it is at its minimum when there is one element that has a probability of one and the rest have a probability of zero.

If we consider the conditional probabilities, we can extend the concept of entropy to *conditional entropy*:

$$H(X \mid y) = \sum_{i=1}^{q} P(x_i \mid y)log_2[1/P(x_i \mid y)] \tag{2.23}$$

Another extension of entropy is the *cross entropy*:

$$H(X, Y) = \sum_X \sum_Y P(X, Y) log_2[P(X, Y)/P(X)P(Y)] \quad (2.24)$$

The cross entropy provides a measure of the mutual information (dependency) between two random variables; it is zero when the two variables are independent.

## 2.5 Additional Reading

Donald Gillies [2] provides a comprehensive account of the different philosophical approaches to probability. An excellent book on probability theory from a logical perspective is [3]. Wasserman [6] gives a concise course on probability and statistics oriented for computer science and engineering students. Other accessible introduction on probability and statistics for computer science is [1]. There are several books on information theory; one that relates it to machine learning and inference is [4].

## 2.6 Exercises

1. If we throw two dices, what is the probability that the outcomes add to exactly seven? Seven or more?
2. If we assume that the height of a group of students follows a normal distribution with a mean of 1.7 m and a standard deviation of 0.1 m, how probable is it that there is a student with a height above 1.9 m?
3. Demonstrate by mathematical induction the chain rule.
4. We have an urn with 10 balls, 4 white and 6 black. What is the probability of getting two white balls and one black ball from the urn if we take three balls from the urn with replacement?
5. In certain country there are 5 states with the following populations: S1, 2000; S2, 1000; S3, 3000; S4, 1000; S5, 3000. In S1 there are 100 poor people, 200 in S2, 500 in S3, 100 in S4 and 300 in S5. What is the poverty percentage in general in the country?
6. Assume that a person has only one of two possible diseases, hepatitis ($H$) or typhoid ($T$). There are two symptoms associated to those diseases: headache ($D$) and fever ($F$), which could be TRUE or FALSE. Given the following probabilities: $P(T) = 0.5$, $P(D \mid T) = 0.7$, $P(D \mid \neg T) = 0.4$, $P(F \mid T) = 0.9$, $P(F \mid \neg T) = 0.5$. Describe the sampling space and complete the partial probability tables.
7. Given the data for the previous problem, and assuming that the symptoms are independent given the disease, obtain the probability that the person has hepatitis given that she does not have a headache and does have a fever.
8. How will you represent graphically that the two symptoms (headache and fever) are independent given the disease?

9. Given the two dimensional probability distribution in the table below, obtain: (a) $P(X_1)$, (b) $P(Y_2)$, and (c) $P(X_1 \mid Y_1)$.

| | $Y_1$ | $Y_2$ | $Y_3$ |
|---|---|---|---|
| $X_1$ | 0.1 | 0.2 | 0.1 |
| $X_2$ | 0.3 | 0.1 | 0.2 |

10. In the previous problem, are $X$ and $Y$ independent?
11. Modify the probabilities of the joint distribution in Problem 9 so that $X$ and $Y$ are independent.
12. In a certain place, the statistics show that in a year the weather behaves in the following way. From 365 days, 200 are sunny, 60 cloudy, 40 rainy, 20 snowy, 20 with thundershowers, 10 with hail, 10 windy and 5 with drizzle. If on each day a message is sent about the weather, what is the information of the message for each type of weather?
13. Considering the information for each type of weather in the previous problem, what is the average information (entropy) of the message?
14. Suppose that a medical expert tells you that the *prevalence* (prior probability) of certain disease in certain population is 10%. Recently they found out from medical records that 12 out of 100 persons in the same population present this disease. What will be your estimate of the probability of the disease under a frequentist interpretation of probability? And under a Bayesian interpretation?
15. *** Investigate the different philosophical interpretations of probability, and discuss the advantages and limitations of each one of them. Which one do you consider the most appropriate? Why?

## References


1. Forsyth, D.: Probability and Statistics for Computer Science. Springer, Switzerland (2018)
2. Gillies, D.: Philosophical Theories of Probability. Routledge, London (2000)
3. Jaynes, E.T.: Probability Theory: The Logic of Science. Cambridge University Press, Cambridge (2003)
4. MacKay, D.J.: Information Theory. Inference and Learning Algorithms. Cambridge University Press, Cambridge (2004)
5. Sucar, L.E., Gillies, D.F., Gillies, D.A.: Objective probabilities in expert systems. Artif. Intell. **61**, 187–208 (1993)
6. Wasserman, L.: All of Statistcs: A Concise Course in Statistical Inference. Springer, New York (2004)

# Chapter 3
# Graph Theory

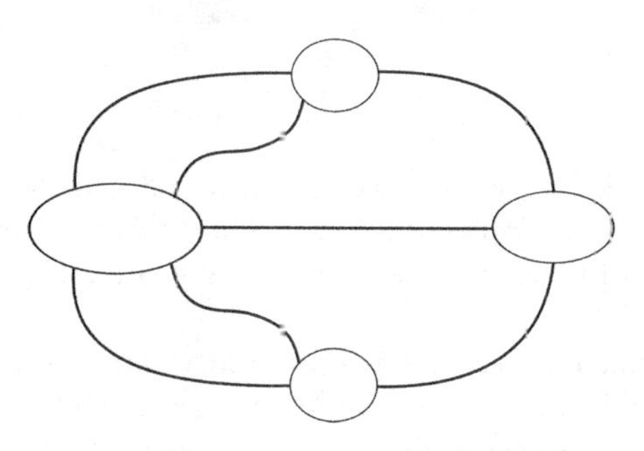

## 3.1 Definitions

A *graph* provides a compact way to represent binary relations between a set of objects. For example, consider a set of cities in a certain region, and the roads that connect these cities. Then a map of the region is essentially a graph, in which the object are the cities and the direct roads between pairs of cities are the relations. Graphs are usually represented graphically. Objects are represented as circles or ovals, and relations as lines or arrows; see Fig. 3.1. There are two basic types of graphs: *undirected graphs* and *directed graphs*. Next we formalize the definitions of directed and undirected graphs.

Given $V$, a non empty set, a binary relation $E \subseteq V \times V$ on $V$ is a set of ordered pairs, $(V_j, V_k)$, such that $V_j, V_k \in V$. A *directed graph* or *digraph* is an ordered pair, $G = (V, E)$, where $V$ is a set of vertices or nodes and $E$ is a set of arcs that represent a binary relation on $V$; see Fig. 3.1b. Directed graphs represent anti-symmetric relations between objects, for instance the "parent" relation.

An *undirected graph* is an ordered pair, $G = (V, E)$, where $V$ is a set of vertices or nodes and $E$ is a set of edges that represent symmetric binary relations: $(V_j, V_k) \in E \rightarrow (V_k, V_j) \in E$; see Fig. 3.1a. Undirected graphs represent symmetric relations between objects, for example, the "brother" relation.



L. E. Sucar, *Probabilistic Graphical Models*, Advances in Computer Vision and Pattern Recognition, https://doi.org/10.1007/978-3-030-61943-5_3

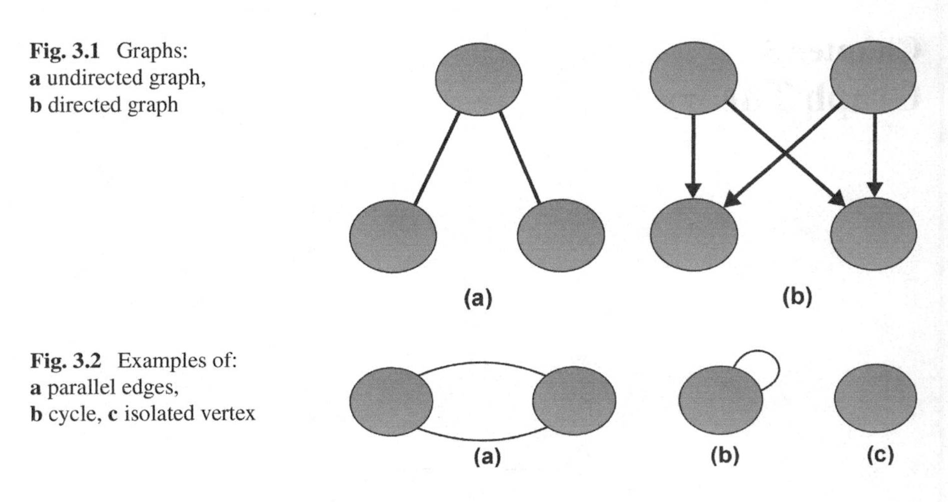


**Fig. 3.1** Graphs: **a** undirected graph, **b** directed graph

**Fig. 3.2** Examples of: **a** parallel edges, **b** cycle, **c** isolated vertex

If there is an edge $E_i(V_j, V_k)$ between nodes $j$ and $k$, then $V_j$ is adjacent to $V_k$. In an undirected graph, the *degree* of a node is the number of edges that are incident in that node. In Fig. 3.1a, the upper node has a degree of 2 and the two lower nodes have a degree of 1.

Two edges associated to the same pair of vertices are said to be *parallel edges*; an edge which is incident on a single vertex is a *cycle*; and a vertex that is not an endpoint to any edge is an *isolated vertex* –it has degree 0. These are illustrated in Fig. 3.2.

In a directed graph, the number of arcs pointing to a node is its *in degree*; the number of edges pointing away from a node is its *out degree*; and the total number of edges that are incident in the node is its *degree* (in degree + out degree). In Fig. 3.1b, the two upper nodes have an in degree of zero and an out degree of two; while the two lower nodes have an in degree of two and an out degree of zero.

Given a graph $G = (V, E)$, a subgraph $G' = (V', E')$ of $G$, is a graph such that $V' \subseteq V$ and $E' \subseteq E$, in which each edge in $E'$ is incident on vertices in $V'$. For example, if we take out the direction of the edges in the graph of Fig. 3.1b (making it an undirected graph), then the graph in Fig. 3.1a is a subgraph of Fig. 3.1b.

## 3.2 Types of Graphs

In addition to the two basic graphs, directed and undirected, there other types of graphs, such as:

Chain graph: a hybrid graph that has directed and undirected edges.

Simple graph: a graph that does not include cycles and parallel arcs.

Multigraph: a graph with several components (subgraphs), such that each component has no edges to the other components, i.e., they are disconnected.

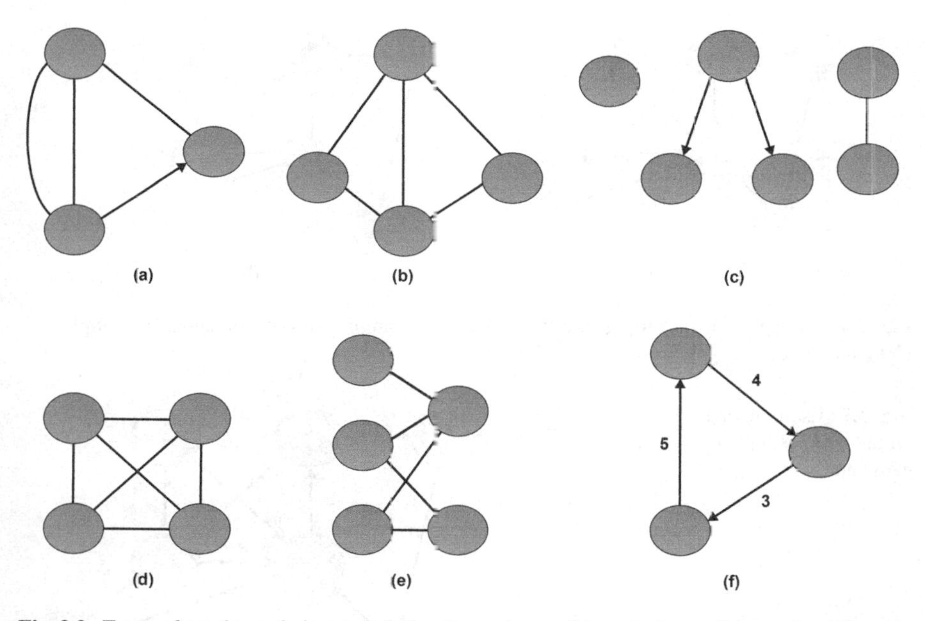


**Fig. 3.3** Types of graphs: **a** chain graph, **b** simple graph, **c** multigraph, **d** complete graph, **e** bipartite graph, **f** weighted graph

Complete graph: a graph that has an edge between each pair of vertices.

Bipartite graph: a graph in which the vertices can be divided in two subsets, $G_1, G_2$, such that all edges connect a vertex in $G_1$ with a vertex in $G_2$; that is, there are no edges between nodes in each subset.

Weighted graph: a graph that has weights associated to its edges and/or vertices.

Examples of these types of graphs are depicted in Fig. 3.3.

## 3.3 Trajectories and Circuits

A *trajectory* is a sequence of edges, $E_1, E_2, ..., E_n$ such that the final vertex of each edge coincides with the initial vertex of the next edge in the sequence (except for the final vertex); that is, $E_i(V_j, V_k)$, $E_{i+1}(V_k, V_l)$, for $i = 1$ to $i = n - 1$. A *simple* trajectory does not include the same edge two o more times; an *elemental* trajectory is not incident on the same vertex more than once. Examples of different trajectories are illustrated in Fig. 3.4.

A graph $G$ is *connected* if there is a trajectory between each pair of distinct vertices in $G$. If a graph $G$ is not connected, each part that is connected is called a *component* of $G$.

A *circuit* is a trajectory such that the final vertex coincides with the initial one, i.e., it is a "closed trajectory". In an analogous way as with trajectories, we can define simple and elemental circuits. Figure 3.5 shows an example of a circuit.

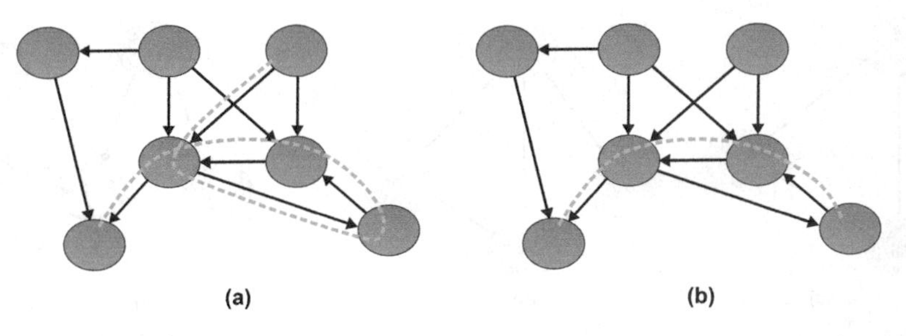


**Fig. 3.4** Examples of trajectories: **a** a trajectory that is simple but not elemental, **b** a simple and elemental trajectory

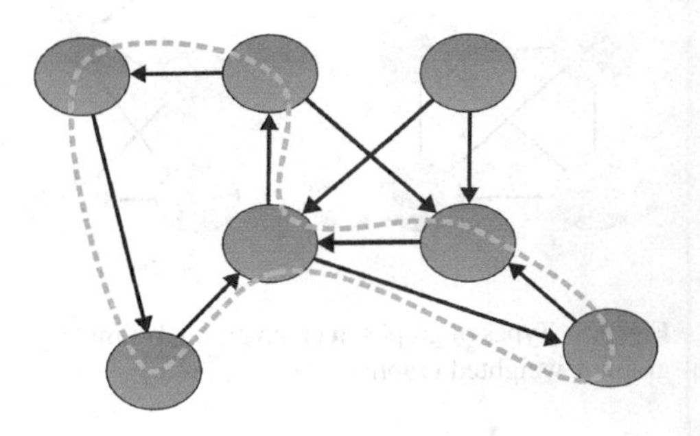

**Fig. 3.5** Example of a circuit that is simple but not elemental

An important type of graph for PGMs is a *Directed Acyclic Graph* (DAG). A DAG is a directed graph that has no directed circuits (a directed circuit is a circuit in which all edges in the sequence follow the directions of the arrows). For instance, Fig. 3.1b is a DAG and Fig. 3.3f is not a DAG.

Some classical problems in graph theory include trajectories and circuits, such as:

- Finding a trajectory that includes all edges in a graph only once (Euler trajectory).
- Finding a circuit that includes all edges in a graph only once (Euler circuit).
- Finding a trajectory that includes all vertices in a graph only once (Hamiltonian trajectory).
- Finding a circuit that includes all vertices in a graph only once (Hamiltonian circuit).
- Finding a Hamiltonian circuit in a weighted graph with minimum cost (Traveling salesman problem).[1]

The solution to these problems is beyond the scope of this book, the interested reader is referred to [2].

[1] In this case the nodes represent cities and the edges roads with an associated distance or time, so the solution will provide a traveling salesman with the "best" (minimum distance or time) route to cover all the cities.

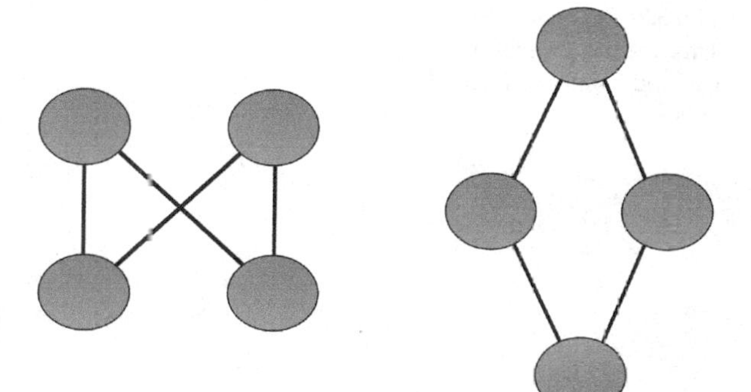

**Fig. 3.6** These two graphs are isomorphic

## 3.4 Graph Isomorphism

Two graphs are isomorphic if there is a one to one correspondence between their vertices and edges, so that the incidences are maintained. Given two graphs, $G_1$ and $G_2$, there are three basic types of isomorphisms:

1. *Graph isomorphism.* Graphs $G_1$ and $G_2$ are isomorphic.
2. *Subgraph isomorphism.* Graph $G_1$ is isomorphic to a subgraph of $G_2$ (or vice versa).
3. *Double subgraph isomorphism.* A subgraph of $G_1$ is isomorphic to a subgraph of $G_2$.

Figure 3.6 shows an example of two graphs that are isomorphic.

Determining if two graphs are isomorphic (type 1) is an NP problem; while the subgraph and double subgraph isomorphism problems (type 2 and 3) are NP-complete. See [1].

## 3.5 Trees

Trees are a type of graph which are very important in computer science in general, and for PGMs in particular. We will discuss two types of trees: undirected and directed.

An undirected tree is a connected graph that does not have simple circuits; Fig. 3.7 depicts an example of an undirected tree. There are two classes of vertices or nodes in an undirected tree: (i) leaf or terminal nodes, with degree one; (ii) internal nodes, with degree greater than one. Some basic properties of a tree are:

- There is a simple trajectory between each pair of vertices.
- The number of vertices, $\mid V \mid$, is equal to the number of edges, $\mid E \mid$ plus one: $\mid V \mid = \mid E \mid +1$.
- A tree with two or more vertices has at least two leaf nodes.

A forest is an undirected graph in which any two vertices are connected by at most one path, or equivalently a disjoint union of trees.

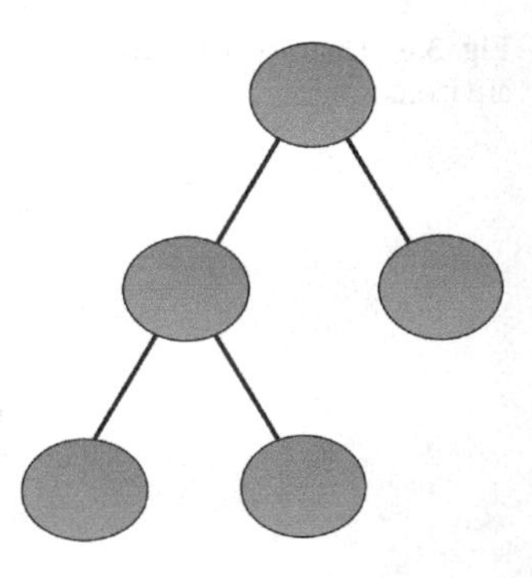

**Fig. 3.7** An undirected tree. This tree has 5 nodes, three leaf nodes and two internal nodes

A directed tree is a connected directed graph such that there is only a single directed trajectory between each pair of nodes (it is also known as a singly connected directed graph). There are two types of directed trees: (i) a rooted tree (or simply a tree), (ii) a polytree. A rooted tree has a single node with an in degree of zero (the root node) and the rest have in degree of one. A polytree might have more than one node with in degree zero (roots), and certain nodes (zero or more) with in degree greater than one (called multi-parent nodes). If we take out the direction of the edges in a polytree, it transforms into an undirected tree. We can think of a tree as a special case of a polytree. Examples of a rooted tree and a polytree are shown in Fig. 3.8

Some relevant terminology for directed trees is the following.

Root: a node with in degree equal to zero.
Leaf: a node with out degree equal to zero.
Internal node: a node with out degree greater than zero (not a leaf node).
Parent / Child: if there is a directed arc from $A$ to $B$, $A$ is parent of $B$ and $B$ is a child of $A$.
Brothers: two or more nodes are brothers if they have the same parent.

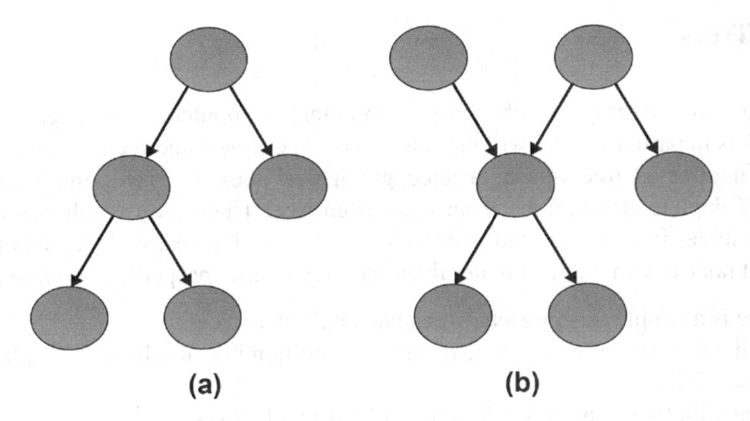


**Fig. 3.8** **a** A rooted tree. **b** A polytree

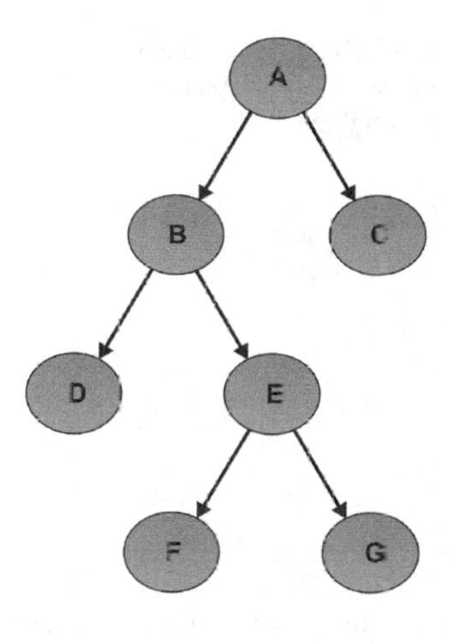


**Fig. 3.9** An example of a directed tree to illustrate the terminology

Ascendants /Descendants: if there is a directed trajectory from $A$ to $B$, $A$ is an ascendant of $B$ and $B$ is a descendant of $A$.

Subtree with root $A$: a subtree with $A$ as its root.

Subtree of $A$: a subtree with a child of $A$ as its root.

K-ary Tree: a tree in which each internal node has at most $K$ children. It is a regular tree if each internal node has $K$ children.

Binary Tree: a tree in which each internal node has at most two children.

For example, in the tree in Fig. 3.9; (i) $A$ is a root node, (ii) $C$, $D$, $F$, $G$ are leaf nodes, (iii) $A$, $B$, $E$ are internal nodes, (iv) $A$ is parent of $B$ and $B$ is child of $A$, (v) $B$ and $C$ are brothers, (vi) $A$ is an ascendant of $F$ and $F$ is a descendant of $A$, (vii) the subtree $B$, $D$, $E$, $F$, $G$ is a subtree with root $B$, (viii) the subtree $E$, $F$, $G$ is a subtree of $B$. The tree of Fig. 3.9 is a regular binary tree.

## 3.6 Cliques

A *complete graph* is a graph, $G_c$, in which each pair of nodes is adjacent; that is, there is an edge between each pair of nodes. Figure 3.3d is an en example of a complete graph. A *complete set*, $W_c$ is a subset of $G$ that induces a complete subgraph of $G$. It is a subset of vertices of $G$ so that each pair of nodes in this subgraph is adjacent. For example, in Fig. 3.3d, each subset of three nodes in the graph is a complete set.

A *clique*, $C$, is a subset of graph $G$ such that it is a complete set that is maximal; that is, there is no other complete set in $G$ that contains $C$. The subsets of three nodes in Fig. 3.3d are not cliques, as these are not maximal; they are contained by the graph which is complete.

The graph in Fig. 3.10 has 5 cliques, one with four nodes, one with three nodes and three with two nodes. Notice that every node in a graph is part of at least one clique; thus, the set of cliques of a graph always covers $V$.

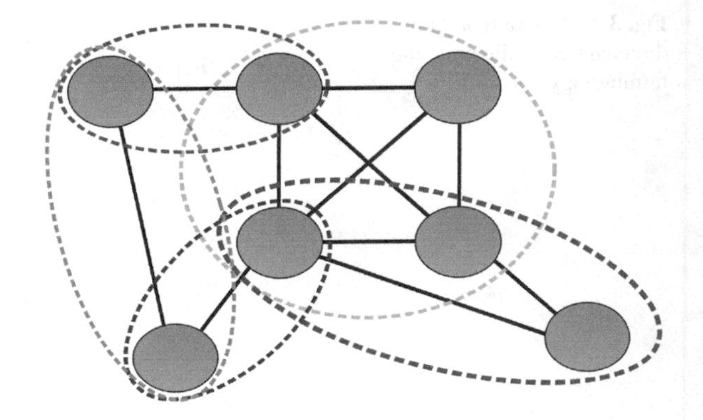

**Fig. 3.10** Cliques: the 5 cliques in the graph are highlighted

The following sections cover some more advanced concepts of graph theory, as these are used by some of the inference algorithms for probabilistic graphical models.

## 3.7 Perfect Ordering

An ordering of the nodes in a graph consists in assigning an integer to each vertex. Given a graph $G = (V, E)$, with $n$ vertices, then $\alpha = [V_1, V_2, ..., V_n]$ is an ordering of the graph; $V_i$ is *before* $V_j$ according to this ordering, if $i < j$.

An ordering $\alpha$ of a graph $G = (V, E)$ is a *perfect ordering* if all the adjacent vertices of each vertex $V_i$ that are before $V_i$, according to this ordering, are completely connected. That is, for every $i$, $Adj(V_i) \cap \{V_1, V_2, ..., V_{i-1}\}$ is a complete subgraph of $G$. $Adj(V_i)$ is the subset of nodes in $G$ that is adjacent to $V_i$. Figure 3.11 depicts an example of a perfect ordering.

Consider the set of cliques $C_1, C_2, ...C_p$ of an undirected connected graph $G$. In an analogous way as an ordering of the nodes, we can define an ordering of the cliques, $\beta = [C_1, C_2, ..., C_p]$. An ordering $\beta$ of the cliques has the *running intersection property*, if all the common nodes of each clique $C_i$ with previous cliques according to this order are contained in a clique $C_j$; $C_j$ is the *parent* of $C_i$. In other words, for every clique $i > 1$ there exists a clique $j < i$ such that $C_i \cap \{C_1, C_2, ..., C_{i-1}\} \subseteq C_j$. It is possible that a clique has more than one parent.

The cliques, $C_1$, $C_2$ and $C_3$ in Fig. 3.11 satisfy the running intersection property. In this example, $C_1$ is the parent of $C_2$, and $C_1$ and $C_2$ are parents of $C_3$.

A graph $G$ is *triangulated* if every simple circuit of length greater than three in $G$ has a chord. A chord is an edge that connects two of the vertices in the circuit and that is not part of that circuit. For example, in Fig. 3.11 the circuit formed by the vertices 1, 2, 4, 3, 1 has a chord that connects nodes 2 and 3. The graph in Fig. 3.11 is triangulated. An example of a graph that is not triangulated is depicted in Fig. 3.12. Although visually this graph might seem triangulated, there is a circuit 1, 2, 5, 4, 1 that does not have any chord.

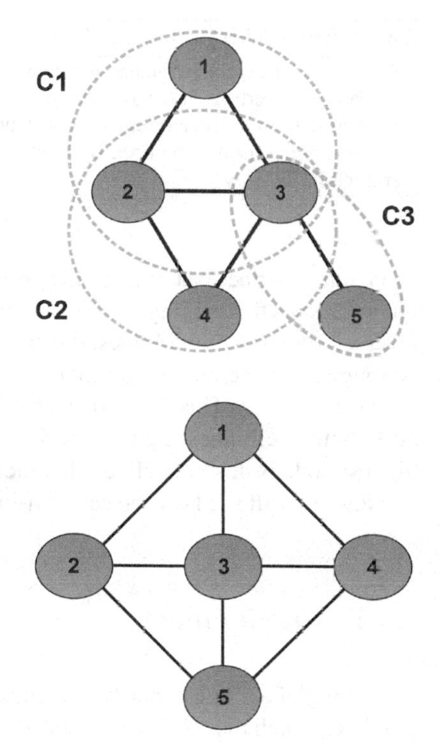


**Fig. 3.11** An example of an ordering of nodes and cliques in a graph. In this case, the nodes have a perfect ordering, and the ordering of the cliques satisfies the running intersection property

**Fig. 3.12** An example of a graph that is not triangulated. The circuit 1,2,5,4,1 does not have a chord

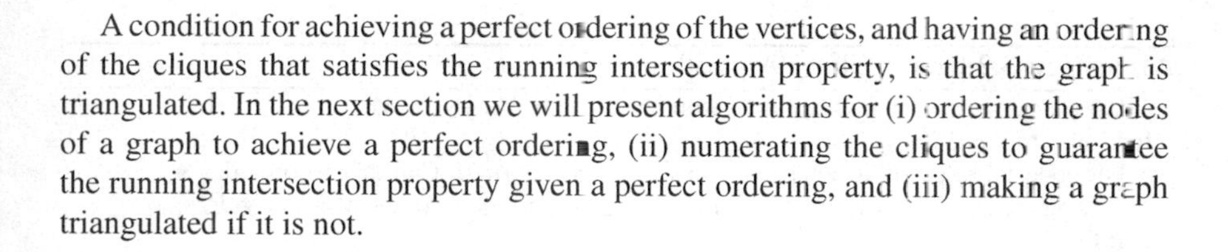

A condition for achieving a perfect ordering of the vertices, and having an ordering of the cliques that satisfies the running intersection property, is that the graph is triangulated. In the next section we will present algorithms for (i) ordering the nodes of a graph to achieve a perfect ordering, (ii) numerating the cliques to guarantee the running intersection property given a perfect ordering, and (iii) making a graph triangulated if it is not.

## 3.8 Ordering and Triangulation Algorithms

### *3.8.1 Maximum Cardinality Search*

Given that a graph is triangulated, the following algorithm, known as *maximum cardinality search*, guarantees a perfect ordering.[2] Given an undirected graph $G = (V, E)$ with $n$ vertices:

[2]The algorithm can be applied to any undirected graph; however, if the graph is not triangulated there is no guarantee of obtaining a perfect ordering.

**Algorithm 3.1** Maximum Cardinality Search Algorithm

Select any vertex from $V$ and assign it number 1.
**while** Not all vertices in $G$ have been numbered **do**
From all the non-labeled vertices, select the one with higher number of adjacent labeled vertices and assign it the next number. Break ties arbitrarily.
**end while**

Given a perfect ordering of the vertices, it is easy to number the cliques so the order satisfies the running intersection property. For this, the cliques are numbered in inverse order. Given $p$ cliques, the clique that has the node with the highest number is assigned $p$; the clique that includes the next highest numbered node is assigned $p - 1$; and so on. This method can be illustrated with the example in Fig. 3.11. The node with the highest number is 5, so the clique that contains it is $C_3$. The next highest node is 4, so the clique that includes it is $C_2$. The remaining clique is $C_1$.

Now we will see how we can "fill-in" a graph to make it triangulated.

### 3.8.2 *Graph Filling*

The filling of a graph consists of adding arcs to an original graph $G$, to obtain a new graph, $G_t$, such that $G_t$ is triangulated. Given an undirected graph $G = (V, E)$, with $n$ nodes, the following algorithm makes the graph triangulated:

**Algorithm 3.2** Graph Filling Algorithm

Order the vertices $V$ with maximum cardinality search: $V_1, V_2, ..., V_n$.
**for** $i = n$ **to** $i = 1$ **do**
For node $V_i$, select all its adjacent nodes $V_j$ such that $j > i$. Call this set of nodes $A_i$.
Add an arc from $V_i$ to $V_k$ if $k > i$ and $k < j \in A_i$ and $V_k \notin A_i$ and $V_i - V_k \notin E$.
**end for**

For example, consider the graph of Fig. 3.12 which is not triangulated. If we apply the previous algorithm, we first order the nodes, generating the labeling illustrated in the figure. Next we process the nodes in inverse order and obtain the sets $A$ for each node:

$A_5$: ∅
$A_4$: 5
$A_3$: 4, 5
$A_2$: 3, 5. An arc is added from 2 to 4.
$A_1$: 2, 3, 4.

The resulting graph has one additional arc $2 - 4$, and we can verify that it is triangulated.

The filling algorithm guarantees that the resulting graph is triangulated, but generally it is not optimal in terms of adding the minimum number of additional arcs, known as the the *minimum triangulation problem* or the minimum fill-in problem. A related problem, of particular interest for inference in PGMs, is the *treewidth problem*, which consists on finding a triangulation where the size of the largest clique is minimized. Both the minimum triangulation and the treewidth problems are NP-hard.

## 3.9 Additional Reading

There are several books in graph theory that cover most of the concepts introduced in this chapter in more detail, including [2–4]. Neapolitan [6], Chap. 3, covers the main graph theory background required for Bayesian networks, including the more advanced concepts. Some of the graph theory techniques from an algorithmic perspective are described in [1], including graph isomorphism. For a survey of minimal triangulations of graphs see [5].

## 3.10 Exercises

1. In the XVIII century, the City of Könisberg (in Prussia, currently part of Russia) was divided in 4 parts connected by 7 bridges. It is said that the residents tried to find a circuit along the entire city so that they cross every bridge only once. Euler transformed the problem to a graph (illustrated at the beginning of the chapter) and established the condition for a circuit in a connected graph that passes through each edge exactly once: all the vertices in the graph must have an even degree. Determine if the residents of Könisberg were able to find an Euler circuit.
2. Prove the condition established by Euler: a graph $G$ has a Euler circuit if and only if all the vertices in $G$ have an even degree.
3. What is the condition for a graph to have a Euler trajectory?
4. Given the graph in Fig. 3.10, determine if it has (a) A Euler circuit (b) A Euler trajectory (c) A Hamiltonian circuit (d) A Hamiltonian trajectory
5. For the tree in Fig. 3.9 list (a) all the parent–child relations, (b) all the brothers relations, (c) all the ascendant–descendant relations, and (d) all the subtrees with root a vertex in the graph.
6. Given the graph in Fig. 3.10, is it triangulated? If it is not triangulated, make it triangulated by applying the graph filling algorithm.
7. Prove that the number of vertices of odd degree in a graph is even.
8. Given the graph in Fig. 3.13, transform it to an undirected graph and order the nodes applying the maximum cardinality search algorithm.
9. Triangulate the graph of the previous problem.

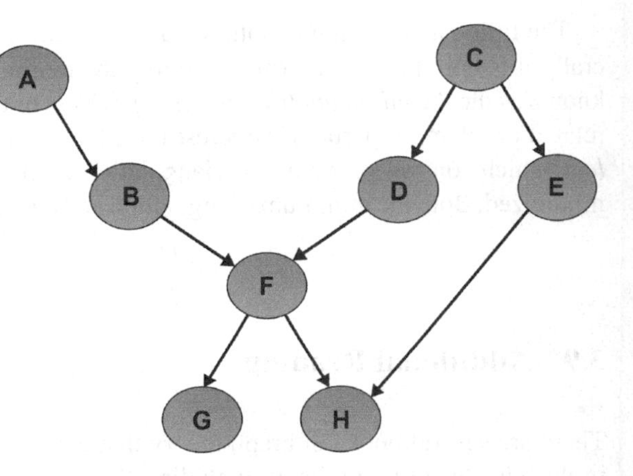


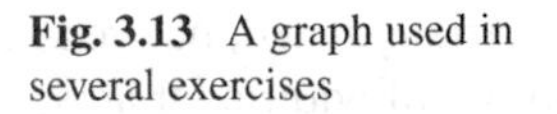
**Fig. 3.13** A graph used in several exercises

10. Given the triangulated graph obtained in the previous problem: (a) find its cliques, (b) order the cliques according to the node ordering and verify that they satisfy the running intersection property, (c) show the resulting tree (or trees) of cliques.
11. Try different orderings of the nodes in the graph in Fig. 3.13 by applying the maximum cardinality search algorithm. Triangulate the graphs resulting from the different orderings and find the cliques. Do the number of additional arcs vary according to the ordering? Do the size of the largest clique changes? Which ordering gives the least number of additional arcs and the smallest largest clique?
12. *** Develop an algorithm to verify if two graphs are isomorphic and try it on different pairs of graphs. What is the computational complexity of your algorithm?
13. *** Investigate alternative algorithms for the minimal triangulation of graphs. What is the computational complexity of the more efficient algorithms?
14. *** Develop a program for generating a tree of cliques given an undirected graph. For this consider (a) ordering the nodes according to maximum cardinality search, (b) triangulating the graph using the graph filling algorithm, (c) finding the cliques of the resulting graph and numbering them. Think of an adequate data structure to represent the graph considering the implementation of the previous algorithms.
15. *** Incorporate to the program of the previous exercise some heuristics for selecting the node ordering such that the size of the biggest clique in the graph is minimized (this is important for the junction tree inference algorithm for Bayesian networks).

# References

1. Aho, A.V., Hopcroft, J.E., Ullman, J.D.: The Design and Analysis of Computer Algorithms. Addison-Wesley, Boston (1974)
2. Golumbic, M.C.: Algorithmic Graph Theory and Perfect Graphs. Elsevier, Netherlands (1994)
3. Gould, R.: Graph Theory. Benjamin/Cummings, Menlo Park (1988)
4. Gross, J.L., Yellen, J.: Graph Theory and its Applications. CRC Press, Boca Raton (2005)
5. Heggernes, P.: Minimal triangulations of graphs: a survey. Discrete Math. **306**(3), 297–317 (2006)
6. Neapolitan, R.: Probabilistic Reasoning in Expert Systems: Theory and Algorithms. Wiley, New York (1990)

# Part II
# Probabilistic Models

The main types of probabilistic graphical models are presented, including Bayesian classifiers, hidden Markov models, Markov random fields, Bayesian networks, and dynamic and temporal Bayesian networks. A chapter is dedicated to each type of model, including representation, inference and learning; and also practical application examples. The models covered in Part II do not include decisions, those are covered in Part III.

# Chapter 4
# Bayesian Classifiers

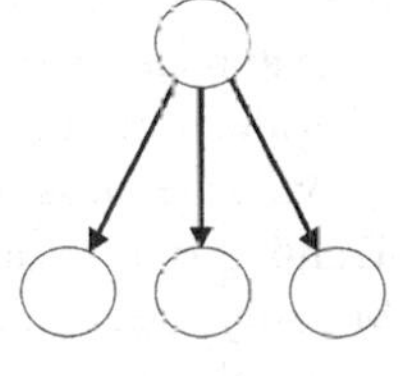

## 4.1 Introduction

Classification consists in assigning classes or labels to objects. There are two basic types of classification problems:

Unsupervised: in this case the classes are unknown, so the problem consists in dividing a set of objects into $n$ groups or clusters, so that a class is assigned to each different group. It is also known as *clustering*.

Supervised: the possible classes or labels are known a priori, and the problem consists in finding a function or rule that assigns one or more classes to each object.

In both cases the objects are usually described by a set of properties or attributes. In this chapter we will focus on supervised classification.

From a probabilistic perspective, the problem of supervised classification consists in assigning to a particular object described by its attributes, $A_1, A_2, \ldots, A_n$, one of $m$ classes, $C = \{c_1, c_2, \ldots, c_m\}$, such that the probability of the class given the attributes is maximized; that is:

$$Arg_C[Max P(C \mid A_1, A_2, \ldots, A_n)] \tag{4.1}$$



L. E. Sucar, *Probabilistic Graphical Models*, Advances in Computer Vision and Pattern Recognition, https://doi.org/10.1007/978-3-030-61943-5_4

If we denote the set of attributes as $\mathbf{A} = \{A_1, A_2, \ldots, A_n\}$, Eq. 4.1 can be written as: $Arg_C[Max P(C \mid \mathbf{A})]$.

There are many ways to build a classifier, including decision trees, rules, neural networks and support vector machines, among others.[1] In this book we will cover classification models based on probabilistic graphical models, in particular Bayesian classifiers. Before we describe the Bayesian classifier, we will present a brief introduction to how classifiers are evaluated.

### 4.1.1 Classifier Evaluation

To compare different classification techniques, we can evaluate a classifier in terms of several aspects. The aspects to evaluate, and the importance given to each aspect, depends on the final application of the classifiers. The main aspects to consider are:

Accuracy: it refers to how well a classifier predicts the correct class for unseen examples (that is, those not considered for learning the classifier).
Classification time: how long the classification process takes to predict the class, once the classifier has been trained.
Training time: how much time is required to learn the classifier from data.
Memory requirements: how much space in terms of memory is required to store the classifier parameters.
Clarity: if the classifier is easily understood by a person.

Usually the most important aspect is accuracy. This can be measured by predicting $N$ unseen data samples, and determining the percentage of correct predictions. Thus, the accuracy in percentage is:

$$Acc = (N_C/N) \times 100 \tag{4.2}$$

Where $N_C$ is the number of correct predictions.

A more detailed way to represent the performance of a classifier is by a *confusion matrix*. Each row of the matrix represents the instances for the predicted class while each column represents the instances for the actual class. In the case of binary classification (with classes positive and negative), there are four cells in the confusion matrix which represent:

- True positive (TP): number of instances of the positive class predicted as positive.
- True negative (TN): number of instances of the negative class predicted as negative.
- False positive (FP): number of instances of the negative class predicted as positive (Type I error).
- False negative (FN): number of instances of the positive class predicted as negative (Type II error).

---

[1]For an introduction and comparison of different types of classifiers we refer the interested reader to [11].

Table 4.1 Confusion matrix for binary classification

| | | Actual Class | |
|---|---|---|---|
| | | Positive | Negative |
| Predicted Class | Positive | TP | FP |
| | Negative | FN | TN |

The confusion matrix for binary classification is depicted in Table 4.1. This can be extended for $N$ classes, in that case it will be an $N \times N$ matrix.

According to the previous notation, the accuracy for a binary classifier is:

$$Acc = \frac{TP + TN}{P + N} = \frac{TP + TN}{TP + TN + FP + FN} \tag{4.3}$$

Where $P$ is the number of positive cases and $N$ the number of negative cases.

When comparing classifiers, in general we want to maximize the classification accuracy; however, this is only *optimal* if the cost of a wrong classification is the same for all the classes. Consider a classifier used for predicting breast cancer. If the classifier predicts that a person has cancer but this is not true (false positive), this might produce an unnecessary treatment. On the other hand, if the classifier predicts no cancer and the person does have breast cancer (false negative), this might cause a delay in the treatment and even death. Clearly the consequences of the several types of errors are different.

When there is imbalance in the costs of misclassification, we must then minimize the expected cost (EC). For two classes this is given by:

$$EC = FN \times P(-)C(- \mid +) + FP \times P(+)C(+ \mid -) \tag{4.4}$$

Where: $FN$ is the false negative rate, $FP$ is the false positive rate, $P(+)$ is the probability of positive, $P(-)$ is the probability of negative, $C(- \mid +)$ is the cost of classifying a positive as negative, and $C(+ \mid -)$ is the cost of classifying a negative as positive. Determining these costs may be difficult for some applications.

## 4.2 Bayesian Classifier

The formulation of the *Bayesian Classifier* is based on the application of the Bayes rule to estimate the probability of each class given the attributes:

$$P(C \mid A_1, A_2, \ldots, A_n) = P(C)P(A_1, A_2, \ldots, A_n \mid C)/P(A_1, A_2, \ldots, A_n) \tag{4.5}$$

Which can be written more compactly as:

$$P(C \mid \mathbf{A}) = P(C)P(\mathbf{A} \mid C)/P(\mathbf{A}) \tag{4.6}$$

The classification problem, based on Eq. 4.6, can be formulated as:

$$Arg_C[Max[P(C \mid \mathbf{A}) = P(C)P(\mathbf{A} \mid C)/P(\mathbf{A})]] \tag{4.7}$$

Equivalently, we can express Eq. 4.7 in terms of any function that varies monotonically with respect to $P(C \mid \mathbf{A})$, for instance:

- $Arg_C[Max[P(C)P(\mathbf{A} \mid C)]]$
- $Arg_C[Max[log(P(C)P(\mathbf{A} \mid C))]]$
- $Arg_C[Max[(log P(C) + log P(\mathbf{A} \mid C)]]$

Note that the probability of the attributes, $P(\mathbf{A})$, does not vary with respect to the class, so it can be considered as a constant for the maximization.

Based on the previous equivalent formulations for solving a classification problem we will require an estimate of $P(C)$, known as the prior probability of the classes, and $P(\mathbf{A} \mid C)$, known as the *likelihood*; $P(C \mid \mathbf{A})$ is the posterior probability. Therefore, to obtain the posterior probability of each class, we just need to multiply its prior probability by the likelihood which depends on the values of the attributes.[2]

The direct application of the Bayes rule results in a computationally expensive problem, as it was mentioned in Chap. 1. This is because the number of parameters in the likelihood term, $P(A_1, A_2, \ldots, A_n \mid C)$, increases exponentially with the number of attributes. This will not only imply a huge amount of memory to store all the parameters, but it will also be very difficult to estimate all the probabilities from the data or with the aid of a domain expert. Thus, the Bayesian classifier can only be of practical use for relatively *small* problems in terms of the number of attributes. An alternative is to consider some independence properties as in graphical models, in particular that all attributes are independent given the class, resulting in the *Naive Bayesian Classifier*.

### 4.2.1 Naive Bayesian Classifier

The naive or simple Bayesian classifier (NBC) is based on the assumption that all the attributes are independent given the class variable; that is, each attribute $A_i$ is conditionally independent of all the other attributes given the class: $P(A_i \mid A_j, C) = P(A_i \mid C), \forall j \neq i$. Under this assumption, Eq. 4.5 can be written as:

$$P(C \mid A_1, A_2, \ldots, A_n) = P(C)P(A_1 \mid C)P(A_2 \mid C) \ldots P(A_n \mid C)/P(\mathbf{A}) \tag{4.8}$$

where $P(\mathbf{A})$ can be considered, as mentioned before, a normalization constant.

[2]The posterior probabilities of the classes will be affected by a constant as we are not considering the denominator in Eq. 4.7, that is, they will not add to one; however, they can be easily *normalized* by dividing each one by the sum of the posterior probabilities for all the classes.

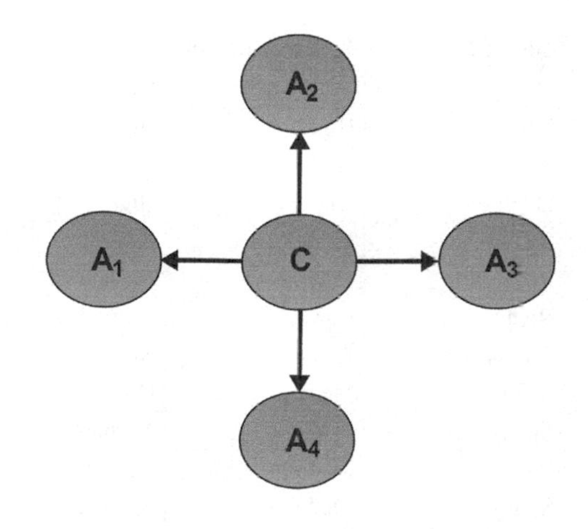


**Fig. 4.1** An example of a naive Bayesian classifier with the class variable, $C$, and four attributes, $A_1, \ldots, A_4$

The naive Bayes formulation drastically reduces the complexity of the Bayesian classifier, as in this case we only require the prior probability (one dimensional vector) of the class, and the $n$ conditional probabilities of each attribute given the class (two dimensional matrices) as the parameters for the model. That is, the space requirement is reduced from exponential to linear in the number of attributes. Also, the calculation of the posterior is greatly simplified, as to estimate it (unnormalized) only $n$ multiplications are required.

A graphical representation of the naive Bayesian classifier is shown in Fig. 4.1. This star like structure depicts the property of conditional independence between all the attributes given the class—as there are not arcs between the attribute nodes.

Learning a NBC consists in estimating the prior probability of the class, $P(C)$, and the conditional probability table (CPT) of each attribute given the class, $P(A_i \mid C)$. These can be obtained via subjective estimates from an expert or from the data by maximum likelihood.[3]

The probabilities can be estimated from data using, for instance, maximum likelihood estimation. The prior probabilities of the class variable, $C$, are given by:

$$P(c_i) \sim N_i/N \tag{4.9}$$

Where $N_i$ is the number of times $c_i$ occurs in the $N$ samples.

The conditional probabilities of each attribute, $A_j$ can be estimated as:

$$P(A_{jk} \mid c_i) \sim N_{jki}/N_i \tag{4.10}$$

Where $N_{jki}$ is the number of times that the attribute $A_j$ takes the value $k$ and it is from class $i$, and $N_i$ is the number of samples of class $c_i$ in the data set.

[3] We will cover parameter estimation in detail in the chapter on Bayesian Networks.

**Table 4.2** Data set for the golf example

| Outlook | Temperature | Humidity | Windy | Play |
|---|---|---|---|---|
| Sunny | High | High | False | No |
| Sunny | High | High | True | No |
| Overcast | High | High | False | Yes |
| Rain | Medium | High | False | Yes |
| Rain | Low | Normal | False | Yes |
| Rain | Low | Normal | True | No |
| Overcast | Low | Normal | True | Yes |
| Sunny | Medium | High | False | No |
| Sunny | Low | Normal | False | Yes |
| Rain | Medium | Normal | False | Yes |
| Sunny | Medium | Normal | True | Yes |
| Overcast | Medium | High | True | Yes |
| Overcast | High | Normal | False | Yes |
| Rain | Medium | High | True | No |

Once the parameters have been estimated, the posterior probability can be obtained just by multiplying the prior by the likelihood for each attribute. Thus, given the values for $m$ attributes, $a_1, \ldots a_m$, for each class $c_i$, the posterior is proportional to:

$$P(c_i \mid a_1, \ldots a_m) \sim P(c_i)P(a_1 \mid c_i) \cdots P(a_m \mid c_i) \tag{4.11}$$

The class $c_k$ that maximizes the previous equation will be selected.[4]

Returning to the *golf* example, Table 4.2 depicts 14 records with 5 variables: four attributes (Outlook, Temperature, Humidity, Windy) and one class (Play). A NBC for the golf example is depicted in Fig. 4.2, including some of the required probability tables.

In summary, the main advantages of the naive Bayesian classifier are:

- The low number of required parameters, which reduces the memory requirements and facilitates learning them from data.
- The low computational cost for inference (estimating the posteriors) and learning.
- The relatively good performance (classification precision) in many domains.
- A simple and intuitive model.

While its main limitations are the following:

- In some domains, the performance is reduced given that the conditional independence assumption is not valid.

[4]This assumes that the misclassification cost is the same for all classes; if these costs are not the same, the class that minimizes the misclassification cost should be selected.

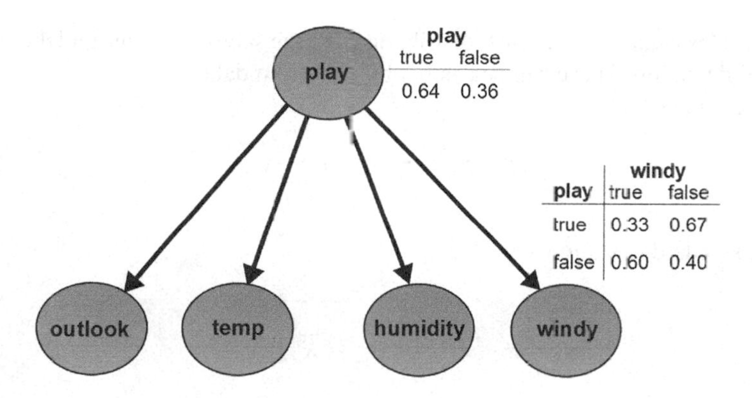


**Fig. 4.2** A naive Bayesian classifier for the golf example. Two of the required probability tables are shown: $P(play)$ and $P(windy \mid play)$. The conditional probabilities of the other three attributes given the class will also be required to complete the parameters of this classifier

- If there are continuous attributes, these need to be discretized (or consider alternative models such as the Gaussian naive Bayes).

In the next section we introduce the Gaussian naive Bayesian classifier. Then we will cover other models that consider certain dependencies between attributes. Later we will describe techniques that can eliminate or join attributes, as another way to overcome the conditional independence assumption.

## 4.3 Gaussian Naive Bayes

In this case, the attributes are continuous random variables, and the class is discrete. Thus, each attribute is modeled by some continuous probability distribution over the range of its values. A common assumption is that within each class, the values of numeric attributes are normally distributed. That is:

$$P(A_i \mid C = c_j) = (1/\sqrt{2\pi\sigma})e^{-\frac{(a_i-\mu)^2}{2c^2}} \tag{4.12}$$

Where $\mu$ and $\sigma$ are the mean and standard deviation, respectively, of the distribution of attribute $A_i$ given class $c_j$. So for each attribute given each class value, the Gaussian naive Bayes (GNB) requires to represent such a distribution in terms of its mean and standard deviation. Additionally, as in the naive Bayes, the prior probability of each class is required.

Learning a GNB consists in estimating the prior probabilities of the classes, and the mean and standard deviation of each attribute given the class. The prior of the class is obtained as for the naive Bayes via maximum likelihood estimation. The maximum likelihood estimates of the mean and standard deviation of each attribute

given the class; that is of a normal distribution, are the sample average and the sample standard deviation. These can be easily obtained from data.
Sample average:

$$\mu_X \sim \overline{X} = (1/N)\sum_{i=1}^{N} x_i \tag{4.13}$$

Sample standard deviation:

$$\sigma_X \sim S_X = \sqrt{\sum_{i=1}^{N}(x_i - \overline{X})/(N-1)} \tag{4.14}$$

In the inference stage, for estimating the posterior probabilities of each class given the values of the attributes, we do the same as for the naive Bayes, multiply the prior probability of the class by the probabilities of each attribute given the class:

$$P(c_i \mid a_1, \ldots a_m) \sim P(c_i)P(a_1 \mid c_i) \cdots P(a_m \mid c_i) \tag{4.15}$$

Where $P(a_j \mid c_i)$ is obtained by calculating it from the normal distribution equation according to the value of $a_j$ and the corresponding parameters (mean and standard deviation).

Note that with this formulation we can easily mix discrete and continuous attributes, estimating the CPTs for the discrete attributes and the mean and standard deviation for the continues attributes (assuming Gaussian distributions); and then simply applying the naive Bayes equation for estimating the posterior probabilities of each class.

## 4.4 Alternative Models: TAN, BAN

The general Bayesian classifier and the naive Bayesian classifier are the two extremes of possible dependency structures for Bayesian classifiers; the former represents the most complex structure with no independence assumptions, while the latter is the simplest structure that assumes that all the attributes are independent given the class. Between these two extremes there is a wide variety of possible models of varying complexities. Two interesting alternatives are the TAN and BAN classifiers.

The *Tree augmented Bayesian Classifier*, or TAN, incorporates some dependencies between the attributes by building a directed tree among the attribute variables. That is, the $n$ attributes form a graph which is restricted to a directed tree that represents the dependency relations between the attributes. Additionally there is an arc between the class variables and each attribute. The structure of a TAN classifier is depicted in Fig. 4.3.

If we take out the limitation of a tree structure between attributes, we obtain the *Bayesian Network augmented Bayesian Classifier*, or BAN, which considers that

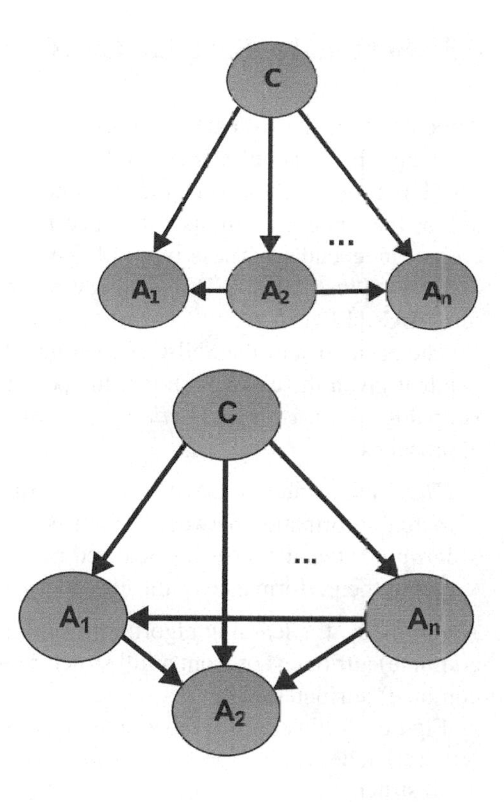


**Fig. 4.3** An example of a TAN classifier

**Fig. 4.4** An example of a BAN classifier

the dependency structure among the attributes constitutes a directed acyclic graph (DAG). As with the TAN classifier, there is a directed arc between the class node and each attribute. The structure of a BAN classifier is depicted in Fig. 4.4.

The posterior probability for the class variable given the attributes can be obtained in a similar way as with the NBC; however, now each attribute not only depends on the class but also on other attributes according to the structure of the graph. Thus, we need to consider the conditional probability of each attribute given the class and its *parent* attributes:

$$P(C \mid A_1, A_2, \ldots, A_n) = P(C)P(A_1 \mid C, Pa(A_1))P(A_2 \mid C, Pa(A_2)) \ldots P(A_n \mid C, Pa(A_n))/P(\mathbf{A}) \tag{4.16}$$

where $Pa(A_i)$ is the set of parent attributes of $A_i$ according to the attribute dependency structure of the TAN or BAN classifier.

The TAN and BAN classifiers can be considered as particular cases of a more general model, that is, Bayesian networks, which will be covered in Chap. 7. In Chaps. 7 and 8, we will cover different techniques for inference and for learning Bayesian networks, which can be applied to obtain the posterior probabilities (inference) and the model (learning) for the TAN and BAN classifiers.

## 4.5 Semi-naive Bayesian Classifiers

Another alternative to deal with dependent attributes is to transform the basic structure of a naive Bayesian classifier, while maintaining a star or tree-structured network. This has as an advantage that the efficiency and simplicity of the NBC is maintained, and at the same time the performance is improved for cases where the attributes are not independent. These types of Bayesian classifiers are known as *Semi-Naive Bayesian Classifiers* (SNBC), and several authors have proposed different variants of SNBCs [12, 19].

The basic idea of the SNBC is to eliminate or *join* attributes which are not independent given the class, such that the performance of the classifier improves. This is analogous to *feature selection* in machine learning, and there are two types of approaches:

*Filter:* the attributes are selected according to a local measure, for instance the mutual information between the attribute and the class.
*Wrapper:* the attributes are selected based on a global measure, usually by comparing the performance of the classifier with and without the attribute.

Additionally, the learning algorithm can start from an empty structure and add (or combine) attributes; or from a full structure with all the attributes, and eliminate (or combine) attributes.

Figure 4.5 illustrates the two alternative operations to modify the structure of a NBC: (i) node elimination, and (ii) node combination, considering that we start from a full structure.

Node elimination consists in simply eliminating an attribute, $A_i$, from the model, this could be because it is not relevant for the class ($A_i$ and $C$ are independent); or because the attribute $A_i$ and another attribute, $A_j$, are not independent given the class (which is a basic assumption of the NBC). The rationale for eliminating one of the dependent attributes is that if the attributes are not independent given the class, one of them is redundant and could be eliminated.

Node combination consists in merging two attributes, $A_i$ and $A_j$, into a new attribute $A_k$, such that $A_k$ has as possible values the cross product of the values of $A_i$ and $A_j$ (assuming discrete attributes). For example, if $A_i = a, b, c$ and $A_j = 1, 2$, then $A_k = a1, a2, b1, b2, c1, c2$. This is an alternative when two attributes are not independent given the class. By merging them into a single combined attribute, the independence condition is not longer relevant.

Thus, when two attributes are not independent given the class there are two alternatives: eliminate one or combine them into a single variable; in principle we should select the alternative that implies a higher improvement in the performance of the classifier, although in practice, this could be difficult to evaluate.

As mentioned before, there are several alternatives for learning a SNBC. A simple greedy scheme is outlined in Algorithm 4.1, where we start from a full NBC with all the attributes [10].

This process is repeated until there are no more superfluous or dependent attributes.

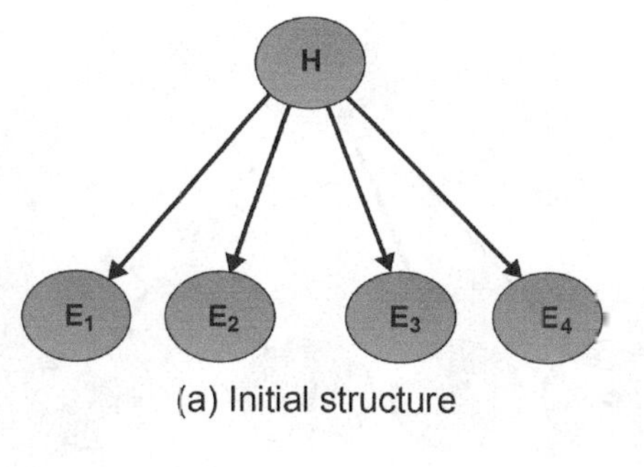


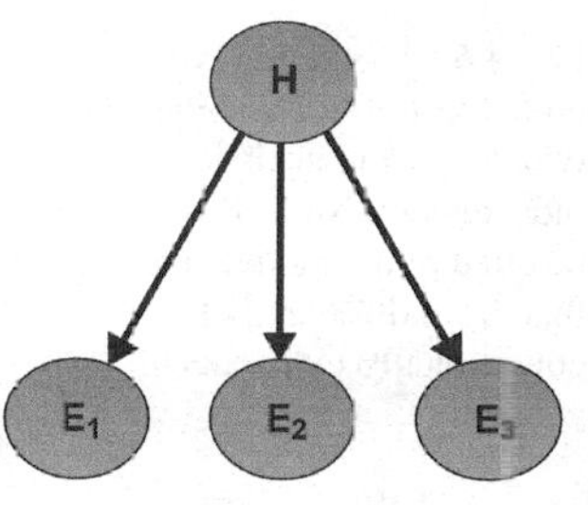


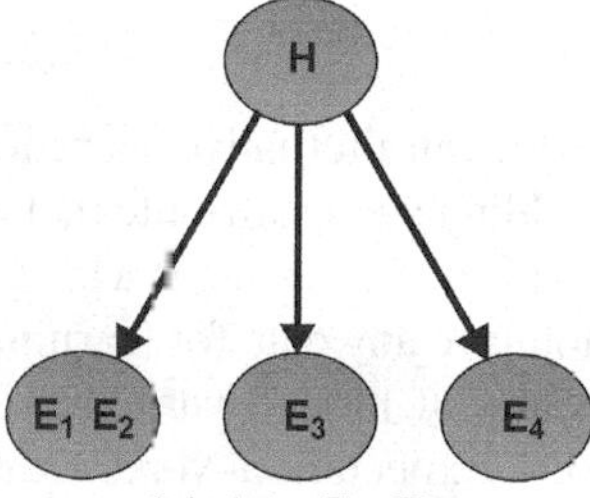


**Fig. 4.5** Structural improvement: **a** initial structure, **b** one attribute is eliminated, **c** two attributes are joined into one variable

---

**Algorithm 4.1** Structural Improvement Algorithm

---

**Require:** $A$, the attributes
**Require:** $C$, the class
/* The dependency between each attribute and the class is estimated (for instance using mutual information).*/
**for all** $a \in$ **A do**
  /* Those attributes below a threshold are eliminated.*/
  **if** $MI(a, C) < \varepsilon$ **then**
    Eliminate $a$
  **end if**
**end for**
/* The remaining attributes are tested to see if they are independent given the class, for example, using conditional mutual information (CMI).*/
**for all** $a \in$ **A do**
  **for all** $b \in$ **A** $- a$ **do**
    /* Those attributes above a threshold are eliminated or combined based on which option gives the best classification performance.*/
    **if** $CMI(a, b|C) > \omega$ **then**
      Eliminate or Combine $a$ and $b$
    **end if**
  **end for**
**end for**

---

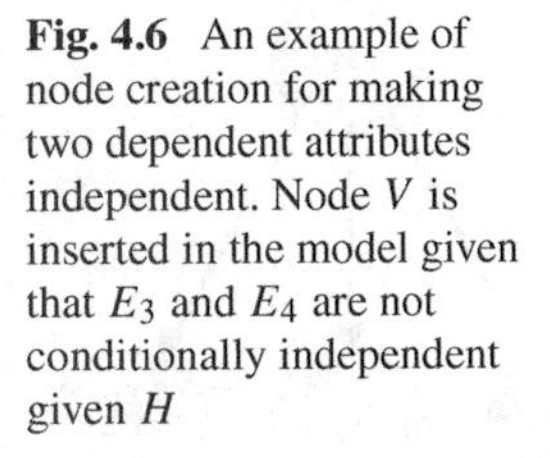

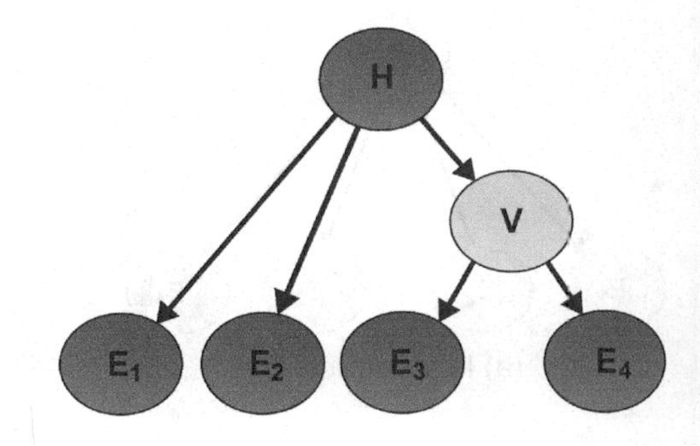


**Fig. 4.6** An example of node creation for making two dependent attributes independent. Node $V$ is inserted in the model given that $E_3$ and $E_4$ are not conditionally independent given $H$

References [9, 19] introduce an alternative operation for modifying the structure of a NBC, which consists in adding a new attribute that makes two dependent attributes independent; see Fig. 4.6. This new attribute is a kind of virtual or hidden node in the model, for which we do not have any data for learning its parameters. An alternative for estimating the parameters of hidden variables in Bayesian networks, such as in this case, is based on the Expectation-Maximization (EM) procedure, and it is described in Chap. 8.

All previous classifiers consider that there is a single class variable; that is, each instance belongs to one and only one class. Next we consider that an instance can belong to several classes, which is known as *multidimensional* classification.

## 4.6 Multidimensional Bayesian Classifiers

Several important problems need to predict several classes simultaneously. For example: text classification, where a document can be assigned to several topics; gene classification, as a gene may have different functions; image annotation, as an image may include several objects; among others. These are examples of *multi-dimensional classification*, in which more than one class can be assigned to an object. Formally, the *multi-dimensional classification* problem corresponds to searching for a function $h$ that assigns to each instance represented by a vector of $m$ features $\mathbf{X} = (X_1, \ldots, X_m)$ a vector of $d$ class values $\mathbf{C} = (C_1, \ldots, C_d)$. The $h$ function should assign to each instance $\mathbf{X}$ the most likely combination of classes, that is:

$$ArgMax_{c_1,\ldots,c_d} P(C_1 = c_1, \ldots, C_d = c_d|\mathbf{X}) \tag{4.17}$$

*Multi-label* classification is a particular case of multi-dimensional classification, where all class variables are binary. In the case of multi-label classification, there are two basic approaches: *binary relevance* and *label power-set* [21]. Binary relevance approaches transform the multi-label classification problem into $d$ independent binary classification problems, one for each class variable, $C_1, \ldots, C_d$. A classifier is independently learned for each class and the results are combined to determine the

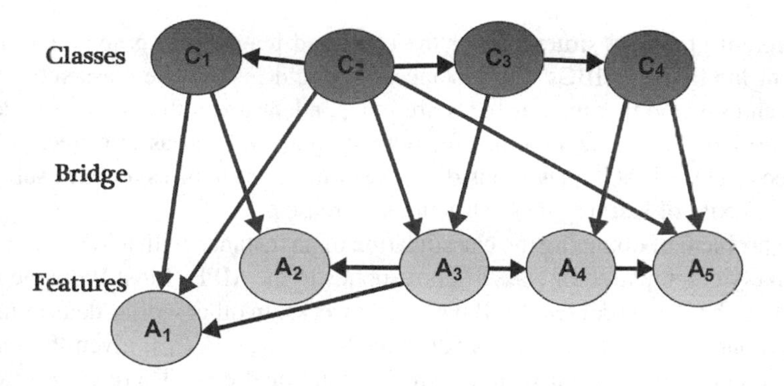


**Fig. 4.7** A multidimensional Bayesian network classifier, showing the three subgraphs: classes, features and bridge

predicted class set; the dependencies between classes are not considered. The label power-set approach transforms the multi-label classification problem into a single-class scenario by defining a new compound class variable whose possible values are all the possible combinations of values of the original classes. In this case the interactions between classes are implicitly considered. It can be an effective approach for domains with only a few class variables; however for many classes this approach is impractical. Essentially, binary relevance can be effective when the classes are relatively independent, and label power-set when there are few class variables.

Under the framework of Bayesian classifiers, we can consider two alternatives to the basic approaches. One is based on Bayesian networks, in which the dependencies between class variables (and also between attributes) are explicitly considered. The other implicitly incorporates the dependencies between classes by adding additional attributes to each independent classifier. Both are described in the following sections.

### *4.6.1 Multidimensional Bayesian Network Classifiers*

A multidimensional Bayesian network classifier (MBC) over a set $\mathbf{V} = \{Z_1, \ldots, Z_n\}$, $n > 1$, of discrete random variables is a Bayesian network with a particular structure.[5] The set $\mathbf{V}$ of variables is partitioned into two sets $\mathbf{V_C} = \{C_1, \ldots, C_d\}, d > 1$, of class variables and $\mathbf{V_X} = \{X_1, \ldots, X_m\}, m \geq 1$, of feature variables $(d + m = n)$. The set $\mathbf{A}$ of arcs is also partitioned into three sets, $\mathbf{A_C}, \mathbf{A_X}, \mathbf{A_{CX}}$, such that $\mathbf{A_C} \subseteq \mathbf{V_C} \times \mathbf{V_C}$ is composed of the arcs between the class variables, $\mathbf{A_X} \subseteq \mathbf{V_X} \times \mathbf{V_X}$ is composed of the arcs between the feature variables, and finally, $\mathbf{A_{CX}} \subseteq \mathbf{V}_C \times \mathbf{V_X}$ is composed of the arcs from the class variables to the feature variables. The corresponding induced sub-graphs are $\mathbf{G_C} = (\mathbf{V_C}, \mathbf{A_C})$, $\mathbf{G_X} = (\mathbf{V_X}, \mathbf{A_X})$ and $\mathbf{G_{CX}} = (\mathbf{V}, \mathbf{A_{CX}})$, called respectively *class*, *feature* and *bridge* sub-graphs (see Fig. 4.7).

[5]Bayesian networks are introduced in Chap. 7.

Different graphical structures for the class and feature sub-graphs may lead to different families of MBCs. For instance, we could restrict the class sub-graph to a tree, and assume that the attributes are independent given the class variables. Or, we could have the same structure for both subgraphs, such as tree-tree, polytree-polytree or DAG-DAG. As we consider more complex structures for each sub-graph, the complexity of learning these structures increases.

The problem of obtaining the classification of an instance with a MBC, that is, the most likely combination of classes, corresponds to the MPE (Most Probable Explanation) or *abduction* problem for Bayesian networks. In other words, determining the most probable values for the class variables $\mathbf{V} = \{C_1, \ldots, C_n\}$, given the features. This is a complex problem with a high computational cost. There are a few ways to try to reduce the time complexity [3], however this approach is still limited to problems with only a limited number of classes.

We will describe how to learn Bayesian networks (MBCs are a particular type of Bayesian networks) and the calculation of the MPE in Chaps. 7 and 8.

### 4.6.2 Chain Classifiers

Chain classifiers are an alternative method for multi-label classification that incorporate class dependencies, while keeping the computational efficiency of the binary relevance approach [13]. A chain classifier consists of $d$ base binary classifiers which are linked in a chain, such that each classifier incorporates the predicted classes by the previous classifiers as additional attributes. Thus, the feature vector for each binary classifier, $L_i$, is extended with the labels (0/1) from all previous classifiers in the chain. Each classifier in the chain is trained to learn the association of label $l_i$ given the features augmented with all previous class labels in the chain, $L_1, L_2, \ldots, L_{i-1}$. For classification, it starts at $L_1$, and propagates the predicted classes along the chain, such that for $L_i \in \mathscr{L}$ (where $\mathscr{L} = \{L_1, L_2, \ldots, L_d\}$) it predicts $P(L_i \mid \mathbf{X}, L_1, L_2, \ldots, L_{i-1})$. As in the binary relevance approach, the class vector is determined by combining the outputs of all the binary classifiers in the chain.

A challenge for chain classifiers is to select the order of the classes in the chain; as the order can affect the performance of the classifier. Next we describe two alternatives for this challenge. The first one, *circular chain classifiers*, which makes that the order does not affect the result. The second one, *Bayesian chain classifiers*, proposes a way to determine a *good* order based on the dependencies between the class variables.

#### 4.6.2.1 Circular Chain Classifiers

In a Circular Chain Classifiers (CCC) the propagation of the classes of the previous binary classifiers is done iteratively in a circular way. In the first cycle, as in the standard chain classifiers, the predictions of the previous classifiers are additional attributes for each classifier in the chain; that is $P(L_2 \mid \mathbf{X}, L_1)$, $P(L_3 \mid \mathbf{X}, L_1, L_2)$, and so on. After the first cycle, the predictions from all other base base classifiers (from the previous cycle) are entered as additional attributes to the first one in the chain, so $P(L_1 \mid \mathbf{X}, L_2, L_3, \ldots, L_d)$. This process continues for all the classifiers in the chain, so that from the second cycle all classifiers receive as additional attributes the predictions of all other classifiers: $P(L_i \mid \mathbf{X}, L_1, L_2, L_{i-1}, L_{i+1} \ldots, L_d)$.

This process is repeated for a prefixed number of cycles or until convergence. Convergence is reached when the posterior probabilities of all classifiers in the chain do not change (above a certain threshold) from one iteration to the next. It has been shown empirically that the CCCs tend to converge in a few iterations and that the order of the classes in the chain does not affect the performance according to several metrics [15].

#### 4.6.2.2 Bayesian Chain Classifiers

*Bayesian chain classifiers* are a type of chain classifier under a probabilistic framework. If we apply the chain rule of probability theory, we can rewrite Eq. (4.17)

$$ArgMax_{C_1,\ldots,C_d} P(C_1|C_2, \ldots, C_d, \mathbf{X}) P(C_2|C_3, \ldots, C_d, \mathbf{X}) \ldots P(C_d|\mathbf{X}) \quad (4.18)$$

If we consider the dependency relations between the class variables, and represent these relations as a directed acyclic graph (DAG), then we can simplify Eq. (4.18) by considering the independencies implied in the graph so that only the *parents* of each class variable are included in the chain, and all other *previous* classes according to the chain order are eliminated. We can rewrite Eq. (4.18) as:

$$ArgMax_{C_1,\ldots,C_d} \prod_{i=1}^{d} P(C_i|\mathbf{Pa}(C_i), \mathbf{X}) \quad (4.19)$$

where $\mathbf{Pa}(C_i)$ are the parents of class $i$ in the DAG that represents the dependencies between the class variables.

Next we make a further simplification by assuming that the most probable joint combination of classes can be approximated by simply concatenating the individual most probable classes. That is, we solve the following set of equations as an approximation of Eq. (4.19):

$$ArgMax_{C_1} P(C_1|\mathbf{Pa}(C_1), \mathbf{X})$$
$$ArgMax_{C_2} P(C_2|\mathbf{Pa}(C_2), \mathbf{X})$$
$$\ldots\ldots\ldots\ldots\ldots\ldots$$
$$ArgMax_{C_d} P(C_d|\mathbf{Pa}(C_d), \mathbf{X})$$

This last approximation corresponds to a Bayesian chain classifier. Thus, a BCC makes two basic assumptions:

1. The class dependency structure given the features can be represented by a DAG.
2. The most probable joint combination of class assignments (total abduction) is approximated by the concatenation of the most probable individual classes.

The first assumption is reasonable if we have enough data to obtain a good approximation of the class dependency structure, and assuming that this is obtained conditioned on the features. With respect to the second assumption, it is well known that the total abduction or most probable explanation is not always equivalent to the maximization of the individual classes. However, the assumption is less strong than that assumed by the binary relevance approach. Bayesian chain classifiers provide an attractive alternative to multi-dimensional classification, as they incorporate in certain ways the dependencies between class variables, and they keep the efficiency of the Binary relevance approach.

For the *base* classifier that belongs to each class we can use any of the Bayesian classifiers presented in the previous sections, for instance a NBC. Assuming that we have a class dependency structure represented as a DAG (this structure can be learned from data, see Chap. 8), each classifier can be learned in a similar way as a NBC, by simply including as additional attributes the class variables according to the class dependency structure. The simplest option is to only include the parent nodes of each class according to this dependency graph. The general idea for building a BCC is illustrated in Fig. 4.8.

For classifying an instance, all the classifiers are applied simultaneously, obtaining the posterior probability of each class.

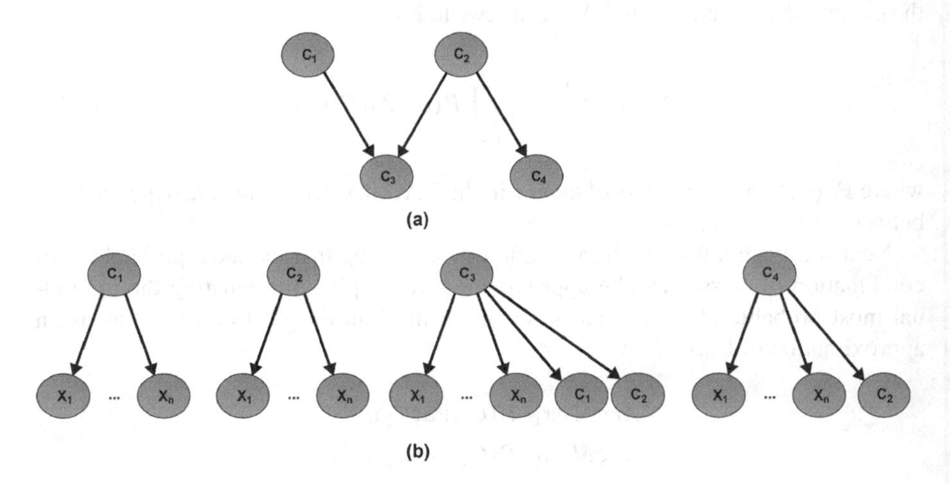


**Fig. 4.8** An example of a BCC. **a** A BN that represents the class dependency structure. **b** Naive Bayesian classifiers, one for each class. Each base classifier defined for $C_i$ includes the set of attributes, $X_1, \ldots, X_n$, plus its parents in the dependency structure as additional attributes

## 4.7 Hierarchical Classification

Hierarchical classification is a type of multi-label classification in which the classes are ordered in a predefined structure, typically a tree, or in general a directed acyclic graph (DAG). By taking into account the hierarchical organization of the classes, the classification performance can be improved. In hierarchical classification, an example that belongs to certain class automatically belongs to all its superclasses; this is known as the *hierarchical constraint*. Hierarchical classification has application in several areas, such as text categorization, protein function prediction, and object recognition.

As in the case of multidimensional classification, there are two basic approaches for hierarchical classification: global classifiers and local classifiers. The global approach builds a classifier to predicts all the classes at once; this becomes too complex computationally for large hierarchies. The local classifiers schemes train several classifiers and combine their outputs. There are three basic approaches. A Local Classifier per hierarchy Level, that trains one multi-class classifier for each level of the class hierarchy. Local Classifier per Node, in which a binary classifier is built for each node (class) in the hierarchy, except the root node. Local Classifier per Parent Node (LCPN), where a multi-class classifier is trained to predict its child nodes.

Local methods commonly use a top-down approach for classification [17]; the classifier at the top level selects certain class, so the other classes are discarded, and then the successors of the selected class are analyzed, and so on. This has the problem that if there is an error at the upper levels of the hierarchy, this can not be recovered and it propagates down to the other levels, known as *error propagation*. An important restriction for hierarchical classification is the *Hierarchical Constraint*: if an instance $z$ is associated to the class $C_i$, the instance $z$ must be associated to all the ancestors of $C_i$.

Next we describe two recent approaches for hierarchical classification. In the first one, to avoid error propagation an alternative is to analyze the paths in the hierarchy, and select the *best* path according to the results of the local classifiers. In the second one, for maintaining the hierarchical constraint, the hierarchal structure is transformed into a Bayesian network.

### *4.7.1 Chained Path Evaluation*

Chained Path Evaluation (CPE) [14] analyzes each possible path from the root to a leaf node in the hierarchy, taking into account the level of the predicted labels to give a score to each path and finally return the one with the best score. Additionally, it considers the relations of each node with its ancestors in the hierarchy, based on chain classifiers. CPE consists of two parts, training and classification.

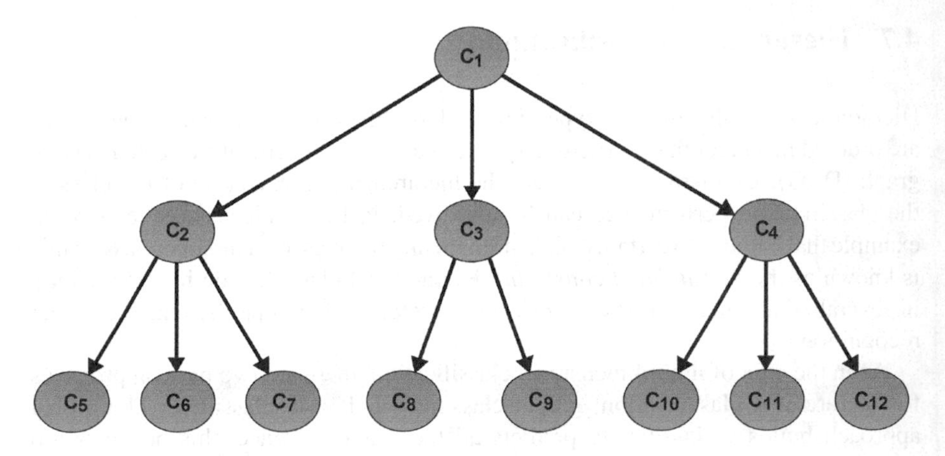


**Fig. 4.9** Example of a hierarchical structure (a tree). For each non-leaf node, a local classifier is trained to predict its child nodes: $C_1$ classifies $C_2, C_3, C_4$; $C_2$, classifies $C_5, C_6, C_7$; and similarly for $C_3$ and $C_4$

#### 4.7.1.1 Training

A local classifier is trained for each node, $C_i$, in the hierarchy, except for the leaf nodes, to classify its child nodes; that is, using the LCPN scheme, see Fig. 4.9. The classifier for each node, $C_i$, for instance a naive Bayesian classifier, is trained considering examples from all it child nodes, as well as some examples of it sibling nodes in the hierarchy. For instance, the classifier $C_2$ in Fig. 4.9 will be trained to classify $C_5, C_6, C_7$; additional examples will be considered from the classes $C_3$ and $C_4$, which represent an *unknown* class for $C_2$.

To consider the relation with other nodes in the hierarchy, the predicted class by the parent (tree structure) or parents (DAG), are included as additional attributes in each local classifier, inspired by Bayesian chain classifiers. For example, the classifier $C_2$ in Fig. 4.9 will have as an additional attribute the predicted class by its parent, $C_1$.

#### 4.7.1.2 Classification

In the classification phase, the probabilities for each class from all local classifiers are obtained based on the input data (features of each instance). After computing the probabilities for each node in the hierarchy, these are combined to obtain a score for each path.

The *score* for each path in the hierarchy is calculated by a weighted sum of the log of the probabilities of all local classifiers in the path:

$$score = \sum_{i=0}^{n} w_{C_i} \times \log(P(C_i|X, Pa(C_i))) \tag{4.20}$$

where $C_i$ are the classes for each LCPN, $X$ is the vector of attributes, $Pa(C_i)$ is the parent predicted class, and $w_{C_i}$ is a weight. The purpose of these weights is to give more importance to the upper levels of the hierarchy, as errors at the upper hierarchy levels (which correspond to more generic concepts) are more *expensive* than those at the lower levels (which correspond to more specific concepts) [14]. Taking the sum of logarithms is used to ensure numerical stability when computing the probability for long paths.

Once the scores for all the paths are obtained, the path with the highest score will be selected as the set of predicted classes for the corresponding instance. For the example in Fig. 4.9, the score will be calculated for each path from the root to each leaf node: Path 1: $C_1 - C_2 - C_5$, Path 2: $C_1 - C_2 - C_6$, etc. In this case there are 8 paths. Suppose that the path with the highest score is Path 4: $C_1 - C_3 - C_8$; then these 3 classes will be returned as the output of the classifier.

### 4.7.2 Hierarchical Classification with Bayesian Networks

The method consists of two levels [2, 16]. In the first level, the hierarchical structure is represented as a Bayesian Network, which represents the data distribution in the nodes while maintaining the hierarchical constraint. In the second level, a local binary classifier is trained for each class in the hierarchy. The hierarchical structure (tree or DAG) is mapped to a Bayesian network (BN), so each node ($y_i$) in the BN corresponds to a class in the hierarchy, and the arcs in the BN correspond to the class–subclass relations in the hierarchy, see Fig. 4.10. An additional node ($q_i$) for each node (except the root) is added to the BN, which represents a local binary classifier for the corresponding node (class). The $y_i$ nodes will have the probability that a new instance is associated to the class $C_i$, while the $q_i$ nodes receive the prediction of the classifiers. The idea is that an initial prediction for each class is obtained via the

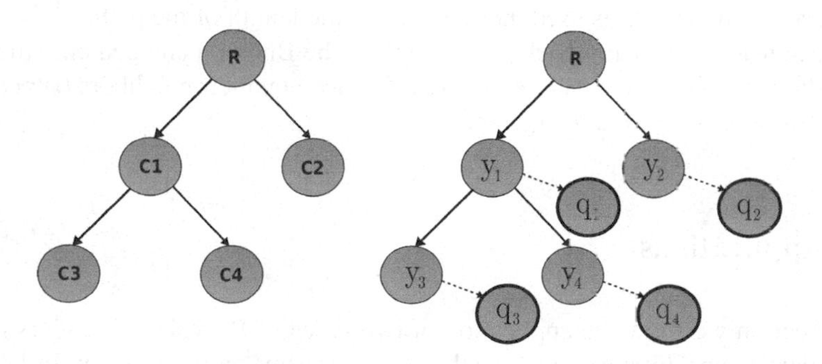


**Fig. 4.10** Example of a hierarchical structure transformed to a Bayesian network. Left: original hierarchical class structure where each node corresponds to a class. Right: Bayesian network, the nodes $y_i$ correspond to each class, and the nodes $q_i$ correspond to the binary classifiers for each class

local classifiers, and then these are *propagated* in the BN which helps to maintain the hierarchical constraint.

#### 4.7.2.1 Training

Once the BN is built, its parameters must be estimated. There are two set of parameters. The first set consists of the probabilities associated to the hierarchy, that is $P(y_i \mid Pa(y_i))$, where $Pa(y_i)$ represents the set of the parents of $y_i$; that is the superclasses of each class in the hierarchy (in the case of a tree hierarchy there will be one parent per node, however there could be multiple parents for DAG hierarchies). These probabilities can be estimate by frequency from the training set.

The second set corresponds to the probabilities of the local classifiers per class, that is $P(q_i \mid y_i)$, that represent how "good" is each local classifier for predicting the corresponding class. These can be estimated using the confusion matrices for each local classifier over validation data.

Once these two sets of parameters are estimated, usually from training data, the BN is completed and we can proceed to the classification stage.

#### 4.7.2.2 Classification

Classification involves three parts. First, an initial probability of each class is estimated based on a set of attributes using the local classifiers, and fed to the BN. Second, once the estimations are received from each local classifier, these are combined by probability propagation in the BN, after which the posterior probabilities for each class (node) are obtained. Third, similarly to the previous method (chained path evaluation), the score for each path in the hierarchy is estimated, and the path with the highest score is selected as the output of the hierarchical classifier. There are different alternatives to score each path, in this case the sum of the probabilities of the classes in the path is used, normalized by the length of the path.

An extension to this approach [16] combines the BN with chained classifiers, so the predictions of each local classifier are influenced by their neighbors (superclass, subclasses or ancestors) in the hierarchy.

## 4.8 Applications

In this section we show the application of two types of Bayesian classifiers in two practical problems. First we illustrate the use of the semi-naive classifier for labeling pixels in an image as skin or not skin. Then we demonstrate the use of multidimensional chain classifiers for HIV drug selection.

### 4.8.1 Visual Skin Detection

Skin detection is a useful pre-processing stage for many applications in computer vision, such as person detection and gesture recognition, among others. Thus it is critical to have a very accurate and efficient classifier. A simple and very fast way to obtain an approximate classification of pixels in an image as *skin* or *not–skin* is based on the color attributes of each pixel. Usually, pixels in a digital image are represented as the combination of three basic (primary) colors: Red (R), Green (G) and Blue (B), in what is known as the $RGB$ model. Each color component can take numeric values in a certain interval, i.e. $0 \dots 255$. There are alternative color models, such as $HSV$, $YIQ$, etc.

A NBC can be built to classify pixels as skin or not-skin using the 3 color values—RGB—as attributes. However, it is possible for a different color model to produce a better classification. Alternatively, we can combine several color models into a single classifier, having as attributes all the attributes from the different models. This last option has the potential of taking advantage of the information provided by different models; however, if we use a NBC the independence assumption is violated—the different models are not independent, one model could be derived from another.

An alternative is to consider a semi-naive Bayesian classifier and *select* the best attributes from the different color models for skin classification by eliminating or joining attributes. For this we used three different color models: RGB, HSV and YIQ, so there are 9 attributes in total. The attributes (color components) were previously discretized into a reduced number of intervals. Then an initial NBC was learned based on data—examples of skin and not-skin pixels taken from several images. This initial classifier obtained a 94% accuracy when applied to other (test) images.

The classifier was then *optimized* using the method described in Sect. 4.5. Starting from the *full* NBC with 9 attributes, the method applies the variable elimination and combination stages until the *simplest* classifier with maximum accuracy is obtained. The sequence of operations and the final structure are depicted in Fig. 4.11. We can observe that initially the algorithm eliminates a number of irrelevant or redundant attributes, it then combines two dependent attributes and subsequently eliminates two more attributes, until it arrives to the final structure with 3 attributes: $RG$, $Y$, $I$ (one is a combination of two original attributes). This final model was evaluated with the same test images and the accuracy improved to 98%. An example of an image with the skin pixels detected by this classifier is shown in Fig. 4.12.

In this example, the SNBC obtains a significant advantage compared to the original NBC, while at the same time producing a simpler model (in terms of the number of variables and parameters required) [10].

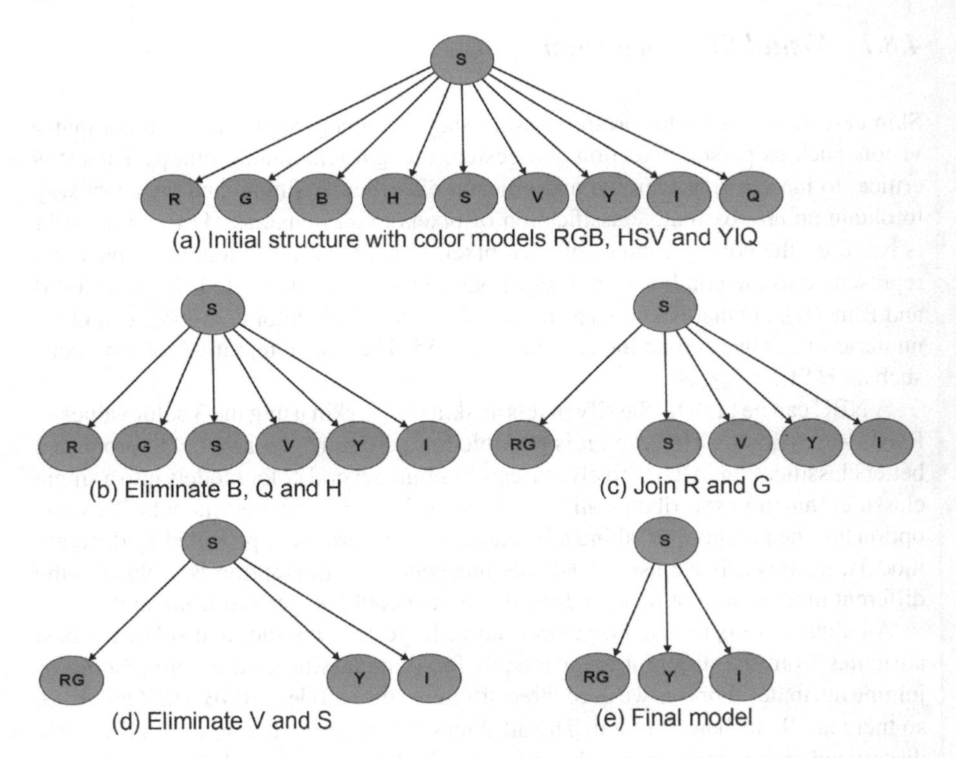


**Fig. 4.11** The figure illustrates the process of optimizing the semi-naive Bayesian classifier for skin detection, from the initial model with 9 attributes (**a**) to the final one with three attributes (**e**)

(a) Original image
(b) Image with skin pixels detected


**Fig. 4.12** An example of an image in which the pixels have been classified as skin or not skin

### 4.8.2 HIV Drug Selection

The Human Immunodeficiency Virus (HIV) is the causing agent of AIDS, a condition in which progressive failure of the immune system allows opportunistic life-threatening infections to occur. To combat HIV infection several antiretroviral (ARV) drugs belonging to different drug classes that affect specific steps in the viral replication cycle have been developed. Antiretroviral therapy (ART) generally consists of combinations of three or four ARV drugs. Selecting a drug combination depends on the patient's condition, which can be characterized according to the mutations of the virus present in the patient. Thus, it is important to select the best drug combination according to the virus' mutations in a patient.

Selecting the best group of antiretroviral drugs for a patient can be seen as an instance of a multi-label classification problem, in which the classes are the different types of antiretroviral drugs and the attributes are the virus mutations. Since multi-label classification is a subcase of multi-dimensional classification, this particular problem can be accurately modeled with a MBC. By applying a learning algorithm we can discover the relations that exist between antiretrovirals and mutations, while also retrieving a model with a high predictive power.

An alternative for learning a MBC is the MB-MBC algorithm [4]. This particular algorithm uses the *Markov blanket* of each class variable to lighten the computational burden of learning the MBC by filtering out those variables that do not improve classification. A Markov blanket of a variable $C$, denoted as $MB(C)$, is the minimal set of variables such that $I(C, \mathbf{S}|MB(C))$ is true for every variable subset $\mathbf{S}$, where $\mathbf{S}$ does not have as members any variables that belong to $MB(C) \cup C$. In other words, the Markov blanket of C is the minimal set of variables under which C is conditionally independent of all remaining variables.

To predict the most likely group of antiretrovirals drugs for a patient given the viral mutations present, the Markov blanket for each antiretroviral is learned. For example, if we consider a group of reverse transcriptase inhibitors (an antiretroviral drug group that attacks the reverse transcriptase phase of the viral HIV lifecycle) as the class variables, and a group of mutations as the attributes, the Markov blanket for the entire set of reverse transcriptase inhibitors is learned to determine the existing relations of type antiretroviral-antiretroviral and antiretroviral-mutation. Learning the Markov blanket of each class variable corresponds to learning an undirected structure for the MBC, i.e. the three subgraphs. Finally, directionality for all three subgraphs is determined in the last step of the MB-MBC algorithm. Figure 4.13 shows the resulting MBC.

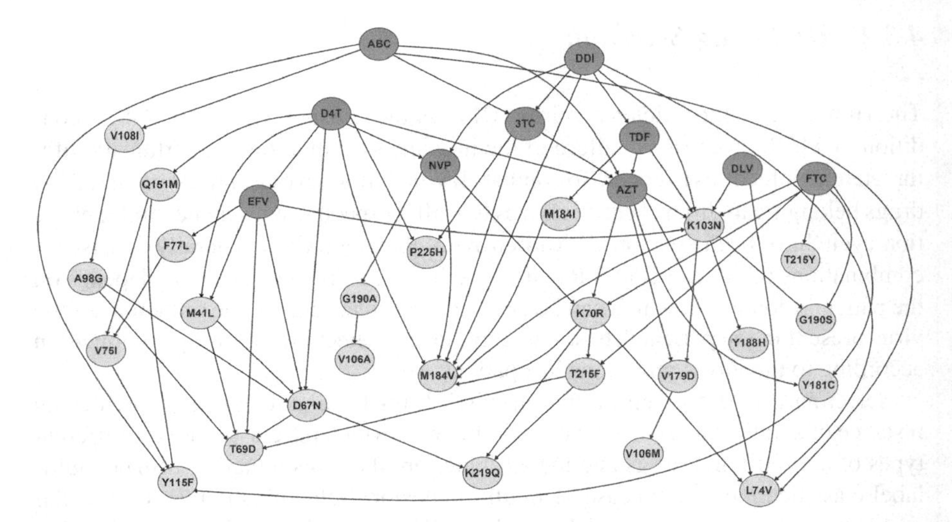


**Fig. 4.13** Multi-dimensional BN classifier for HIV drug selection. A graphical structure for the multi-dimensional Bayesian network classifier learned with the MB-MBC algorithm for a set of reverse transcriptase inhibitors (green) and a set of mutations (yellow) (Figure based on [4])

## 4.9 Additional Reading

For a general introduction and comparison of different classification approaches see [11]. The consideration of classification costs is described in [6]. The TAN and BAN classifiers are described in [7]; a comparison of different BN classifiers is presented in [5]. The semi-naive approach was initially introduced in [19], and later extended by [12]. Reference [21] presents an overview of multidimensional classifiers. Different alternatives for MBC are presented in [3]. Chain classifiers were introduced in [13], and Bayesian chain classifiers in [20]. A review of different approaches for hierarchical classification and their applications is presented in [18].

## 4.10 Exercises

1. Given the data for the *golf* example in Table 4.2, complete the CPTs for the NBC using maximum likelihood estimation.
2. Obtain the class with the maximum probability for the *golf* NBC of the previous problem, considering all the combinations of values of the attributes.
3. Based on the results of the previous problem, design a set of classification rules that are equivalent to the NBC for determining play/no-play based on the attribute values.
4. Consider that we transform the NBC for *golf* to a TAN with the following dependency structure for the attributes: $outlook \rightarrow temperature$, $outlook \rightarrow$

**Table 4.3** Data set for the golf example with discrete and continuous attributes

| Outlook | Temperature | Humidity | Windy | Play |
|---|---|---|---|---|
| Sunny | 30.5 | 90 | False | No |
| Sunny | 29 | 82 | True | No |
| Overcast | 28.2 | 95 | False | Yes |
| Rain | 21 | 86 | False | Yes |
| Rain | 5.3 | 60 | False | Yes |
| Rain | 8.8 | 63 | True | No |
| Overcast | 0 | 55 | True | Yes |
| Sunny | 17 | 80 | False | No |
| Sunny | 7 | 65 | False | Yes |
| Rain | 15.4 | 70 | False | Yes |
| Sunny | 16 | 59 | True | Yes |
| Overcast | 19.5 | 85 | True | Yes |
| Overcast | 31 | 55 | False | Yes |
| Rain | 20 | 88 | True | No |

$humidity, temperature \rightarrow wind$. Using the same dataset, estimate the CPTs for this TAN.

5. Given the data set for the *golf* example, estimate the mutual information between the class and each attribute. Build a semi-naive Bayesian classifier by eliminating those attributes that have a low mutual information with the class (define a threshold).
6. Extend the previous problem by now estimating the mutual information between each pair of attributes given the class variable. Eliminate or join those attributes that are not conditionally independent given the class according to a predefined threshold. Show the structure and parameters of the resulting classifiers.
7. Table 4.3 shows data for the golf example with discrete and continuous attributes. Estimate the parameters for a hybrid classifier based on these data, that is the CPTs for the discrete attributes and the conditional mean and standard deviation for the continuous ones.
8. Based on the parameters estimated for the hybrid golf data, calculate the posterior probabilities of the class (Play) given: (a) Outlook = sunny, Temperature = 22, Humidity = 80, Windy = yes; (b) Outlook = overcast, Temperature = 15, Humidity = 60, Windy = no.
9. Repeat problems 5 and 6 considering the hybrid golf data.
10. Consider that we transform the *golf* example to a multidimensional problem such that there are two classes, play and outlook, and three attributes, temperature, humidity and windy. Consider that we build a multidimensional classifier based on binary relevance—an independent classifier for each class variable. Given that each classifier is a NBC, what will the structure of the resulting classifier be? Obtain the parameters for this classifier based on the same data in Table 4.2.

11. For the previous problem, consider that we now build a NBC based on the power set approach. What will the structure and parameters of the resulting model be? Use the same dataset.
12. Given the hierarchical structure in Fig. 4.9, assume that there are local classifiers for each class (except the root) and they provide the following posterior probabilities: $C_2 = 0.7$, $C_3 = 0.5$, $C_4 = 0.8$, $C_5 = 0.9$, $C_6 = 0.3$, $C_7 = 0.2$, $C_8 = 0.3$, $C_9 = 0.9$, $C_{10} = 0.7$, $C_{11} = 0.5$, $C_{12} = 0.7$. Estimate the score of each path using (a) the score used by chain path evaluation, (b) the normalized sum of probabilities. Which is the most probable path according to each method?
13. *** Compare the structure, complexity (in terms of the number of parameters) and classification accuracy of different Bayesian classifiers—NBC, TAN, BAN—using several datasets (use, for example, the WEKA [8] implementation of the Bayesian classifiers; and some of the data sets from the UCI repository [1]). Do TAN or BAN always outperform the naive Bayesian classifier? Why?
14. *** A hierarchical classifier is a particular type of multidimensional classifier in which the classes are arranged in a hierarchy; for instance, an animal hierarchy. A restriction of a hierarchy is that an instance that belongs to a certain class, must belong to all its superclasses in the hierarchy (hierarchical constraint). How can a multidimensional classifier be designed to guarantee the hierarchical constraint? Extend the Bayesian chain classifier for hierarchical classification.
15. *** Implement the circular chain classifier. Using different multidimensional datasets, analyze experimentally: (a) if the order of the classes influences the results; (b) how many iterations does it take for the results to converge.

## References


1. Bache, K., Lichman, M.: UCI machine learning repository. School of Information and Computer Science, University of California, Irvine, CA (2013). Available at: http://archive.ics.uci.edu/ml. Accessed 22 Sept 2014
2. Barutcuoglu, Z., Decoro, C.: Hierarchical shape classification using Bayesian aggregation. In: IEEE International Conference on Shape Modeling and Applications (2006)
3. Bielza, C., Li, G., Larrañaga, P.: Multi-dimensional classification with Bayesian networks. Int. J. Approx. Reason. (2011)
4. Borchani, H., Bielza, C., Toro, C., Larrañaga, P.: Predicting human immunodeficiency virus inhibitors using multi-dimensional Bayesian network classifiers. Artif. Intell. Med. **57**, 219–229 (2013)
5. Cheng, J., Greiner, R.: Comparing Bayesian network classifiers. In: Proceedings of the 15th Conference on Uncertainty in Artificial Intelligence, pp. 101–108 (1999)
6. Drummond, C., Holte, R.C.: Explicitly representing expected cost: an alternative to the ROC representation. In: Proceedings of the 6th ACM SIGKDD International Conference on Knowledge Discovery and Data Mining, pp. 198–207 (2000)
7. Friedman, N., Geiger, D., Goldszmidt, M.: Bayesian network classifiers. Mach. Learn. **29**, 131–163 (1997)
8. Hall, M., Frank, E., Holmes, G., Pfahringer, B., Reutemann, P., Witten, I.H.: The WEKA data mining software: an update. ACM SIGKDD Explor. Newsl. (ACM) 10–18 (2009)

9. Kwoh, C.K, Gillies, D.F.: Using hidden nodes in Bayesian networks. Artif. Intell. (Elsevier, Essex) **88**, 1–38 (1996)
10. Martinez, M., Sucar, L.E.: Learning an optimal naive bayes classifier. In: International Conference on Pattern Recognition (ICPR), vol. 3, pp. 1236–1239 (2006)
11. Michie, D., Spiegelhalter, D.J., Taylor, C.C.: Machine Learning, Neural and Statistical Classification. Ellis Howard, England (2004)
12. Pazzani, M.J.: Searching for attribute dependencies in Bayesian classifiers. Preliminary Papers of the Intelligence and Statistics, pp. 424–429 (1996)
13. Read, J., Pfahringer, B., Holmes, G., Frank, E.: Classifier chains for multi-label classification. In: Proceedings ECML/PKDD, pp. 254–269 (2009)
14. Ramírez, M., Sucar, L.E., Morales, E.: Path evaluation for hierarchical multi-label classification. In: Proceedings of the 27th International Florida Artificial Intelligence Research Society Conference (FLAIRS), pp. 502–507 (2014)
15. Rivas, J.J., Orihuela-Espina, F., Sucar, L.E.: Circular chain classifiers. In: Proceedings of Machine Learning Research, vol. 72, pp. 380–391 (2018)
16. Serrano-Pérez, J., Sucar, L.E.: Hierarchical classification with Bayesian networks and chained classifiers. In: Proceedings of the 32nd International Florida Artificial Intelligence Research Society Conference (FLAIRS-32), AAAI, pp. 488–493 (2019)
17. Silla-Jr., C.N., Freitas, A.A.: Novel top-down approaches for hierarchical classification and their application to automatic music genre classification. In: IEEE International Conference on Systems, Man, and Cybernetics, October pp. 3499–3504 (2009)
18. Silla-Jr., C.N., Freitas, A.A.: A survey of hierarchical classification across different application domains. Data Min. Knowl. Discov. **22**(1–2), 31–72 (2011)
19. Sucar, L.E., Gillies, D.F., Gillies, D.A.: Objective probabilities in expert systems. Artif. Intell. **61**, 187–208 (1993)
20. Sucar, L.E., Bielza, C., Morales, E., Hernandez, P., Zaragoza, J., Larrañaga, P.: Multi-label classification with Bayesian network-based chain classifiers. Pattern Recognit. Lett. **41**, 14–22 (2014)
21. Tsoumakas, G., Katakis, I.: Multi-label classification: an overview. Int. J. Data Warehous. Min. **3**, 1–13 (2007)
22. van der Gaag L.C., de Waal, P.R.: Multi-dimensional Bayesian network classifiers. In: 3rd European Conference on Probabilistic Graphic Models, Prague, Czech Republic, pp. 107–114 (2006)

# Chapter 5
# Hidden Markov Models

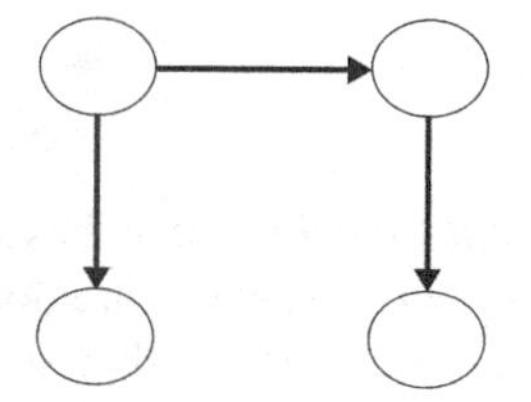

## 5.1 Introduction

Markov Chains are a class of PGMs that represent dynamic processes, in particular how the state of a process changes with time. For instance, consider that we are modeling how the weather in a particular place changes over time. In a very simplified model, we assume that the weather is constant throughout the day, and can have three possible states: *sunny, cloudy, raining*. Additionally, we assume that the weather on a certain day only depends on the previous day. Thus, we can think of this simple weather model as a Markov chain in which there is a state variable per day, with 3 possible values; these variables are linked in a *chain*, with a directed arc from one day to the next, (see Fig. 5.1). This implies what is known as the *Markov property*, the state of the weather for the next day, $S_{t+1}$, is independent of all previous days given the present weather, $S_t$, i.e., $P(S_{t+1} \mid S_t, S_{t-1}, ...) = P(S_{t+1} \mid S_t)$. Thus, in a Markov chain the main parameter required is the probability of a state given the previous one.

The previous model assumes that we can measure the weather with precision each day, that is, the state is *observable*. However, this is not necessarily true. In many applications we cannot observe the state of the process directly, so we have what is called a *Hidden Markov Model*, where the state is hidden. In this case, in addition to the probability of the next state given the current state, there is another parameter which models the uncertainty about the state, represented as the probability of the *observation* given the state, $P(O_t \mid S_t)$. This type of model is more powerful than the simple Markov chain and has many applications, for example in speech and gesture recognition.



L. E. Sucar, *Probabilistic Graphical Models*, Advances in Computer Vision and Pattern Recognition, https://doi.org/10.1007/978-3-030-61943-5_5

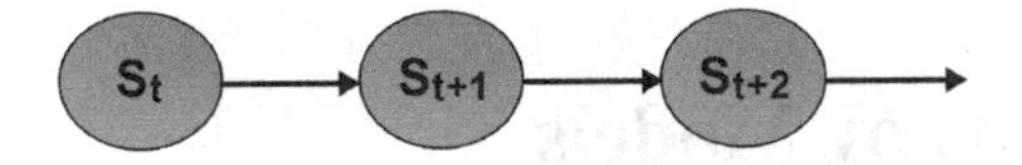


**Fig. 5.1** The figure illustrates a Markov chain where each node represents the state at certain point in time

After a brief introduction to Markov chains, in the following sections we will discuss hidden Markov models in detail, including how the computations of interest for this type of model are solved. Then we will discuss several extensions to standard HMMs, and we conclude the chapter with two application examples.

## 5.2 Markov Chains

A Markov chain (MC) is a *state machine* that has a discrete number of states, $q_1, q_2, ..., q_n$, and the transitions between states are non-deterministic; i.e. there is a probability of transiting from a state $q_i$ to another state $q_j$: $P(S_{t+1} = q_j \mid S_t = q_i)$. Time is also discrete, such that the chain can be at a certain state $q_i$ for each time step $t$. It satisfies the Markov property, that is, the probability of the next state only depends on the current state.

Formally, a Markov chain is defined by:

Set of states: $Q = \{q_1, q_2, ..., q_n\}$
Vector of prior probabilities: $\Pi = \{\pi_1, \pi_2, ..., \pi_n\}$, where $\pi_i = P(S_0 = q_i)$
Matrix of transition probabilities: $A = \{a_{ij}\}$, $i = [1..n], j = [1..n]$, where $a_{ij} = P(S_{t+1} = q_j \mid S_t = q_i)$

where $n$ is the number of states, and $S_0$ is the initial state. In a compact way, a MC is represented as $\lambda = \{A, \Pi\}$.

A (first order) Markov chain satisfies the following properties:

1. Probability axioms: $\sum_i \pi_i = 1$ and $\sum_j a_{ij} = 1$
2. Markov property: $P(S_{t+1} = q_j \mid S_t = q_i, S_{t-1} = q_k, ...) = P(S_{t+1} = q_j \mid S_t = q_i)$
3. Stationary process: $P(S_t = q_j \mid S_{t-1} = q_i) = P(S_{t+1} = q_j \mid S_t = q_i)$

For example, consider the previous simple weather model with three states: $q_1 =$ *sunny*, $q_2 =$ *cloudy*, $q_3 =$ *raining*. In this case to specify a MC we will require a vector with three prior probabilities (see Table 5.1) and a $3 \times 3$ matrix of transition probabilities (see Table 5.2).

The transition matrix can be represented graphically with what is called a *state transition diagram* or simply a *state diagram*. This diagram is a directed graph,

**Table 5.1** Prior probabilities for the weather example

| Sunny | Cloudy | Raining |
|---|---|---|
| 0.2 | 0.5 | 0.3 |

**Table 5.2** Transition probabilities for the weather example. Each row represents the transition probabilities from a present state (e.g., sunny) to the future states (sunny, cloudy, raining); thus the probabilities per row must add to one

| | Sunny | Cloudy | Raining |
|---|---|---|---|
| Sunny | 0.8 | 0.1 | 0.1 |
| Cloudy | 0.2 | 0.6 | 0.2 |
| Raining | 0.3 | 0.3 | 0.4 |

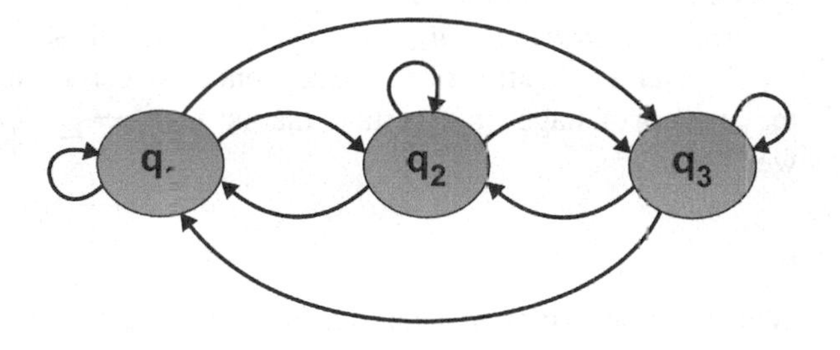


**Fig. 5.2** The figure illustrates the state transition diagram for the weather example

where each node is a state and the arcs represent possible transitions between states. If an arc between state $q_i$ and $q_j$ does not appear in the diagram, it means that the corresponding transition probability is zero. An example of a state diagram for the weather example is depicted in Fig. 5.2.[1]

Given a Markov chain model, there are three basic questions that we can ask:

- What is the probability of a certain state sequence?
- What is the probability that the chain remains in a certain state for a period of time?
- What is the expected time that the chain will remain in a certain state?

Next we will see how we can answer these questions and we will illustrate them using the weather example.

The probability of a sequence of states given the model is basically the product of the transition probabilities of the sequence of states:

$$P(q_i, q_j, q_k, ...) = a_{0i} a_{ij} a_{jk} .... \quad (5.1)$$

where $a_{0i}$ is the transition to the initial state in the sequence, which could be its prior probability ($\pi_i$) or the transition from the previous state (if this is known).

For example, in the weather model, we might want to know the probability of the following sequence of states: $Q =$ *sunny, sunny, rainy, rainy, sunny, cloudy, sunny*. Assuming that *sunny* is the initial state in the MC, then:

[1] Do not confuse a state diagram, where a node represents each state—a specific value of a random variable- and the arcs transitions between states, with a graphical model diagram, where a node represents a random variable and the arcs represent probabilistic dependencies.

$P(Q) = \pi_1 a_{11} a_{13} a_{33} a_{31} a_{12} a_{21} = (0.2)(0.8)(0.1)(0.4)(0.3)(0.1)(0.2) = 3.84 \times 10^{-5}$

The probability of staying $d$ time steps in a certain state, $q_i$, is equivalent to the probability of a sequence in this state for $d-1$ time steps and then transiting to a different state. That is:

$$P(d_i) = a_{ii}^{d-1}(1 - a_{ii}) \quad (5.2)$$

Considering the weather model, what is the probability of 3 cloudy days? This can be computed as follows:

$P(d_2 = 3) = a_{22}^2(1 - a_{22}) = 0.6^2(1 - 0.6) = 0.144$

The average duration of a state sequence in a certain state is the expected value of the number of stages in that state, that is: $E(D) = \sum_i d_i P(d_i)$. Substituting Eq. 5.2 we obtain:

$$E(d_i) = \sum_i d_i a_{ii}^{d-1}(1 - a_{ii}) \quad (5.3)$$

Which can be written in a compact form as:

$$E(d_i) = 1/(1 - a_{ii}) \quad (5.4)$$

For instance, what is the expected number of days that the weather will remain cloudy? Using the previous equation:

$E(d_2) = 1/(1 - a_{22}) = 1/(1 - 0.6) = 2.5$

### 5.2.1 Parameter Estimation

Another important question is how to determine the parameters of the model, what is known as *parameter estimation*. For a MC, the parameters can be estimated simply by counting the number of times that the sequence is in a certain state, $i$; and the number of times there is a transition from a state $i$ to a state $j$. Assume there are $N$ sequences of observations. $\gamma_{0i}$ is the number of times that the state $i$ is the initial state in a sequence, $\gamma_i$ is the number of times that we observe state $i$, and $\gamma_{ij}$ is the number of times that we observe a transition from state $i$ to state $j$. The parameters can be estimated with the following equations.

Initial probabilities:

$$\pi_i = \gamma_{0i}/N \quad (5.5)$$

Transition probabilities:

$$a_{ij} = \gamma_{ij}/\gamma_i \quad (5.6)$$

Note that for the last observed state in a sequence we do not observe the next state, so the last state for all the sequences is not considered in the counts.

For instance, consider that for the weather example we have the following 4 observation sequences:

**Table 5.3** Calculated prior probabilities for the weather example

| Sunny | Cloudy | Raining |
|---|---|---|
| 0.25 | 0.5 | 0.25 |

**Table 5.4** Calculated transition probabilities for the weather example

| | Sunny | Cloudy | Raining |
|---|---|---|---|
| Sunny | 0 | 0.33 | 0.67 |
| Cloudy | 0.285 | 0.43 | 0.285 |
| Raining | 0.18 | 0.18 | 0.64 |

$q_2, q_2, q_3, q_3, q_3, q_3, q_1$
$q_1, q_3, q_2, q_3, q_3, q_3, q_3$
$q_3, q_3, q_2, q_2$
$q_2, q_1, q_2, q_2, q_1, q_3, q_1$
Given these four sequences, the corresponding parameters can be estimated as depicted in Tables 5.3 and 5.4.

### 5.2.2 Convergence

An additional interesting question is: if a sequence transits from one state to another a large number of times, $M$, what is the probability in the limit (as $M \rightarrow \infty$) of each state, $q_i$?

Given an initial probability vector, $\Pi$, and transition matrix, $A$, the probability of each state, $P = \{p_1, p_2, ..., p_n\}$ after $M$ *iterations* is:

$$P = \pi A^M \tag{5.7}$$

What happens when $M \rightarrow \infty$? The solution is given by the Perron–Frobenius theorem, which says that when the following two conditions are satisfied:

1. Irreducibility: from every state $i$ there is a probability $a_{ij} > 0$ of transiting to any state $j$.
2. Aperiodicity: the chain does not form *cycles* (a subset of states in which the chain remains once it arrives to one of these state).

Then as $M \rightarrow \infty$, the chain converges to an invariant distribution $P$, such that $P \times A = P$, where $A$ is the transition probability matrix. The rate of convergence is determined by the second *eigen-value* of matrix $A$.

For example, consider a MC with three states and the following transition probability matrix:

$$A = \begin{matrix} 0.9 & 0.075 & 0.025 \\ 0.15 & 0.8 & 0.05 \\ 0.25 & 0.25 & 0.5 \end{matrix}$$

It can be shown that in this case the steady state probabilities converge to $P = \{0.625, 0.3125, 0.0625\}$.

An interesting application of this convergence property of Markov chains for ranking web pages is presented in the applications section. Next we discuss hidden Markov models.

## 5.3 Hidden Markov Models

A Hidden Markov model (HMM) is a Markov chain where the states are not directly observable. For example, if we consider the weather example, the *weather* cannot be directly measured; in reality, the weather is estimated based on a series of sensors—temperature, pressure, wind velocity, etc. So in this, as in many other phenomena, states are not directly observable, and HMMs provide a more appropriate and powerful modeling tool. Another way of thinking about a HMM is that it is a double stochastic process: (i) a hidden stochastic process that we cannot directly observe, (ii) and a second stochastic process that produces the sequence of observations given the first process.

For instance, consider that we have two unfair or "biased" coins, $M_1$ and $M_2$. $M_1$ has a higher probability of *heads*, while $M_2$ has a higher probability of *tails*. Someone sequentially flips these two coins, however we do not know which one. We can only observe the outcome, *heads* or *tails*:

$H, T, T, H, T, H, H, H, T, H, T, H, T, T, T, H, T, H, H, \ldots$

Assume the person flipping the coins selects the first coin in the sequence (prior probabilities) and the next coin to flip given the previous one (transition probabilities) with equal probability. Aside from the prior and transition probabilities for the states (as with a MC), in a HMM we need to specify the *observation* probabilities, in this case the probabilities of $H$ or $T$ given each coin (the state). Lets assume that $M_1$ has an 80% probability for *heads* and $M_2$ has an 80% probability for *tails*. Then we have specified all the required parameters for this simple example, which are summarized in Table 5.5.

The state diagram for the coins example is depicted in Fig. 5.3, with two state variables and two possible observations, that depend on the state.

**Table 5.5** The prior probabilities ($\Pi$), transition probabilities ($A$) and observation probabilities ($B$) for the unfair coins example

$\Pi =$

| $M_1$ | $M_2$ |
|---|---|
| 0.5 | 0.5 |

$A =$

| | $M_1$ | $M_2$ |
|---|---|---|
| $M_1$ | 0.5 | 0.5 |
| $M_2$ | 0.5 | 0.5 |

$B =$

| | $M_1$ | $M_2$ |
|---|---|---|
| $H$ | 0.8 | 0.2 |
| $T$ | 0.2 | 0.8 |

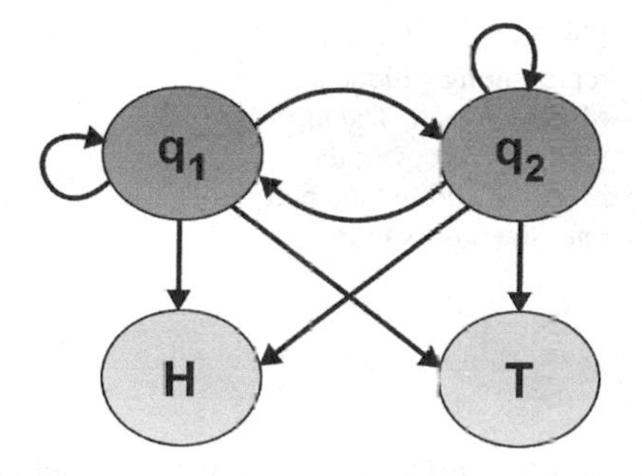


**Fig. 5.3** State diagram for the HMM coins example. The two states, $q_1$ and $q_2$, and two observations, $H$ and $T$ are shown, with arcs representing the transitions and observation probabilities

Formally, a hidden Markov model is defined by:

Set of states: $Q = \{q_1, q_2, ..., q_n\}$
Set of observations: $O = \{o_1, o_2, ..., o_m\}$
Vector of prior probabilities: $\Pi = \{\pi_1, \pi_2, ..., \pi_n\}$, where $\pi_i = P(S_0 = q_i)$
Matrix of transition probabilities: $A = \{a_{ij}\}$, $i = [1..n], j = [1..n]$, where $a_{ij} = P(S_{t+1} = q_j \mid S_t = q_i)$
Matrix of observation probabilities: $B = \{b_{ij}\}$, $i = [1..n], j = [1..m]$, where $b_{ik} = P(O_t = o_k \mid S_t = q_i)$

where $n$ is the number of states and $m$ the number of observations; $S_0$ is the initial state. Compactly, a HMM is represented as $\lambda = \{A, B, \Pi\}$.

A (first order) HMM satisfies the following properties:

Markov property: $P(S_{t+1} = q_j \mid S_t = q_i, S_{t-1} = q_k, ...) = P(S_{t+1} = q_j \mid S_t = q_i)$
Stationary process: $P(S_t = q_j \mid S_{t-1} = q_i) = P(S_{t+1} = q_j \mid S_t = q_i)$ and $P(O_{t-1} = o_k \mid S_{t-1} = q_j) = P(O_t = o_k \mid S_t = q_j), \forall(t)$
Independence of observations: $P(O_t = o_k \mid S_t = q_i, S_{t-1} = q_j, ...) = P(O_t = o_k \mid S_t = q_i)$

As in the case of a MC, the Markov property implies that the probability of the next state only depends on the current state, and it is independent of the rest of the history. The second property says that the transition and observation probabilities do not change over time; i.e. the process is stationary. The third property specifies that the observations only depend on the current state. There are extensions to the *basic* HMM that relax some of these assumptions; some of these will be discussed in the next section and in further chapters.

According to the previous properties, the graphical model of a HMM is shown in Fig. 5.4, which includes two series of random variables, the state at time $t$, $S_t$, and the observation at time $t$, $O_t$.

Given a HMM representation of a certain domain, there are three basic questions that are of interest in most applications [7]:

1. *Evaluation*: given a model, estimate the probability of a sequence of observations.

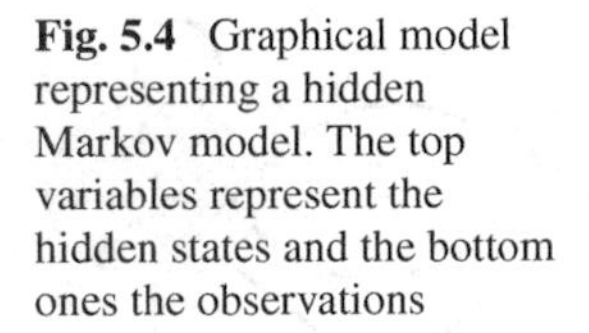

**Fig. 5.4** Graphical model representing a hidden Markov model. The top variables represent the hidden states and the bottom ones the observations

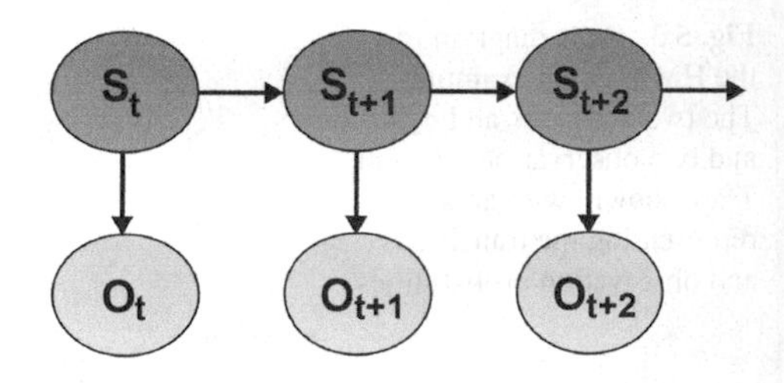


2. *Optimal Sequence*: given a model and a particular observation sequence, estimate the most probable state sequence that produced the observations.
3. *Parameter learning*: given a number of sequence of observations, adjust the parameters of the model.

Algorithms for solving these questions, assuming a *standard HMM* with finite number of states and observations, are described next.

## 5.3.1 Evaluation

Evaluation consists in determining the probability of an observation sequence, $O = \{o_1, o_2, o_3, ...\}$, given a model, $\lambda$, that is, estimating $P(O \mid \lambda)$. We present two methods. First we present the direct method, a *naive* algorithm that motivates the need for a more efficient one, which is then described.

### 5.3.1.1 Direct Method

A sequence of observations, $O = \{o_1, o_2, o_3, ...o_T\}$, can be generated by different state sequences, $Q_i$, as the states are unknown for HMMs. Thus, to calculate the probability of an observation sequence, we can estimate it for a certain state sequence, and then add the probabilities for all the possible state sequences:

$$P(O \mid \lambda) = \sum_i P(O, Q_i \mid \lambda) \tag{5.8}$$

To obtain $P(O, Q_i \mid \lambda)$ we simply multiply the probability of the initial state, $q_1$, by the transition probabilities for a state sequence, $q_1, q_2, ...$ and the observation probabilities for an observation sequence, $o_1, o_2, ...$:

$$P(O, Q_i \mid \lambda) = \pi_1 b_1(o_1) a_{12} b_2(o_2) ... a_{(T-1)T} b_T(o_T) \tag{5.9}$$

Thus, the probability of $O$ is given by a summation over all the possible state sequences, $Q$:

$$P(O \mid \lambda) = \sum_Q \pi_1 b_1(o_1) a_{12} b_2(o_2) ... a_{(T-1)T} b_T(o_T) \quad (5.10)$$

For a model with $N$ states and an observation length of $T$, there are $N^T$ possible state sequences. Each term in the summation requires $2T$ operations. As a result, the evaluation requires a number of operations in the order of $2T \times N^T$.

For example, if we consider a model with five states, $N = 5$, and an observation sequence of length $T = 100$, which are common parameters for HMM applications, the number of required operations is in the order of $10^{72}$. A more efficient method is required!

#### 5.3.1.2 Iterative Method

The basic idea of the iterative method, also known as *Forward*, is to estimate the probabilities of the states/observations per time step. That is, calculate the probability of a partial sequence of observations until time $t$ (starting from $t = 1$), and based on this partial result, calculate it for time $t + 1$, and so on.

First, we define an auxiliary variable called *forward*:

$$\alpha_t(i) = P(o_1, o_2, ..., o_t, S_t = q_i \mid \lambda) \quad (5.11)$$

That is, the probability of the partial sequence of observations until time $t$, being in state $q_i$ at time $t$. $\alpha_t(i)$, for certain time $t$, can be calculated iteratively based on the values from the previous time, $\alpha_{t-1}(i)$:

$$\alpha_t(j) = [\sum_i \alpha_{t-1}(i) a_{ij}] b_j(O_t) \quad (5.12)$$

The iterative algorithm consists of three main parts: initialization, induction and termination. In the initialization phase, the $\alpha$ variables for all states at the initial time are obtained. The induction phase consists in calculating $\alpha_{t+1}(i)$ in terms of $\alpha_t(i)$; this is repeated from $t = 2$ to $t = T$. Finally, $P(O \mid \lambda)$ is obtained by adding all the $\alpha_T$ in the termination phase. The procedure is shown in Algorithm 5.1.

We now analyze the time complexity of the iterative method. Each iteration requires $N$ multiplications and $N$ additions (approx.), so for the $T$ iterations, the number of operations is in the order of $N^2 \times T$. Thus, the time complexity is reduced from exponential in $T$ for the direct method to linear in $T$ and quadratic in $N$ for the iterative method, a significant reduction in complexity; note that in most applications $T >> N$.

Returning to the example where $N = 5$ and $T = 100$, now the number of operations is approx. 2500, which can be carried efficiently with a standard personal computer.

The iterative procedure just described is the basis for solving the other two questions for HMMs. These are described next.

**Algorithm 5.1** The Forward Algorithm

**Require:** HMM, $\lambda$; Observation sequence, $O$; Number of states, $N$; Number of observations, $T$
  **for** $i = 1$ **to** $N$ **do**
    $\alpha_1(i) = P(O_1, S_1 = q_i) = \pi_i b_i(O_1)$ (Initialization)
  **end for**
  **for** $t = 2$ **to** $T$ **do**
    **for** $j = 1$ **to** $N$ **do**
      $\alpha_t(j) = [\sum_i \alpha_{t-1}(i) a_{ij}] b_j(O_t)$ (Induction)
    **end for**
  **end for**
  $P(O) = \sum_i \alpha_T(i)$ (Termination)
  **return** $P(O)$

### 5.3.2 State Estimation

Finding the most probable sequence of states for an observation sequence, $O = \{o_1, o_2, o_3, ...\}$, can be interpreted in two ways: (i) obtaining the most probable state, $S_t$ at each time step $t$, (ii) obtaining the most probable sequence of states, $s_0, s_1, ..., s_T$. Notice that the concatenation of the most probable states for each time step, for $t = 1...T$, is not necessarily the same as the most probable sequence of states.[2] First we solve the problem of finding the most probable or *optimum* state for a certain time $t$, and then the problem of finding the *optimum* sequence.

We first need to define some additional auxiliary variables. The *backward* variable is analogous to the forward one, but in this case we start from the end of the sequence, that is:

$$\beta_t(i) = P(o_{t+1}, o_{t+2}, ..., o_T, S_t = q_i \mid \lambda) \tag{5.13}$$

That is, the probability of the partial sequence of observations from $t + 1$ to $T$, being in state $q_i$ at time $t$. In a similar way to $\alpha$, it can be calculated iteratively but now backwards:

$$\beta_t(i) = \sum_j \beta_{t+1}(j) a_{ij} b_j(o_{t+1}) \tag{5.14}$$

For $t = T - 1, T - 2, ..., 1$.The $\beta$ variables for $T$ are defined as $\beta_T(j) = 1$.

Thus, we can also solve the evaluation problem of the previous section using $\beta$ instead of $\alpha$, starting from the end of the observation sequence and iterating backwards through time. Or we can combine both variables, and iterate forward and backward, *meeting* at some intermediate time $t$; that is:

$$P(O, s_t = q_i \mid \lambda) = \alpha_t(i)\beta_t(i) \tag{5.15}$$

Then:

---

[2]This is a particular case of the *Most Probable Explanation* or MPE problem, which will be discussed in Chap. 7.

$$P(O \mid \lambda) = \sum_i \alpha_t(i)\beta_t(i) \tag{5.16}$$

Now we define an additional variable, $\gamma$, that is the conditional probability of being in a certain state $q_i$ given the observation sequence:

$$\gamma_t(i) = P(s_t = q_i \mid O, \lambda) = P(s_t = q_i, O \mid \lambda)/P(O \mid \lambda) \tag{5.17}$$

Which can be written in terms of $\alpha$ and $\beta$ as:

$$\gamma_t(i) = \alpha_t(i)\beta_t(i)/\sum_i \alpha_t(i)\beta_t(i) \tag{5.18}$$

This variable, $\gamma$, provides the answer to the first subproblem, the most probable state (MPS) at a time $t$; we just need to find for which state it has the maximum value. That is:

$$MPS(t) = ArgMax_i \gamma_t(i) \tag{5.19}$$

Lets now solve the second subproblem—the most probable state sequence $Q$ given the observation sequence $O$, such that we want to maximize $P(Q \mid O, \lambda)$. By Bayes rule: $P(Q \mid O, \lambda) = P(Q, O \mid \lambda)/P(O)$. Given that $P(O)$ does not depend on $Q$, this is equivalent to maximizing $P(Q, O \mid \lambda)$.

The method for obtaining the optimum state sequence is known as the *Viterbi* algorithm, which in an analogous way as the forward algorithm, solves the problem iteratively. Before we see the algorithm, we need to define an additional variable, $\delta$. This variable gives the maximum value of the probability of a subsequence of states and observations until time $t$, being at state $q_i$ at time $t$; that is:

$$\delta_t(i) = MAX[P(s_1, s_2, \ldots s_t = q_i, o_1, o_2, \ldots, o_t \mid \lambda)] \tag{5.20}$$

Which can also be obtained in an iterative way:

$$\delta_{t+1}(i) = [MAX\, \delta_t(i) a_{ij}] b_j(o_{t+1}) \tag{5.21}$$

The Viterbi algorithm requires four phases: initialization, recursion, termination and backtracking. It requires an additional variable, $\psi_t(i)$, that stores for each state $i$ at each time step $t$ the previous state that gave the maximum probability. This is used to reconstruct the sequence by backtracking after the termination phase. The complete procedure is depicted in Algorithm 5.2.

With the Viterbi algorithm we can obtain the most probable sequence of states given the observation sequence, even if these are hidden for HMMs.

**Algorithm 5.2** The Viterbi Algorithm

**Require:** HMM, $\lambda$; Observation sequence, $O$; Number of states, $N$; Number of observations, $T$
  **for** $i = 1$ **to** $N$ **do**
    (Initialization)
    $\delta_1(i) = \pi_i b_i(O_1)$
    $\psi_1(i) = 0$
  **end for**
  **for** $t = 2$ **to** $T$ **do**
    **for** $j = 1$ **to** $N$ **do**
      (Recursion)
      $\delta_t(j) = MAX_i[\delta_{t-1}(i)a_{ij}]b_j(O_t)$
      $\psi_t(j) = ARGMAX_i[\delta_{t-1}(i)a_{ij}]$
    **end for**
  **end for**
  (Termination)
  $P^* = MAX_i[\delta_T(i)]$
  $q_T^* = ARGMAX_i[\delta_T(i)]$
  **for** $t = T$ **to** 2 **do**
    (Backtracking)
    $q_{t-1}^* = \psi_t(q_t^*)$
  **end for**

### *5.3.3 Learning*

Finally, we will see how we can learn a HMM from data via the *Baum-Welch* algorithm. First, we should note that this method assumes that the *structure* of the model is known: the number of states and observations is previously defined; therefore it only estimates the parameters. Usually the observations are given by the application domain, but the number of states, which are hidden, are not so easy to define. Sometimes the number of hidden states can be defined based on domain knowledge; other times it is done experimentally through trial and error: test the performance of the model with different numbers of states (2, 3, ...) and select the number that gives the best results. It should be noted that there is a tradeoff in this selection, as a larger number of states tends to produce better results but also implies additional computational complexity.

The Baum–Welch algorithm determines the parameters of a HMM, $\lambda = \{A, B, \Pi\}$, given a number of observation sequences, $\mathbf{O} = O_1, O_2, ...O_K$. For this it maximizes the probability of the observations given the model: $P(\mathbf{O} \mid \lambda)$. For a HMM with $N$ states and $M$ observations, we need to estimate $N + N^2 + N \times M$ parameters, for $\Pi$, $A$ and $B$, respectively.

We need to define one more auxiliary variable, $\xi$, the probability of a transition from a state $i$ at time $t$ to a state $j$ at time $t + 1$ given an observation sequence $O$:

$$\xi_t(i, j) = P(s_t = q_i, s_{t+1} = q_j \mid O, \lambda) = P(s_t = q_i, s_{t+1} = q_j, O \mid \lambda)/P(O) \quad (5.22)$$

In terms of $\alpha$ and $\beta$:

$$\xi_t(i,j) = \alpha_t(i)a_{ij}b_j(o_{t+1})\beta_{t+1}(j)/P(O) \tag{5.23}$$

Writing $P(O)$ also in terms of $\alpha$ and $\beta$:

$$\xi_t(i,j) = \alpha_t(i)a_{ij}b_j(o_{t+1})\beta_{t+1}(j)/\sum_i\sum_j \alpha_t(i)a_{ij}b_j(o_{t+1})\beta_{t+1}(j) \tag{5.24}$$

$\gamma$ can also be written in terms of $\xi$: $\gamma_t(i) = \sum_j \xi_t(i,j)$

By adding $\gamma_t(i)$ for all time steps, $\sum_t \gamma_t(i)$, we obtain an estimate of the number of times that the chain is in state $i$; and by accumulating $\xi_t(i,j)$ over $t$, $\sum_t \xi_t(i,j)$, we estimate the number of transitions from state $i$ to state $j$.

Based on the previous definitions, the Baum–Welch procedure for parameter estimation for HMMs is summarized in Algorithm 5.3.

**Algorithm 5.3** The Baum–Welch Algorithm

1. Estimate the prior probabilities—the number of times being in state $i$ at time $t = 1$.
   $\pi_i = \gamma_1(i)$
2. Estimate the transition probabilities—the number of transitions from state $i$ to $j$ between the number of times in state $i$.
   $a_{ij} = \sum_{t=1}^{T-1} \xi_t(i,j)/\sum_{t=1}^{T-1} \gamma_t(i)$
3. Estimate the observation probabilities—the number of times being in state $j$ and observing $k$ between the number of times in state $j$.
   $b_{jk} = \sum_{t=1}^{T} \gamma_t(j)/\sum_{t=1}^{T} \gamma_t(j)$, iff $O(t) = k$

Notice that the calculation of the $\gamma$ and $\xi$ variables is done in terms of $\alpha$ and $\beta$, which require the parameters of the HMM, $\Pi, A, B$. So we have encountered a "chicken and egg" problem—we need the model parameters for the Baum–Welch algorithm, which estimates the model parameters! The solution to this problem is based on the EM (for expectation-maximization) principle.

The idea is to start with some initial parameters for the model (E-step), $\lambda = \{A, B, \Pi\}$, which can be initialized randomly or based on some domain knowledge. Then, via the Baum–Welch algorithm, these parameters are re-estimated (M-step). This cycle is repeated until convergence; that is, until the difference between the parameters for the model from one step to the next is below a certain threshold.

The EM algorithm provides what is known as a *maximum-likelihood* estimator, that does not guarantee an optimal solution—it depends on the initial conditions. However, this estimator tends to work well in practice.[3]

[3] If we have some domain knowledge this could provide a *good* initialization for the parameters; otherwise, we can set them to uniform probabilities.

### 5.3.4 Gaussian Hidden Markov Models

In many applications the observations are continuous; in this case an alternative to discretization is to work directly with the continuous distributions, which are assumed Gaussian. A Gaussian hidden Markov model (GHMM) is a type of HMM where the observation probability distribution is the normal distribution. The initial state probability vector and the transition probability matrix are as in the discrete HMM. The observation probabilities given the state are modeled as Gaussian distribution. Considering a single Gaussian per state:

$$P(O_j \mid S_j) = N(\mu_j, \sigma_j^2) \tag{5.25}$$

Where $\mu_j$ is the mean and $\sigma_j^2$ the variance for state $j$.

Sometimes the observations can not be described by a single Gaussian, in this case we can use a *Gaussian Mixture Model* (GMM) [9]. A GMM consists of a number of Gaussian distributions that are combined to represent the desired distribution. The parameters required include a vector of means and a covariance matrix:

$$P(O_j \mid S_j) = N(\mu_{\mathrm{j}}, \Sigma_j) \tag{5.26}$$

The algorithms for solving the three basic problems (evaluation, optimal sequence and parameter estimation) are essentially the same as for discrete HMMs, just considering that the observations are modeled as a Gaussian distribution or a GMM.

### 5.3.5 Extensions

Several extensions to standard HMMs have been proposed to deal with particular issues in several applications [2]. Next we briefly describe some of these extensions, whose graphical models are depicted in Fig. 5.5.

Parametric HMMs (PHMMs) represent domains that involve variations in the models. In PHMMs, observation variables are conditioned to the state variable and one or more parameters that account for such variations, see Fig. 5.5b. Parameter values are known and constant on training. On testing, values that maximize the likelihood of the PHMM are recovered via a tailored expectation-maximization algorithm.

Coupled HMMs (CHMMs) join HMMs by introducing conditional dependencies between state variables—see Fig. 5.5c. These models are suitable to represent influences between sub-processes that occur in parallel.

Input-Output HMMs (IOHMMs) consider an extra *input* parameter that affects the states of the Markov chain, and optionally, the observation variables. This type of models is illustrated in Fig. 5.5d. The input variable corresponds to the observations. The output signal of IOHMMs is the class of model (e.g., a phoneme in speech

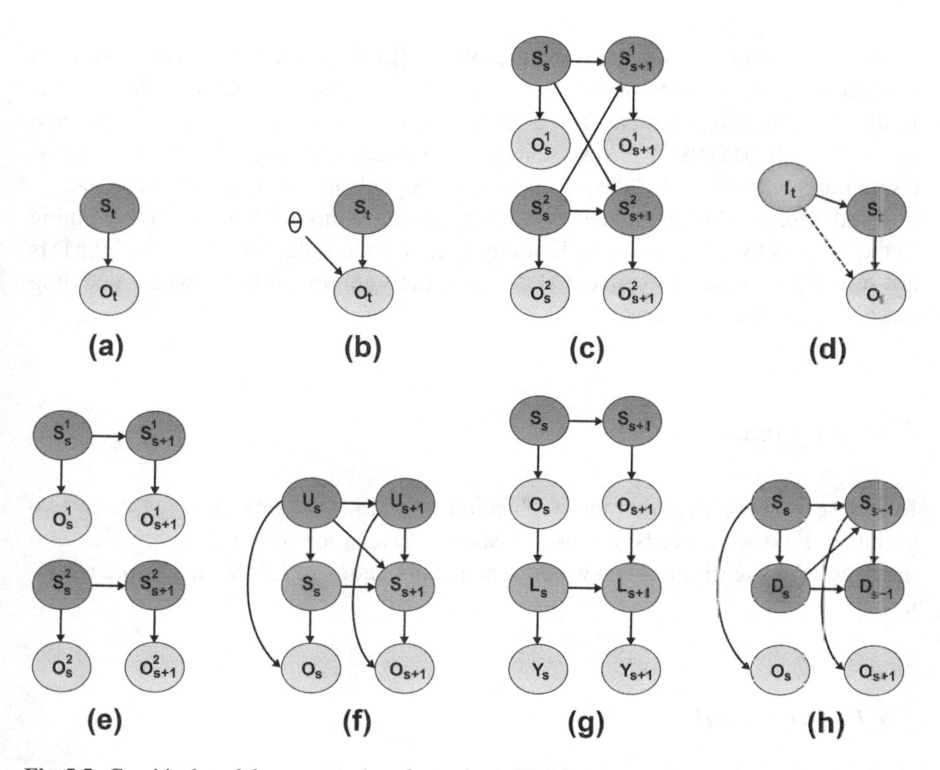


**Fig. 5.5** Graphical model representations for the basic HMM and several extensions. **a** Basic model. **b** Parametric HMMs. **c** Coupled HMMs. **d** Input-Output HMMs. **e** Parallel HMMs. **f** Hierarchical HMMs. **g** Mixed-state dynamic Bayesian networks. **h** Hidden semi-Markov models

recognition or a particular movement in gesture recognition) that is being executed. A single IOHMM can describe a complete set of classes.

Parallel HMMs (PaHMMs) require fewer HMMs than CHMMs for composite processes by assuming mutual independence between HMMs, see Fig. 5.5e. The idea is to construct independent HMMs for two (or more) independent parallel processes (for example, the possible motions of each hand in gesture recognition), and combine them by multiplying their individual likelihoods. PaHMMs with the most probable joint likelihood define the desired class.

Hierarchical hidden Markov models (HHMMs) arrange HMMs into layers at different levels of abstraction; Fig. 5.5f. In a two-layer HHMM, the lower layer is a set of HMMs that represents sub-model sequences. The upper layer is a Markov chain that governs the dynamics of these sub-models. Layering allows us to re-use the basic HMMs simply by changing upper layers.

Mixed-state dynamic Bayesian networks (MSDBNs)[4] combine discrete and continuous state spaces into a two-layer structure. MSDBNs are composed by a HMM

[4]Hidden Markov models, including these extensions, are particular types of dynamic Bayesian networks, a more general model that is described in Chap. 9.

in the upper layer and a linear dynamic system (LDS) in the lower layer. The LDS is used to model transitions between real-valued states. The output values of the HMM drive the linear system. The graphical representation of MSDBNs is depicted in Fig. 5.5g. In MSDBNs, HMMs can describe discrete high-level concepts, such as a grammar, while the LDS describes the input signals in a continuous-state space.

Hidden semi-Markov models (HSMMs) exploit temporal knowledge belonging to the process by defining an explicit duration on each state, see Fig. 5.5h. HSMMs are suitable to avoid an exponential decay of the state probabilities when modeling large observation sequences.

## 5.4 Applications

In this section we illustrate the application of Markov chains and HMMs in two domains. First we describe the use of Markov chains for ordering web pages with the PageRank algorithm. Then we present an application of HMMs in gesture recognition.

### *5.4.1 PageRank*

We can think of the World Wide Web (WWW) as a very large Markov chain, such that each web page is a state and the hyperlinks between web pages correspond to state transitions. Assume that there are $N$ web pages. A particular web page, $w_i$, has $m$ outgoing hyperlinks. If someone is at web page $w_i$, she can select any of the hyperlinks in this page to go to another page. A reasonable assumption is that each outgoing link can be selected with equal probability; thus, the transition probability from $w_i$ to any of the web pages with which it has hyperlinks, $w_j$, is $A_{ij} = 1/m$. For the other web pages for which it has no outgoing links, the transition probability is zero. In this way, according to the structure of the WWW, we can obtain the transition probability matrix, $A$, for the corresponding Markov chain. The state diagram of a small example with three web pages is shown in Fig. 5.6.

Given the transition probability matrix of the WWW, we can obtain the convergence probabilities for each state (web page) according to the Perron–Frobenius theorem (see Sect. 5.2). The convergence probability of a certain web page can be thought to be equivalent to the probability of a person, who is navigating the WWW, visiting this web page. Intuitively, web pages that have more ingoing links, from web pages with more ingoing links, will have a higher probability of being accessed.

Based on the previous ideas, L. Page et al. developed the *PageRank* algorithm which is the basis of how web pages are ordered when we make a search in *Google* [6]. The web pages retrieved by the search algorithm are presented to the user according to their convergence probabilities. The idea is that more relevant (important) web pages will tend to have a higher convergence probability.

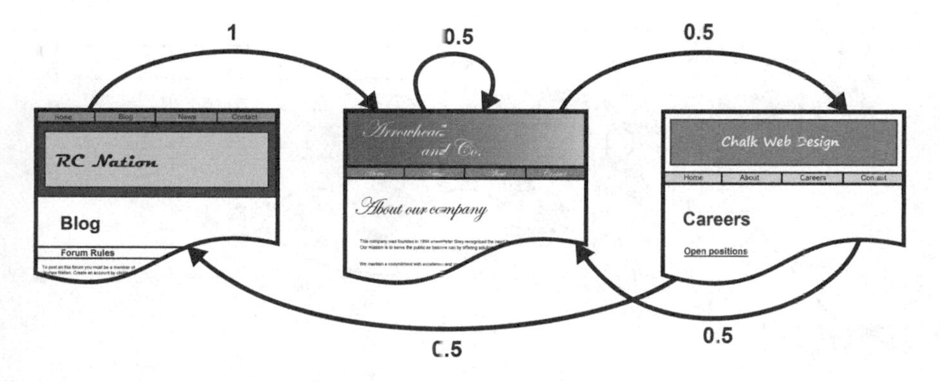


Fig. 5.6 A small example of a WWW with 3 pages

## 5.4.2 Gesture Recognition

Gestures are essential for human-human communication, so they are also important for human-computer interaction. For example, we can use gestures to command a service robot. We will focus on dynamic hand gestures, those that consist in movements by the hand/arm of a person. For example, Fig. 5.7, depicts some frames of a person executing a *stop* gesture.

For recognizing gestures, a powerful option is a hidden Markov model [1, 10]. HMMs are appropriate for modeling sequential processes, and are robust to the temporal variations common in the execution of dynamic gestures. Before we can apply HMMs to model and recognize gestures, the images in the video sequence need to be processed and a set of features extracted from them; these will constitute the observations for the HMM.

Image processing consists in detecting the person in the image, detecting their hand, and then tracking the hand in the sequence of images. From the image sequence, the position of the hand (*XYZ*) is extracted from each image. Additionally, some other regions of the body could be detected, such as the head and torso, which are used to obtain posture features as described below.

Alternatives to describe gestures can be divided in: (a) motion features, (b) posture features, and (c) posture-motion features. Motion features describe the motion of the

Fig. 5.7 A video sequence depicting a person performing a *stop* gesture with his right hand. A few key frames are shown

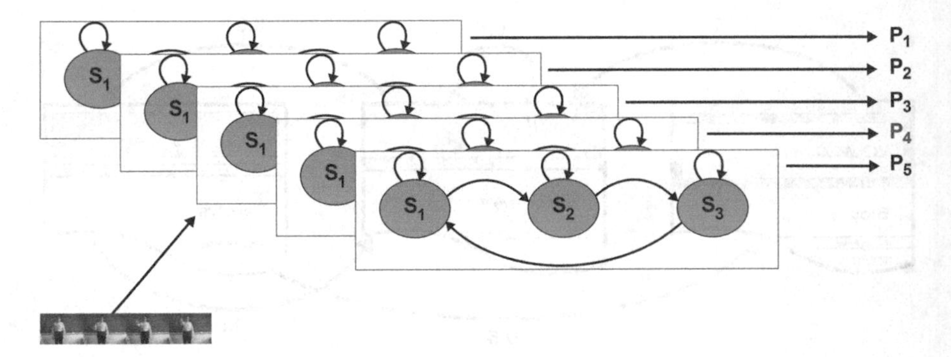


**Fig. 5.8** Gesture recognition using HMMs. The features extracted from the video sequence are fed as observations to each HMM, one per gesture class, and the probability of each model is obtained. The model with highest probability defines the recognized gesture

person's hand in the Cartesian space *XYZ*. Posture features represent the position of the hand with respect to other parts of the person's body, such as the head or torso. These motion and posture features are usually codified in a finite number of code words that provide the observations for the HMMs. For example, if we consider three values to describe each motion coordinate and two binary posture features (e.g., hand above the head and hand over torso), there will be $m = 3^5 \times 2^2 = 972$ possible observations. These are obtained for each frame (or each $n$ frames) in the video sequence of the gesture.

To recognize $N$ different gestures, we need to train $N$ HMMs, one for each gesture. The first parameter that needs to be defined is the number of hidden states for each model. As mentioned before, this can be set based on domain knowledge, or obtained via cross-validation by experimentally evaluating different amounts of states. In the case of dynamic gestures, we can think of the states as representing the different stages in the movement of the hand. For instance, the *stop* gesture can be thought of as having three phases: moving the arm up and forward, extending the hand, and moving the arm down; this implies 3 hidden states. Experimentally, it has been found that using 3 to 5 states usually produces good results.

Once the number of states for each gesture model are defined (these could be different for each model), the parameters are obtained using the Baum–Welch procedure. For this, a set of $M$ training sequences for each gesture class are required, and the more samples the better. Thus, $N$ HMMs are obtained, one for each gesture type.

For recognition, the features are extracted from the video sequence. These are the observations, $O$, for the $N$ HMMs models $\lambda_i$, one for each gesture type. The probability of each model given the observation sequence, $P(O \mid \lambda_i)$, are obtained using the Forward algorithm. The model with the highest probability, $\lambda_k^*$, is selected as the recognized gesture. Figure 5.8 illustrates the recognition process, considering 5 classes of gestures.

## 5.5 Additional Reading

A general introduction to Markov chains is provided in [4]. Rabiner [7] presents an excellent introduction to HMMs and their application to speech recognition; [8] provides a more general overview of speech recognition. A review of several extensions of HMMs in the context of gesture recognition is given in [2]. [5] analyzes search engines and the PageRank algorithm. The application of HMMs to gesture recognition is described by [10] and [1] Open software for HMMs is available in [3].

## 5.6 Exercises

1. For the Markov chain weather model: (i) determine the probability of the state sequence: *cloudy*, *raining*, *sunny*, *sunny*, *sunny*, *cloudy*, *raining*, *raining*, (ii) what is the probability of four continuous rainy days?, (iii) what is the expected number of days that it will continue raining?
2. Given the following sequences of observations, estimate the parameters for the weather MC (s=sunny, c=cloudy, r=raining):
   c,c,r,c,r,c,s,s
   s,c,r,r,r,s,c,c
   s,s,s,c,r,r,r,c
   r,r,c,r,c,s,s,s
3. Given the following transition matrix for a Markov chain: $A = \begin{matrix} 0.9 & 0.075 & 0.025 \\ 0.15 & 0.8 & 0.05 \\ 0.25 & 0.25 & 0.5 \end{matrix}$, obtain the convergence probabilities for each state.
4. For the small web page example of Fig. 5.6, determine: (i) if the convergence conditions are satisfied, and if so, (ii) the order in which the three web pages will be presented to the user.
5. What are the three basic assumptions in standard HMMs? Express them mathematically.
6. Consider the unfair coin example. Given the parameters in Table 5.5, obtain the probability of the following observation sequence: *HHTT* using (i) the direct method, (ii) the forward algorithm.
7. For the previous problem, what is the number of operations for each of the two methods?
8. For problem 6, obtain the most probable state sequence using Viterbi's algorithm.
9. Consider we extend the weather model to a HMM with 3 hidden states (q1=sunny, q2=cloudy, q3=raining), and 3 observations that correspond to the amount (in millimeters) of registered rain in a day: o1=0, o2=5, o3=10. Given the following observation sequences:
   0,0,5,10,5,0,5,10
   5,10,10,10,5,0,0,0
   10,5,5,0,0,10,5,10
   0,5,10,5,0,0,5,10
   Estimate the parameters of the HMM.

10. Based on the HMM's parameters from the previous problem, obtain the most probable state sequence using Viterbi's algorithm given the following observations: 0,10,5,10
11. We want to model the weather example as a Gaussian HMM, considering a single Gaussian per state. Given the following parameters for the GHMM: $\Pi = 0.2, 0.5, 0.3$; $A = \begin{matrix} 0.7 & 0.2 & 0.1 \\ 0.2 & 0.6 & 0.2 \\ 0.3 & 0.2 & 0.5 \end{matrix}$; $b(q_1) = 2, 0.5$, $b(q_2) = 5, 1$, $b(q_3) = 10, 0.5$, where $b(q_i) = \mu, \sigma^2$ represents the mean and variance of the Gaussian distribution per state. Obtain the probability of the following observation: 5.7, 1.1, 3.5, 9.9.
12. Assume there are two HMMs that represent two phonemes: *ph*1 and *ph*2. Each model has two states and two observations with the following parameters: Model 1: $\Pi = [0.5, 0.5]$, $A = [0.5, 0.5 \mid 0.5, 0.5]$, $B = [0.8, 0.2 \mid 0.2, 0.8]$ Model 2: $\Pi = [0.5, 0.5]$, $A = [0.5, 0.5 \mid 0.5, 0.5]$, $B = [0.2, 0.8 \mid 0.8, 0.2]$ Given the following observation sequence: "o1,o1,o2,o2", which is the most probable phoneme?
13. We want to develop a head gesture recognition system and we have a vision system that can detect the following movements of the head: (1) up, (2) down, (3) left, (4) right, (5) stable. The vision system provides a number for each type of movement (1 to 5) each second, which is the input (observation) to the gesture recognition system. The gesture recognition system should recognize four classes of head gestures: (a) affirmative action, (b) negative action, (c) turn right, (d) turn left. (i) Specify a model that is adequate for this recognition problem, including the structure and required parameters. (ii) Indicate which algorithms are appropriate for learning the parameters of the model, and for recognition.
14. *** Develop a program to solve the previous problem.
15. *** An open problem for HMMs is establishing the *optimum* number of states for each model. Develop a search strategy for determining the optimum number of states for each model for the head gesture recognition system that maximizes the recognition rate. Consider that the data set (examples for each class of gesture) is divided in three sets: (a) training–to estimate the parameters of the model, (b) validation–to compare the different models, (c) test–for testing the final models.

## References

1. Aviles, H., Sucar, L.E., Mendoza C.E.: Visual recognition of similar gestures. In: 18th International Conference on Pattern Recognition, pp. 1100–1103 (2006)
2. Aviles, H., Sucar, L.E., Mendoza, C.E., Pineda, L.A.: A comparison of dynamic Naive bayesian classifiers and hidden markov models for gesture recognition. J. Appl. Res. Technol. **9**(1), 81–102 (2011)

3. Kanungo, T.: Hidden markov models software. http://www.kanungo.com/. Accessed 26 May 2008
4. Kemeny, J.K., Snell, L.: Finite Markov Chains. Van Nostrand, New York (1965)
5. Langville, N., Carl, D., Meyer, C.D.: Google's PageRank and Beyond: The Science of Search Engine Rankings. Princeton University Press, Princeton (2012)
6. Page, L., Brin, S., Motwani, R., Winograd T.: The PageRank Citation Ranking: Bringing Order to the Web, Stanford Digital Libraries Working Paper (1998)
7. Rabiner, L.E.: A tutorial on hidden markov models and selected applications in speech recognition. In: Waibel A., Lee, K. (eds.) Readings in speech recognition, Morgan Kaufmann, pp. 267–296 (1990)
8. Rabiner, L., Juang, B.H.: Fundamentals on Speech Recognition. Prentice-Hall Signal Processing Series. Prentice-Hall, Englewood Cliffs, NJ (1993)
9. Gaussian, R.D.: Models, Mixture: In: Li, S.Z., Jain, A. (eds.) Encyclopedia of Biometrics. Springer, Boston, MA (2009)
10. Wilson, A., Bobick, A.: Using hidden markov models to model and recognize gesture under variation. Int. J. Pattern Recogn. Artif. Intell. Special Issue Hidden Markov Models Comput. Vis. **15**(1), 123–160 (2000)

# Chapter 6
# Markov Random Fields

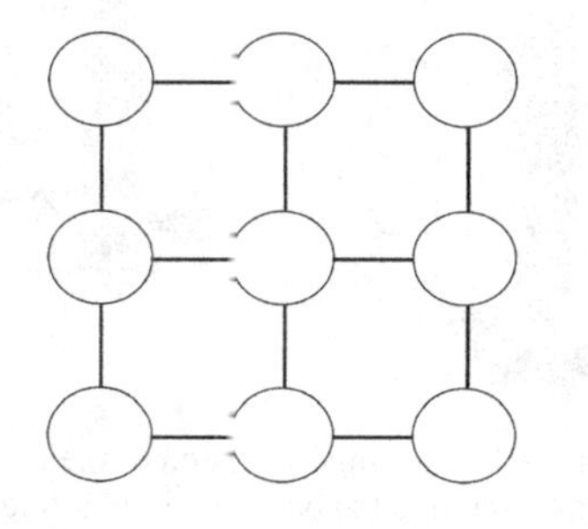

## 6.1 Introduction

Certain processes, such as a ferromagnetic material under a magnetic field, or an image, can be modeled as a series of *states* in a chain or a regular grid. Each state can take different values and is influenced probabilistically by the states of its neighbors. These models are known as *Markov random fields* (MRFs).

MRFs originated to model ferromagnetic materials in what is known as the *Ising* model [2]. In an Ising model, there are a series of random variables in a line; each random variable represents a dipole that could be in two possible states, *up* (+) or *down* (−). The state of each dipole depends on an external field and the state of its neighbor dipoles in the line. A simple example with four variables is illustrated Fig. 6.1. A *configuration* of a MRF is a particular assignment of values to each variable in the model; in the case of the model in Fig. 6.1, there are 16 possible configurations: $++++, +++-, ++-+, ..., ----$.

A MRF is represented as an undirected graphical model, such as in the previous example. An important property of a MRF is that the state of a variable is independent of all other variables in the model given its neighbors in the graph. For instance, in the example in Fig. 6.1, $q_1$ is independent of $q_3$ and $q_4$ given $q_2$. That is $P(q_1 \mid q_2, q_3, q_4) = P(q_1 \mid q_2)$.



L. E. Sucar, *Probabilistic Graphical Models*, Advances in Computer Vision and Pattern Recognition, https://doi.org/10.1007/978-3-030-61943-5_6

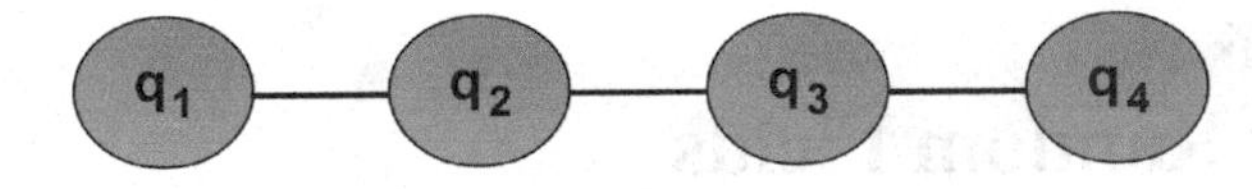


**Fig. 6.1** An example of an Ising (MRF) model with four variables

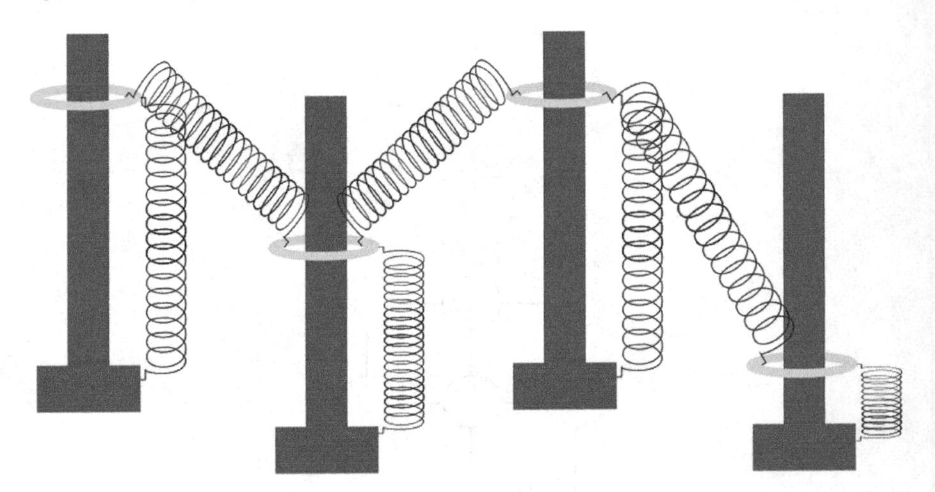

**Fig. 6.2** Physical analogy of a MRF. The rings will tend to the configuration of minimum energy according to the springs that attach them to the base (external influence) and its neighbors (internal influence)

The central problem in a MRF is to find the configuration of maximum probability. Usually, the probability of a configuration depends on the combination of an *external* influence (e.g., a magnetic field in the Ising model) and the *internal* influence of its neighbors. More generally, the posterior probability of a configuration depends on the prior knowledge or context, and the data or likelihood.

Using a physical analogy, a MRF can be thought of as a series of rings in poles, where each ring represents a random variable, and the height of a ring in a pole corresponds to its state. The rings are arranged in a line, see Fig. 6.2. Each ring is attached to its neighbors with a spring, this corresponds to the internal influences; and it is also attached to the base of its pole with another spring, representing the external influence. The relation between the springs' constants defines the relative weight between the internal and external influences. If the rings are left loose, they will stabilize to a configuration of minimum energy, that corresponds to the configuration with maximum probability in a MRF.

Markov random fields, also known as *Markov networks*, are formally defined in the next section.

## 6.2 Markov Random Fields

A *random field* (RF) is a collection of $S$ random variables, $\mathbf{F} = F_1, \ldots, F_s$, indexed by sites. Random variables can be discrete or continuous. In a discrete RF, a random variable can take a value $l_i$ from a set of $m$ possible values or labels $L = \{l_1, l_2, \ldots, l_m\}$. In a continuous RF, a random variable con take values from the real numbers, $R$, or from an interval of them.

A *Markov random field* or Markov network (MN) is a random field that satisfies the locality property, also known as the Markov property: a variable $F_i$ is independent of all other variables in the field given its neighbors, $Nei(F_i)$. That is:

$$P(F_i \mid \mathbf{F_c}) = P(F_i \mid Nei(F_i)) \tag{6.1}$$

where $\mathbf{F_c}$ is the set of all random variables in the field except $F_i$.

Graphically, a MRF is an undirected graphical model which consists of a set of random variables, $\mathbf{V}$, and a set of undirected edges, $\mathbf{E}$. These form an undirected graph that represents the independency relations between the random variables according to the following criteria. A subset of variables $\mathbf{A}$ is independent of the subset of variables $\mathbf{C}$ given $\mathbf{B}$, if the variables in $\mathbf{B}$ *separate* $\mathbf{A}$ and $\mathbf{C}$ in the graph. That is, if the nodes in $\mathbf{B}$ are removed from the graph, then there are no trajectories between $\mathbf{A}$ and $\mathbf{C}$.

Figure 6.3 depicts an example of a MRF with 5 variables, $q_1, \ldots, q_5$. For instance, in this example, $q_1, q_4$ (**A**) are independent of $q_3$ (**C**) given $q_2, q_5$ (**B**).

The joint probability of a MRF can be expressed as the product of local functions on subsets of variables in the model. These subsets should include, at least, all the *cliques* in the network. For the MRF of Fig. 6.3, the joint probability distribution can be expressed as:

$$P(q_1, q_2, q_3, q_4, q_5) = (1/k)P(q_1, q_4, q_5)P(q_1, q_2, q_5)P(q_2, q_3, q_5) \tag{6.2}$$

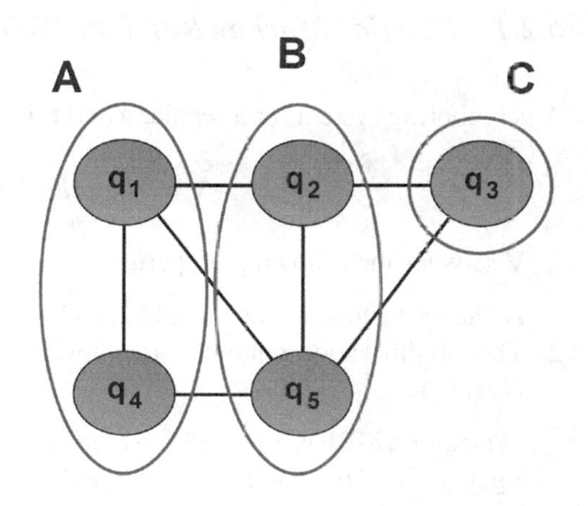


**Fig. 6.3** Example of a MRF, in which $q_1, q_4$ $(A)$ are independent of $q_3$ $(C)$ given $q_2, q_5$ $(B)$

where $k$ is a normalizing constant. For practical convenience, other subsets of variables can also be considered for the joint probability calculation. If we also include subsets of size two, then the joint distribution for the previous example can be written as:

$$P(q_1, q_2, q_3, q_4, q_5) = (1/k)P(q_1, q_4, q_5)P(q_1, q_2, q_5)P(q_2, q_3, q_5)$$
$$P(q_1, q_2)P(q_1, q_4), P(q_1, q_5)P(q_2, q_3)P(q_2, q_5)P(q_3, q_5)P(q_4, q_5) \quad (6.3)$$

Formally, a MRF is a set of random variables, $\mathbf{X} = X_1, X_2, \ldots, X_n$ that are indexed by $V$, such that $G = (V, E)$ is an undirected graph, that satisfies the Markov property: a variable $X_i$ is independent of all other variables given its neighbors, $Nei(X_i)$:

$$P(X_i \mid X_1, ...X_{i-1}, X_{i+1}, ..., X_n) = P(X_i \mid Nei(X_i)) \quad (6.4)$$

The neighbors of a variable are all the variables that are directly connected to it in the graph.

Under certain conditions (if the probability distribution is strictly positive), the the joint probability distribution of a MRF can be factorized over the cliques of the graph:

$$P(\mathbf{X}) = (1/k) \prod_{C \in Cliques(G)} \phi_C(X_C) \quad (6.5)$$

where $k$ is a normalizing constant and $\phi_C$ is a local function over the variables in the corresponding clique $C$.

A MRF can be categorized as *regular* or *irregular*. When the random variables are in a lattice it is considered regular; for instance, they could represent the pixels in an image; if not, they are irregular. Next we will focus on regular Markov random fields.

### 6.2.1 Regular Markov Random Fields

A neighboring system for a regular MRF $\mathbf{F}$ is defined as:

$$\mathbf{V} = \{Nei(F_i) \mid \forall_i \in \mathbf{F_i}\} \quad (6.6)$$

$\mathbf{V}$ satisfies the following properties:

1. A site in the field is not a neighbor to itself.
2. The neighborhood relations are symmetric, that is, if $F_j \in Nei(F_i)$ then $F_i \in Nei(F_j)$.

Typically, a MRF is arranged as a regular grid. An example of a 2D grid is depicted in Fig. 6.4. For a regular grid, a neighborhood of order $i$ is defined as:

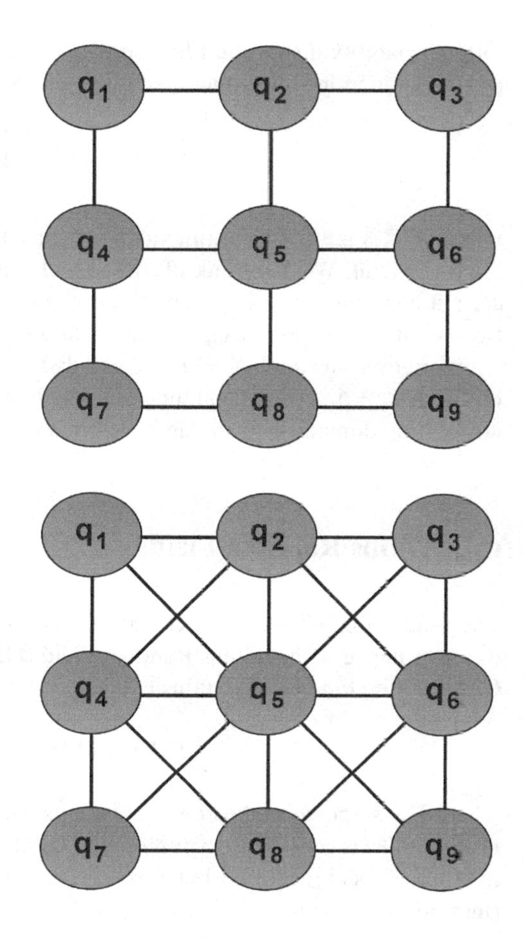


**Fig. 6.4** A regular 2D MRF with a first order neighborhood

**Fig. 6.5** A regular 2D MRF with a second order neighborhood

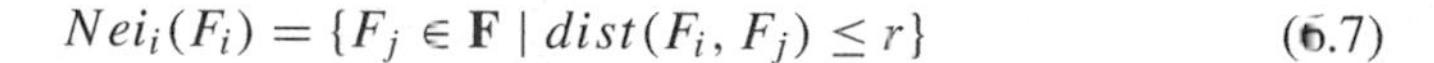

$$Nei_i(F_i) = \{F_j \in \mathbf{F} \mid dist(F_i, F_j) \leq r\} \tag{6.7}$$

where $dist(x, y)$ is the Euclidean distance between $x$ and $y$, considering a unit distance as the vertical and horizontal distance between sites in the grid. The radius, $r$, is defined for each order. For example, $r = 1$ for order one, each interior site has 4 neighbors; $r = \sqrt{2}$ for order two, each interior site has 8 neighbors; $r = 2$ for order three, each interior site has 12 neighbors; and so on. Figure 6.4 shows and example of a neighborhood of order one, and Fig. 6.5 of a neighborhood of order two.

Once the structure of the MRF is specified based on the neighborhood order, its parameters must be defined. The parameters of a regular MRF are specified by a set of local functions. These functions correspond to joint probability distributions of subsets of completely connected variables in the graph. It is sufficient to include all the cliques in the graph, but other completely connected subsets can also be included. For instance, in the case of a first order MRF, there are subsets of 2 variables; in the case of a second order MRF, there are subsets of 2, 3 and 4 variables. In general, the joint probability distribution for the whole field can be expressed as the product of the local functions for different subsets of variables:

$$P(\mathbf{F}) = (1/k) \prod_i f(X_i) \tag{6.8}$$

where $f(X_i)$ are the local functions for subsets of variables $X_i$ and $k$ is a normalizing constant. We can think of these local functions as *constraints* that will favor certain configurations. For example, in the case of the Ising model, if we consider two neighboring variables, $X, Y$, the local function will favor (higher probabilities) configurations in which $X = Y$ and will disfavor (lower probabilities) configurations in which $X \neq Y$. These local functions can be defined subjectively depending on the application domain, or they can be learned from data.

## 6.3 Gibbs Random Fields

The joint probability of a MRF can be expressed in a more convenient way given its equivalence with a Gibbs Random Field (GRM), according to the Hammersley–Clifford theorem [4]. Given this equivalence, we can rewrite Eq. 6.8 as:

$$P(\mathbf{F}) = (1/z) \exp(-\mathbf{U_F}) \tag{6.9}$$

where $\mathbf{U_F}$ is known as the *energy*, given its analogy with physical energy. So maximizing $P(\mathbf{F})$ is equivalent to minimizing $\mathbf{U_F}$. The energy function can also be written in terms of local functions, but as this is an exponent, it is the sum of these functions (instead of a product):

$$\mathbf{U_F} = \sum_i U_i(X_i) \tag{6.10}$$

Considering a regular MRF of order $n$, the energy function can be expressed in terms of functions of subsets of completely connected variables of different sizes, $1, 2, 3, \ldots$:

$$\mathbf{U_F} = \sum_i U_1(F_i) + \sum_{i,j} U_2(F_i, F_j) + \sum_{i,j,k} U_3(F_i, F_j, f_k) + \cdots \tag{6.11}$$

where $U_i$ are the local energy functions, known as *potentials*, for subsets of size $i$. Note that potentials are the inverse of probabilities, so low potentials are equivalent to high probabilities.

Given the Gibbs equivalence, the problem of finding the configuration of maximum probability for a MRF is transformed to find the configuration of minimum energy.

In summary, to specify a MRF we must define:

- A set of random variables, **F**, and their possible values, $L$.
- The dependency structure, or in the case of a regular MRF a neighborhood scheme.
- The potentials for each subset of completely connected nodes (at least the cliques).

## 6.4 Inference

As we mentioned before, the more common application of MRFs consists in finding the most probable configuration; that is, the value for each variable that maximizes the joint probability. Given the Gibbs equivalence, this is the same as minimizing the energy function, expressed as a sum of local functions.

The set of all possible configurations of a MRF is usually very large, as it increases exponentially with the number of variables in **F**. For the discrete case with $m$ possible labels, the number of possible configurations is $m^N$, where $N$ is the number of variables in the field. If we consider a MRF representing a binary image of $100 \times 100$ pixels (a small image), then the number of configurations is $2^{10,000}$! Thus, it is impossible to calculate the energy (potential) for every configuration, except in the case of very small fields.

Finding the most probable configuration is usually posed as a stochastic search problem. Starting from an initial, random assignment of each variable in the MRF, this configuration is *improved* via local operations, until a configuration of minimum energy is obtained. In general, the minimum is a local minimum in the energy function; it is difficult to guarantee a global optimum.

A general stochastic search procedure for finding a configuration of minimum energy is outlined in Algorithm 6.1. After initializing all the variables with a random value, each variable is changed to an alternative value and its new energy is estimated. If the new energy is lower than the previous one, the value is changed; otherwise, the value may also change with a certain probability—this is done to avoid local minima–. The function $random$ generates a pseudo-random number (with a uniform distribution) which is multiplied by the energy difference and compared to a threshold to decide if the value is changed or not. There are different variants of this algorithm according on how this threshold is defined, as explained below. This process is repeated for a number of iterations (or until convergence).

There are several variants of this general algorithm according to variations on different aspects. One is the way in which the *optimal* configuration is defined, for which there are two main alternatives: MAP and MPM. In the *Maximum A posteriori Probability* or MAP, the optimum configuration is taken as the configuration at the end of the iterative process. In the case of *Maximum Posterior Marginals* or MPM, the most frequent value for each variable in all the iterations is taken as the optimum configuration.

**Algorithm 6.1** Stochastic Search Algorithm

**Require:** MRF, F; Energy function, $U_F$; Number of iterations, $N$; Number of variables, $S$; Probability threshold, $T$;
**for** $i = 1$ **to** $S$ **do**
  $F(i) = l_k$ (Initialization)
**end for**
**for** $i = 1$ **to** $N$ **do**
  **for** $j = 1$ **to** $S$ **do**
    $t = l_{k+1}$ (An alternative value for variable F(i))
    **if** $U(t) < U(F(i))$ **then**
      $F(i) = t$ (Change value of F(i) if the energy is lower)
    **else**
      **if** $random \times \mid U(t) - U(F(i)) \mid < T$ **then**
        $F(i) = t$ (With certain probability change F(i) if the energy is higher)
      **end if**
    **end if**
  **end for**
**end for**
**return F*** (Return final configuration)

Regarding the optimization process, there are three main variations:

Iterative Conditional Modes (ICM): it always selects the configuration of minimum energy.
Metropolis: with a fixed probability, $P$, it selects a configuration with a higher energy.
Simulated annealing (SA): with a variable probability, $P(T)$, it selects a configuration with higher energy; where $T$ is a parameter known as *temperature*. The probability of selecting a value with higher energy is determined based on the following expression: $P(T) = e^{-\delta U/T}$; where $\delta U$ is the energy difference. The algorithm starts with a *high* value for $T$ and this is reduced with each iteration. This makes the probability of going to higher energy states high initially, and it subsequently decreases tending to zero at the end of the process.

ICM provides the most efficient approach, and can be used if the optimization *landscape* is known to be convex or relatively "simple". SA is the most robust approach against local minima, nevertheless it tends to be the most costly computationally. Metropolis provides a middle ground between ICM and SA. The choice of which variation to use depends on the application, in particular the size of the model, the complexity of the *landscape* and efficiency considerations.

## 6.5 Parameter Estimation

The definition of a Markov random field includes several aspects:

- The structure of the model—in the case of a regular MRF the neighborhood system.
- The form of the local probability distribution functions—for each complete set in the graph.
- The parameters of the local functions.

In some applications, the previous aspects can be defined subjectively; however, this is not always easy. Different choices for the structure, distribution and parameters can have a significant impact on the results of applying the model to specific problems. Thus, it is desirable to learn the model from data, which can have several levels of complexity. The simplest case, which is non trivial, is when we know the structure and functional form, and we only need to estimate the parameters given a *clean* realization (without noise) of the MRF, $f$. It becomes more complex if the data is noisy, and even more difficult if we want to learn the functional form of the probability distribution and the order of the neighborhood system. Next we will cover the basic case, learning the parameters of a MRF from data.

### 6.5.1 Parameter Estimation with Labeled Data

The set of parameters, $\theta$, of a MRF, $F$, are estimated from data, $f$, assuming no noise. Given $f$, the maximum likelihood (ML) estimator maximizes the probability of the data given the parameters, $P(f \mid \theta)$; thus the optimum parameters are:

$$\theta^* = ArgMax_\theta P(f \mid \theta) \tag{6.12}$$

When the prior distribution of the parameters, $P(\theta)$, is known, we can apply a Bayesian approach and maximize the posterior density obtaining the MAP estimator:

$$\theta^* = ArgMax_\theta P(\theta \mid f) \tag{6.13}$$

where:

$$P(\theta \mid f) \sim P(\theta)P(f \mid \theta) \tag{6.14}$$

The main difficulty in the ML estimation for a MRF is that it requires the evaluation of the normalizing partition function $Z$, in the Gibbs distribution, since it involves summing over all possible configurations. Remember that the likelihood function is given by:

$$P(f \mid \theta) = (1/Z)\exp(-U(f \mid \theta)) \tag{6.15}$$

where the partition function is:

$$Z = \sum_{f \in F} \exp(-U(f \mid \theta)) \tag{6.16}$$

Thus, the computation of $Z$ is intractable even for MRFs of moderate size. So approximations are used for solving this problem efficiently.

One possible approximation is based on the conditional probabilities of each variable in the field, $f_i$, given its neighbors, $N_i$: $P(f_i \mid f_{N_i})$, and assuming that these are independent, we obtain what is known as the *pseudo-likelihood* (PL) [1]. Then the energy function can be written as:

$$U(f) = \sum_i U_i(f_i, f_{N_i}) \tag{6.17}$$

Assuming a first order regular MRF, only single and pairs of nodes are considered, so:

$$U_i(f_i, f(N_i)) = V_1(f_i) + \sum_{f_j \in f(N_i)} V_2(f_i, f_j) \tag{6.18}$$

where $V_1$ and $V_2$ are the single and pair potentials, respectively; and $f_j$ are the neighbors of $f_i$.

The pseudo-likelihood (PL) is defined as the simple product of the conditional likelihoods:

$$PL(f) = \prod_i P(f_i \mid f_{N_i}) = \prod_i \frac{[\exp(-U_i(f_i, f_{N_i}))]}{[\sum_{f_i} \exp(-U_i(f_i, f_{N_i}))]} \tag{6.19}$$

Given that $f_i$ and $f_{N_i}$ are not independent, the PL is not the *true* likelihood; however, it has been proven that in the large lattice limit it converges to the truth with probability one [3].

Using the PL approximation, and given a particular structure and form of the local functions, we can estimate the parameters of a MRF model based on data.

Assuming a discrete MRF and given several realizations (examples), the parameters can be estimated using histogram techniques. Assume there are $N$ distinct sets of instances of size $k$ in the dataset, and that a particular configuration $(f_i, f_{N_i})$ occurs $H$ times, then an estimate of the probability of this configuration is $P(f_i, f_{N_i}) = H/N$. Note that the potentials are inversely proportional to the probabilities, so an option is simply to make them the inverse: $U(f_i, f_{N_i}) = 1/P(f_i, f_{N_i})$

## 6.6 Conditional Random Fields

A limitation of MRFs (and HMMs) is that it is usually assumed that the observations are independent given each state variable. For example, in a hidden Markov model, an observation $O_t$ is conditionally independent of all other observations and states

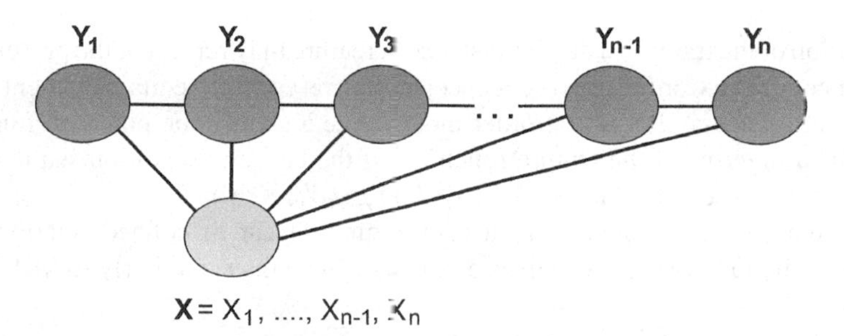


**Fig. 6.6** Graphical representation of a chain–structured conditional random field

given $S_t$. In traditional MRFs, it is also assumed that each observation only depends on a single variable, and it is conditionally independent of the other variables in the field (see Sect. 6.7). There are applications in which these independence assumptions are not appropriate, for example labeling the words in a sentence in natural language; in which there could be long-range dependencies between observations (words).

HMMs and *traditional* MRFs are *generative* models; which represent the joint probability distribution as the product of local functions based on the independence assumptions. If these conditional independence assumptions are removed, the models become intractable. One alternative that does not require these assumptions are *conditional random fields* (CRF) [10, 11].

Generative models explicitly attempt to model a joint probability distribution $P(Y, X)$ over inputs and outputs. Although this approach has advantages, it also has important limitations. Not only the dimensionality of $X$ can be very large, but the features may have complex dependencies. Modeling the dependencies among inputs can lead to intractable models, but ignoring them can lead to reduced performance. A solution to this problem is the discriminative approach, that models the conditional distribution $P(Y|X)$ directly, which is all that is needed for classification. This is the approach taken by conditional random fields.

A conditional random field is an undirected graphical model globally conditioned on $\mathbf{X}$, the random variable representing observations [8]. Conditional models are used to label an observation sequence $\mathbf{X}$ by selecting the label sequence $\mathbf{Y}$ that maximizes the conditional probability $P(\mathbf{Y}|\mathbf{X})$. The conditional nature of such models means that no effort is wasted on modeling the observations, and one is free from having to make unnecessary independence assumptions. The simplest structure of a CRF is that in which the nodes corresponding to elements of $Y$ form a simple first-order chain, as illustrated in Fig. 6.6.

In an analogous way to MRFs, the joint distribution is factorized on a set of potential functions, where each potential function is defined over a set of random variables whose corresponding vertices form a maximal clique of $G$, the graphical structure of the CRF. In the case of a chain-structured CRF each potential function will be specified on pairs of adjacent label variables, $Y_i$ and $Y_{i+1}$.

The potential functions can be defined in terms of *feature functions* [8], which are based on real-valued features that express some characteristic of the empirical distribution of the training data. For instance, a feature may represent the presence (1) or absence (0) of a word in a text sequence; or the presence of a certain element (edge, texture) in an image. For a first-order model (like a chain), the potential functions are defined in terms of the feature functions of the entire observation sequence, $\mathbf{X}$, and a pair of consecutive labels, $Y_{i-1}, Y_i$: $U(Y_{i-1}, Y_i, \mathbf{X}, f)$.

Considering a first-order chain, the energy function can be defined in terms of the potential function of single variables and pairs of variables, similarly to MRFs:

$$E = \sum_j \lambda_j t_j(Y_{i-1}, Y_i, \mathbf{X}, f) + \sum_k \mu_k g_k(Y_i, \mathbf{X}, f), \tag{6.20}$$

where $\lambda_j$ and $\mu_k$ are parameters which weigh the contribution of the variable pairs (internal influence) and the single variables (external influence) respectively; these could be estimated from training data. The main difference with MRFs is that these potentials are conditioned on the entire observation sequence. Parameter estimation and inference are performed in a similar way as for MRFs.

## 6.7 Applications

Markov random fields have been applied to several tasks in image processing and computer vision. For example, MRFs are used for image smoothing, image restoration, segmentation, image registration, texture synthesis, super-resolution, stereo matching, image annotation and information retrieval. We describe two applications: image smoothing and improving image annotation.

### *6.7.1 Image Smoothing*

Digital images are usually corrupted by high frequency noise. For reducing the noise a *smoothing* process can be applied to the image. For this, there are several alternatives; one is to use a MRF.

We can define a MRF associated to a digital image, in which each pixel corresponds to a random variable. Considering a first order MRF, each interior variable is connected to its 4 neighbors. Additionally, each variable is also connected to an *observation* variable that has the value of the corresponding pixel in the image, see Fig. 6.7.

Once the structure of the MRF is defined, we need to specify the local potential functions. A property of natural images is that, in general, they have certain continuity, that is, neighboring pixels will tend to have similar values. Thus, we can propose a restriction that forces neighboring pixels to have similar values, by punishing (higher energy) configurations in which neighbors have different values. At the same time, it is desirable for each variable in the MRF to have a value similar to the one in the

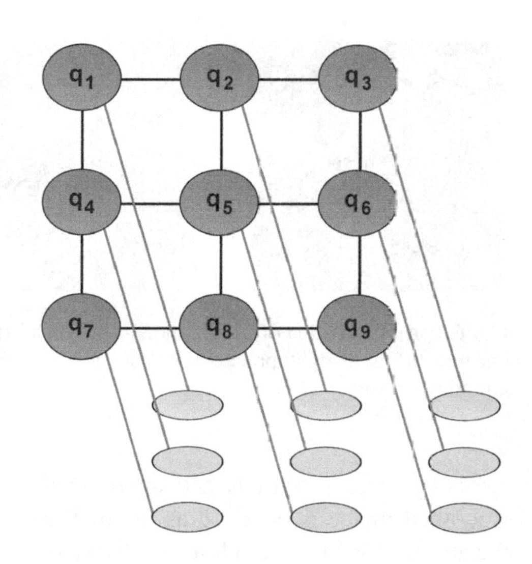


**Fig. 6.7** An example of a MRF associated to an image. The upper part depicts a first order 2D MRF for a 3 × 3 image. The lower part represents the image pixels, each one connected to a corresponding variable in the field

original image; so we also punish configurations in which the variables have different values to their corresponding observations. So the solution will be a compromise between these two types of restrictions, similarity to neighbors and similarity to observations.

The energy function, in this case, can be expressed as the sum of two types of potentials: one associated to pairs of neighbors, $U_c(f_i, f_j)$; and the other for each variable and its corresponding observation, $U_o(f_i, g_i)$. Thus, the energy will be the summation of these two types of potentials:

$$\mathbf{U_F} = \sum_{i,j} U_c(f_i, f_j) + \lambda \sum_{i} U_o(f_i, g_i) \tag{6.21}$$

where $\lambda$ is a parameter which controls which aspect is given more importance, the observations ($\lambda > 1$) or the neighbors ($\lambda < 1$); and $g_i$ is the observation variable associated to $f_i$.

Depending on the desired behavior for each type of potential, these can be defined to penalize the difference with the neighbors or the observations. Thus, a reasonable function is the quadratic difference. Then, the neighbors potential is:

$$U_c(f_i, f_j) = (f_i - f_j)^2 \tag{6.22}$$

And the observation potential is:

$$U_o(f_i, g_i) = (f_i - g_i)^2 \tag{6.23}$$

**Fig. 6.8** An illustration of image smoothing with a MRF. Left: original image; center: processed image with $\lambda = 1$; right: processed image with $\lambda = 0.5$. It can be observed that a smaller value of $\lambda$ produces a smoother image

Using these potentials and applying the stochastic optimization algorithm, a smoothed image is obtained as the final configuration of $\mathbf{F}$. Figure 6.8 illustrates the application of a smoothing MRF to a digital image varying the value of $\lambda$.

### 6.7.2 Improving Image Annotation

Automatic image annotation is the task of automatically assigning annotations or labels to images or segments of images, based on their local features. Image annotation is frequently performed by automatic systems; it is a complex task due to the difficulty of extracting adequate features which allow to generalize and distinguish an object of interest from others with similar visual properties. Erroneous labeling of regions is a common consequence of the lack of a good characterization for the classes by low-level features.

When labeling a segmented image, we can incorporate additional information to improve the annotation of each region of the image. The labels of each region of an image are usually not independent; for instance in an image of animals in the jungle, we will expect to find a sky region *above* the animal, and trees or plants *below* or *near* the animal. Thus, the *spatial relations* between the different regions in the image can help to improve the annotation [5].

We can use Markov random fields to represent the information about the spatial relations among the regions in an image, such that the probability of occurrence of a certain spatial relation between each pair of labels could be used to obtain the most probable label for each region, i.e., the most probable configuration of labels for the entire image. Thus, using a MRF we can combine the information provided by the visual features for each region (external potential) and the information from the spatial relations with other regions in the image (internal potential). By combining both aspects in the potential function, and applying the optimization process, we can obtain a configuration of labels that *best* describe the image.

The procedure is basically the following (see Fig. 6.9):

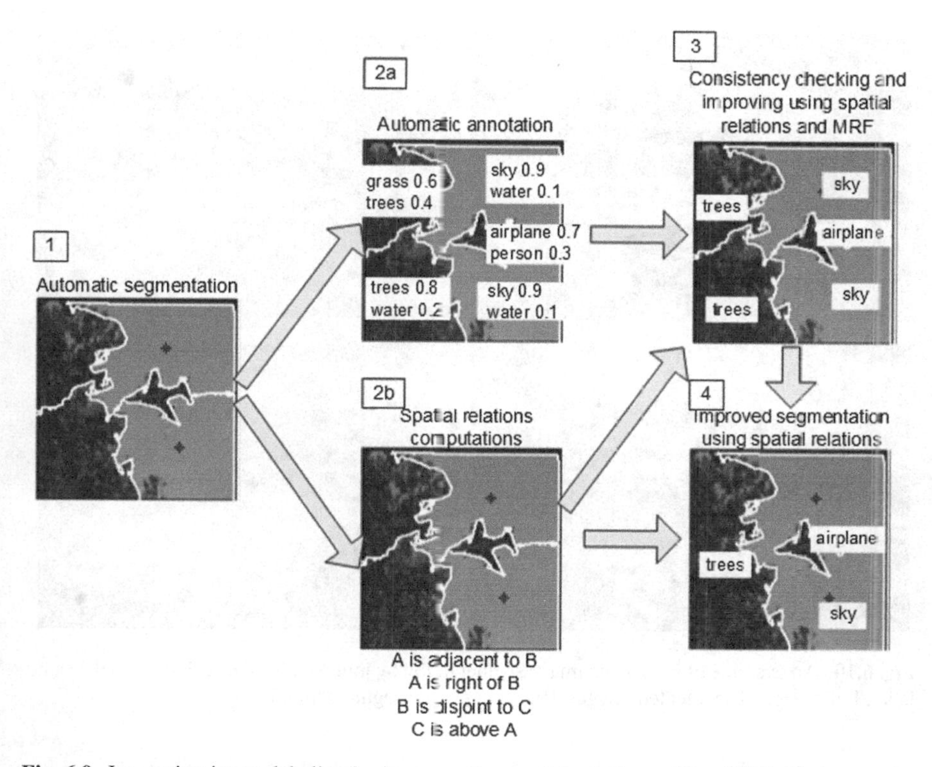


**Fig. 6.9** Improving image labeling by incorporating spatial relations with a MRF. (1) Automatic image segmentation. (2a) Initial region labeling. (2b) Spatial relations among regions are obtained. (3) Improved labeling with a MRF. (4) Improved segmentation [5]

1. An image is automatically segmented (using Normalized cuts).
2. The obtained segments are assigned a list of labels and their corresponding probabilities based on their visual features using a classifier.
3. Concurrently, the spatial relations among the same regions are computed.
4. The MRF is applied, combining the original labels and the spatial relations, resulting in a new labeling for the regions by applying simulated annealing.
5. Adjacent regions with the same label are joined.

The energy function to be minimized combines the information provided by the classifiers (labels' probabilities) with the spatial relations (relations' probabilities). In this study, spatial relations are divided in three groups: topological relations, horizontal relations and vertical relations. Thus, the energy function contains four terms, one for each type of spatial relation and one for the initial labels. So the energy function is:

$$U_p(f) = \alpha_1 V_T(f) + \alpha_2 V_H(f) + \alpha_3 V_V(f) + \lambda \sum_o V_o(f) \tag{6.24}$$

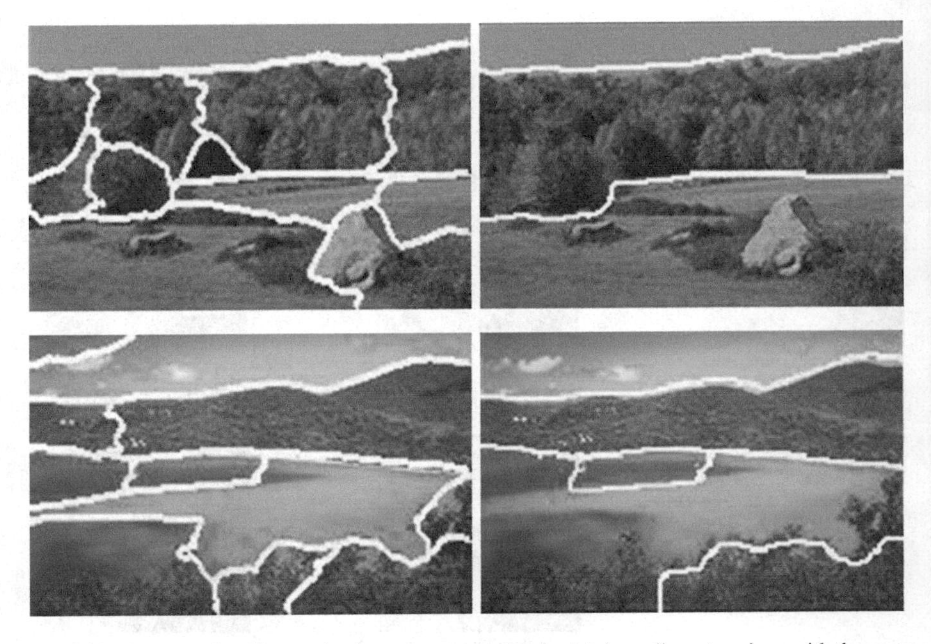

**Fig. 6.10** An example of improving image segmentation by joining adjacent regions with the same label. Left: original segmented images. Right: improved segmentation [5]

where $V_T$ is the potential for topological relations, $V_H$ for horizontal relations, and $V_V$ for vertical relations; $\alpha_1, \alpha_2, \alpha_3$ are the corresponding constants for giving more or less weight to each type of relation. $V_o$ is the classification (label) potential weighted by the $\lambda$ constant. These potentials can be estimated from a set of labeled training images. The potential for a certain type of spatial relation between two regions of classes $A$ and $B$ is inversely proportional to the probability (frequency) of that relation occurring in the training set.

By applying this approach, a significant improvement can be obtained over the initial labeling of an image [6]. In some cases, by using the information provided by this new set of labels, we can also improve the initial image segmentation as illustrated in Fig. 6.10.

## 6.8 Additional Reading

An introduction to Markov random fields and their applications is given in [7]. A comprehensive coverage of MRFs for image processing is presented in [9]. General introductions to conditional Markov random fields are given in [10, 11].

## 6.9 Exercises

1. For the MRF in Fig. 6.3: (a) determine the cliques in the graph, (b) express the joint probability as a product of clique potentials, (c) assuming all the variables are binary, define the required parameters for this model.
2. Given the following data for the MRF of Fig. 6.3, considering that all variables are binary (the table gives the number of cases for each value of the variables):

| Variable | Value 0 | Value 1 |
|---|---|---|
| q1 | 25 | 75 |
| q2 | 50 | 50 |
| q3 | 10 | 90 |
| q4 | 80 | 20 |
| q5 | 60 | 40 |

   Make the simplifying assumption that the variables in each clique are independent (this is strictly not true), to estimate the joint probability of each clique in the graph, and then the joint probability of the MRF.
3. Given the first order MRF of Fig. 6.4, specify the minimum *Markov blanket* for each variable. The Markov blanket of a variable, $q_i$, is a set of variables that make it independent from the rest of the variables in the graph.
4. Repeat the previous problem for the second order MRF of Fig. 6.5.
5. Given a regular MRF of $4 \times 4$ sites with a first order neighborhood, consider that each site can take one of two values, 0 and 1. Consider that we use the smoothing potentials as in the image smoothing application, with $\lambda = 4$, giving more weight to the observations. Given the initial configuration $F$ and the observation $G$, obtain the MAP configuration using the ICM variant of the stochastic simulation algorithm.

$$F: \begin{matrix} 0\,0\,0\,0 \\ 0\,0\,0\,0 \\ 0\,0\,0\,0 \\ 0\,0\,0\,0 \end{matrix} \qquad G: \begin{matrix} 0\,0\,0\,0 \\ 0\,1\,1\,0 \\ 0\,1\,0\,0 \\ 0\,0\,1\,0 \end{matrix}$$

6. Repeat the previous problem using the Metropolis version of stochastic simulation, with $P = 0.5$. Hint: you can use a coin flip to decide if a higher energy configuration is kept or not.
7. Repeat problem 5 with the simulated annealing approach, using as initial value $T = 10$, and reducing it by 50% in each iteration.
8. Solve the previous three problems using MPM instead of MAP.
9. Given data on binary images, the following statistics regarding neighboring pixels were obtained: $0-0$, 45%; $0-1$, 5%; $1-0$, 5%; $1-1$ 45%. Based on these values, estimate the the *internal* potentials for the smoothing MRF.
10. An edge in an image is where there is an abrupt change in the values of the neighboring pixels (a high value of the first derivative if we consider the image as a two dimensional function). Specify the potentials for a first order MRF that emphasizes the edges in an image, considering that each site is binary, where 1

indicates an edge and 0 no edge. The observation is a gray level image in which each pixel varies from 0 to 255.

11. What is the time complexity of the stochastic simulation algorithm for its different variants?
12. Consider we want to model a sequence of words (sentence), $y_1, y_2, \ldots, y_n$, as a linear CRF to determine the part of speech for each word (noun, verb ...). Given an observation vector, $\mathbf{X}$, for each sentence, (i) specify the parameters required for this model; (ii) given an observation vector, how will you estimate the probability of each possible sentence, assuming a fixed length?
13. *** Implement the image smoothing algorithm using a first order regular MRF. Vary the parameter $\lambda$ and observe the effects on the processed image. Repeat considering a second order MRF.
14. *** Implement a program to generate a super-resolution image using MRFs. For example, generate an image that doubles the dimensions $2n \times 2m$ of the original $n \times m$ image.
15. *** Use a deep convolutional neural network to label different objects (regions) in an image, and then implement the MRF based on spatial relations described in Sect. 6.7.2 to improve the original labels.

## References

1. Besag, J.: Statistical analysis of non-lattice data. Statistician **24**(3), 179–195 (1975)
2. Binder, K.: Ising Model. Michiel, Encyclopedia of Mathematics. Springer, Berlin (2001)
3. Geman, D., Geman, S., Graffigne, C.: Locating Object and Texture Boundaries. Pattern Recognition Theory and Applications. Springer, Heidelberg (1987)
4. Hammersley, J.M., Clifford, P.: Markov fields on finite graphs and lattices. Unpublished Paper. http://www.statslab.cam.ac.uk/~grg/books/hammfest/hamm-cliff.pdf (1971). Accessed 14 Dec. 2014
5. Hernández–Gracidas, C., Sucar, L.E.: Markov Random Fields and Spatial Information to Improve Automatic Image Annotation. Advances in Image and Video Technology, Lecture Notes in Computer Science, vol. 4872, pp. 879–892. Springer, Berlin (2007)
6. Hernández-Gracidas, C., Sucar, L.E., Montes, M.: Improving image retrieval by using spatial relations. J. Multimed. Tools Appl. **62**, 479–505 (2013)
7. Kindermann, R., Snell, J.L.: Markov Random Fields and Their Applications, vol. 1. American Mathematical Society, Providence (1980)
8. Lafferty, J., McCallum, A., Pereira, F.: Conditional random fields: probabilistic models for segmenting and labeling sequence data. In: International Conference on Machine Learning (2001)
9. Li, S.Z.: Markov Random Field Modeling in Image Analysis. Springer, London (2009)
10. Sutton, C., McCallum, A.: An introduction to conditional random fields for relational learning. In: Geetor, L., Taskar, B. (eds.) Introduction to Statistical Relational Learning. MIT Press, Cambridge (2006)
11. Wallach, H.M.: Conditional random fields: an introduction. Technical report MS-CIS-04-21, University of Pennsylvania, Philadelphia (2004)

# Chapter 7
# Bayesian Networks: Representation and Inference

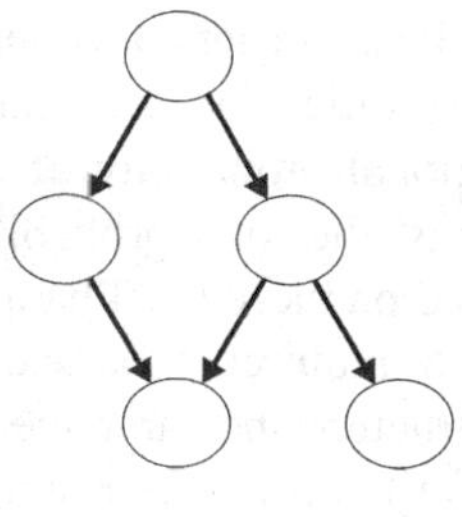

## 7.1 Introduction

In contrast to Markov networks, which are undirected graphical models, Bayesian networks are directed graphical models that represent the joint distribution of a set of random variables. Some of the techniques that we have revised in previous chapters, such as Bayesian classifiers and hidden Markov models, are particular cases of Bayesian networks. Given their importance and the amount of research done for this topic in recent years, we have devoted two chapters to Bayesian networks. In this chapter we will cover the representational aspects and inference techniques. The next chapter will discuss learning; specifically, structure and parameter learning.

An example of a hypothetical medical Bayesian network is shown in Fig. 7.1. In this graph, the nodes represent random variables and the arcs direct dependencies between variables. The structure of the graph encodes a set of conditional independence relations between the variables. For instance, the following conditional independencies can be inferred from the example:

- *Fever* is independent of *Body ache* given *Flu* (common cause).
- *Fever* is independent of *Unhealthy food* given *Typhoid* (indirect cause).
- *Typhoid* is independent of *Flu* when *Fever* is NOT known (common effect). Knowing Fever makes Typhoid and Flu dependent—for example, if we know that someone has Typhoid and Fever, this *diminishes* the probability of having the Flu.



L. E. Sucar, *Probabilistic Graphical Models*, Advances in Computer Vision and Pattern Recognition, https://doi.org/10.1007/978-3-030-61943-5_7

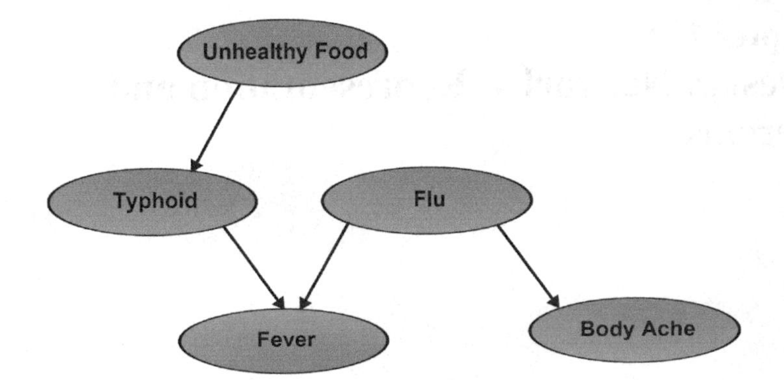


**Fig. 7.1** A simple, hypothetical example of a medical Bayesian network

In addition to the structure, a Bayesian network considers a set of local parameters, which are the conditional probabilities for each variable given its parents in the graph. For example, the conditional probability of Fever given Flu and Typhoid, $P(Fever \mid Typhoid, Flu)$. Thus, the joint probability of all the variables in the network can be represented based on these local parameters; this usually implies an important saving in the number of required parameters.

Given a Bayesian network (structure and parameters) we can answer several probabilistic queries. For instance, for the previous example: What is the probability of Fever given Flu? Which is more probable, Typhoid or Flu, given Fever and Unhealthy food?

In the next section we formalize the representation of a Bayesian network, and then we present several algorithms to answer different types of probabilistic queries.

## 7.2 Representation

A Bayesian network (BN) represents the joint distribution of a set of $n$ (discrete) variables, $X_1, X_2, \ldots, X_n$, as a directed acyclic graph (DAG) and a set of conditional probability tables (CPTs). Each node, that corresponds to a variable, has an associated CPT that contains the probability of each state of the variable given its parents in the graph. The structure of the network implies a set of conditional independence assertions, which give power to this representation.

Figure 7.2 depicts an example of a simple BN. The structure of the graph implies a set of conditional independence assertions for this set of variables. For example, $R$ is conditionally independent of $C, G, F, D$ given $T$, that is:

$$P(R \mid C, T, G, F, D) = P(R \mid T) \tag{7.1}$$

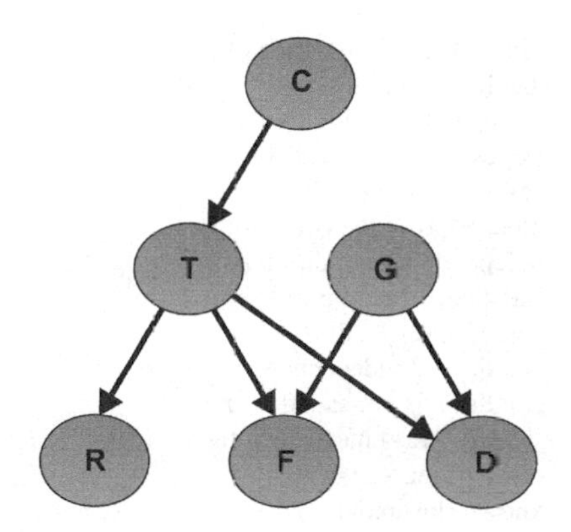


**Fig. 7.2** A Bayesian network. The nodes represent random variables and the arcs direct dependencies

## 7.2.1 *Structure*

The conditional independence assertions implied by the structure of a BN should correspond to the conditional independence relations of the joint probability distribution, and vice versa. These are usually represented using the following notation. If $X$ is conditionally independent of $Z$ given $Y$:

- In the probability distribution: $P(X|Y, Z) = P(X|Y)$ or $I(X, Y, Z)$.
- In the graph: $I < X \mid Y \mid Z >$.

Conditional independence assertions can be verified directly from the structure of a BN using a criteria called *D–separation*. Before we define it, we consider the 3 basic BN structures for 3 variables and 2 arcs:

- Sequential: $X \rightarrow Y \rightarrow Z$.
- Divergent: $X \leftarrow Y \rightarrow Z$.
- Convergent: $X \rightarrow Y \leftarrow Z$.

In the first two cases, $X$ and $Z$ are conditionally independent given $Y$, however in the third case this is not true. This last case, called *explaining away*, corresponds intuitively to having two *causes* with a common *effect*; knowing the effect and one of the causes, alters our belief in the other cause. These cases can be associated to the separating node, $Y$, in the graph. Thus, depending on the case, $Y$ is sequential, divergent or convergent.

### 7.2.1.1 D-Separation

Given a graph $G$, a set of variables $A$ is conditionally independent of a set $B$ given a set $C$, if there is no trajectory in $G$ between $A$ and $B$ such that:

1. All convergent nodes are or have descendants in $C$.
2. All other nodes are outside $C$.

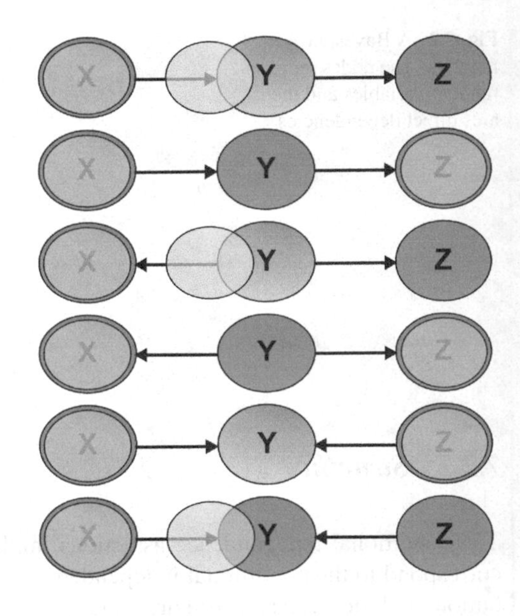


**Fig. 7.3** An illustration of the different cases for D-Separation using the Bayes ball procedure. In the cases where the ball is blocked by $Y$ the conditional independence condition is satisfied; when the ball passes (over $Y$) the conditional independence condition is not satisfied. $Y$ shaded (gray) means that the value of the variable is known (instantiated)

For instance, for the BN in Fig. 7.2, $R$ is independent of $C$ given $T$, but $T$ and $G$ are not independent given $F$.

Another way to verify D-Separation is by using an algorithm known as the *Bayes ball*. Considering that there is a path from node $X$ to $Z$ with $Y$ in the middle (see Fig. 7.3); $Y$ is shaded if it is known (instantiated), otherwise it is not shaded. We *throw a ball* from $X$ to $Z$, if the ball arrives to $Z$ then $X$ and $Z$ are NOT independent given $Y$ according to the following rules:

1. If $Y$ is sequential or divergent and is not shaded, the ball goes through.
2. If $Y$ is sequential or divergent and it is shaded, the ball is blocked.
3. If $Y$ is convergent and not shaded, the ball is blocked.
4. If $Y$ is convergent and shaded, the ball goes through.

To verify if a variable or subset of variables $X$ is independent of $Z$ given $Y$, we need to analyze all the paths between $X$ and $Z$, only if all the paths are blocked, then $X$ and $Z$ are conditionally independent given $Y$.

According to the previous definition of D-separation, any node $X$ is conditionally independent of all nodes in $G$ that are not descendants of $X$ given its parents in the graph, $Pa(X)$. This is known as the *Markov assumption*. The structure of a BN can be specified by the parents of each variable; thus the set of parents of a variable $X$ is known as the *contour* of $X$. For the example in Fig. 7.2, its structure can be specified as:

1. Pa(C) = ∅
2. Pa(T) = C
3. Pa(G) = ∅
4. Pa(R) = T

5. Pa(F) = T,G
6. Pa(D) = T,G

Given this condition and using the chain rule, we can specify the joint probability distribution of the set of variables in a BN as the product of the conditional probability of each variable given its parents:

$$P(X_1, X_2, ..., X_n) = \prod_{i=1}^{n} P(X_i|Pa(X_i)) \quad (7.2)$$

For the example in Fig. 7.2:

$$P(C, T, G, R, F, D) = P(C)P(G)P(T \mid C)P(R \mid T)P(F \mid T, G)P(D \mid T, G)$$

The *Markov Blanket* of a node $X$, $MB(X)$, is a set of nodes that make it independent of all the other nodes in $G$, that is $P(X \mid G - X) = P(X \mid MB(X))$. For a BN, the Markov blanket of $X$ is:

- the parents of $X$,
- the sons of $X$,
- and other parents of the sons of $X$.

For instance, in the BN of Fig. 7.2, the Markov blanket of $R$ is $T$ and the Markov blanket of $T$ is $C, R, F, D, G$.

#### 7.2.1.2 Mappings

Given a probability distribution $P$ of $\mathbf{X}$, and its graphical representation $G$, there must be a correspondence between the conditional independence in $P$ and in $G$; this is called a *mapping*. There are three basic types of mappings:

D-Map: all the conditional independence relations in $P$ are satisfied (by D-Separation) in $G$.
I-Map: all the conditional independence relations in $G$ are true in $P$.
P-Map: or perfect map, it is a D-Map and an I-Map.

In general, it is not always possible to have a *perfect mapping* of the independence relations between the graph $(G)$ and the distribution $(P)$[1], so we settle for what is called a *Minimal I–Map*: all the conditional independence relations implied by $G$ are true in $P$, and if any arc is deleted in $G$ this condition is lost [16].

[1] For instance, the conditional independence relations $I(X, WV, Y)$ and $I(W, XY, V)$ can not be represented by a BN.

#### 7.2.1.3 Independence Axioms

Given some conditional independence relations between subsets of random variables, we can derive other conditional independence relations axiomatically, that is, without the need to estimate probabilities or independence measures. There are some basic rules to derive new conditional independence relations from other conditional independence relations, known as the *independence axioms*:

Symmetry: $I(X, Z, Y) \rightarrow I(Y, Z, X)$
Decomposition: $I(X, Z, Y \cup W) \rightarrow I(X, Z, Y) \wedge I(X, Z, W)$
Weak Union: $I(X, Z, Y \cup W) \rightarrow I(X, Z \cup W, Y)$
Contraction: $I(X, Z, Y) \wedge I(X, Z \cup Y, W) \rightarrow I(X, Z, Y \cup W)$
Intersection: $I(X, Z \cup W, Y) \wedge I(X, Z \cup Y, W) \rightarrow I(X, Z, Y \cup W)$

Graphical examples of the application of the independence axioms are illustrated in Fig. 7.4.

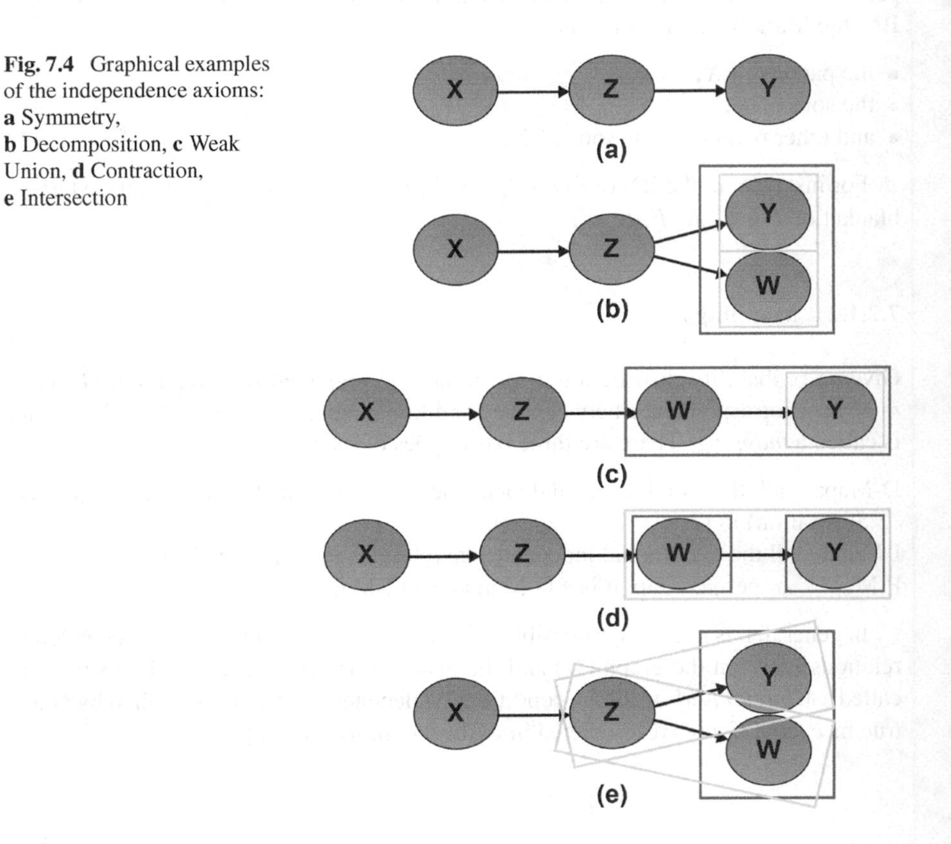


**Fig. 7.4** Graphical examples of the independence axioms: **a** Symmetry, **b** Decomposition, **c** Weak Union, **d** Contraction, **e** Intersection

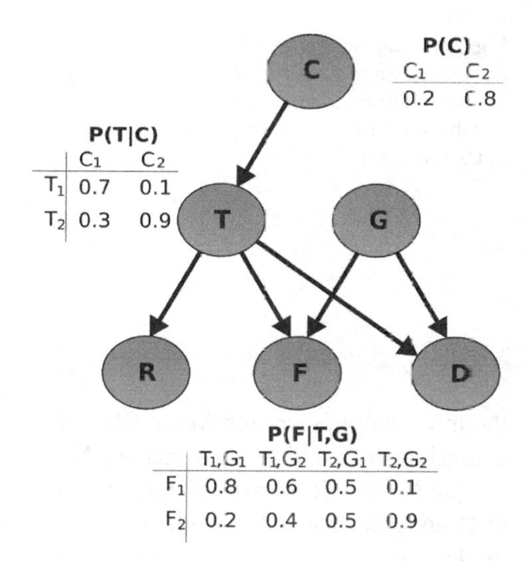


**Fig. 7.5** Parameters for the BN in Fig. 7.2. It shows the CPTs for some of the variables in the example: $P(C)$; $P(T \mid C)$; and $P(F \mid T, G)$. We assume in this case that all variables are binary

## 7.2.2 *Parameters*

To complete the specification of a BN, we need to define its parameters. In the case of a BN, these parameters are the conditional probabilities of each node given its parents in the graph. If we consider discrete variables:

- Root nodes: vector of marginal probabilities.
- Other nodes: conditional probability table (CPT) of the variable given its parents in the graph.

Figure 7.5 shows some of the CPTs of the BN in Fig. 7.2. In case of continuous variables we need to specify a function that relates the density function of each variable to the density of its parents (for example, Kalman filters consider Gaussian distributed variables and linear functions).

In the case of discrete variables, the number of parameters in a CPT increases exponentially with the number of parents of a node. This can become problematic when there are *many* parents. The memory requirements can become very large, and it is also difficult to estimate so many parameters. Two main alternatives have been proposed to overcome this issue, one is based on *canonical models* and the other on graphical representations of CPTs. Next we briefly present both schemes.

### 7.2.2.1 Canonical Models

Canonical models represent the relations between a set of random variables for particular interactions using few parameters. It can be applied when the probabilities of a random variable in a BN conform to certain *canonical* relations with respect to the configurations of its parents. There are several classes of canonical models,

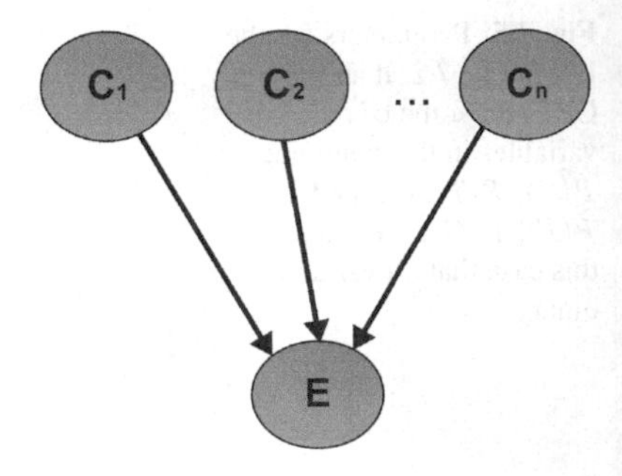


**Fig. 7.6** Graphical representation of a Noisy OR structure. The $n$ cause variables (C) are the parents of the effect variable (E)

the most common are the *Noisy OR* and *Noisy AND* for binary variables, and their extensions for multivalued variables, *Noisy Max* and *Noisy Min*, respectively.

The Noisy OR is basically an extension of the OR relation in logic. Consider an OR logic gate, in which the output is $True$ if any of its inputs are $True$. The Noisy OR model is based on the concept of the logic OR; the difference is that there is a certain (small) probability that the variable is not $True$ even if one or more of its parents are $True$. In an analogous way, the Noisy And model is related to the logical AND. These models apply only when all the variables are binary, however there are extensions for multivalued variables, which consider a set of *ordered* values for each variable. For example, consider a variable that represents a disease, $D$. In the case of the binary canonical models it has two values, $True$ and $False$. For a multivalued model, it could be defined as $D \in \{False, Mild, Intermediate, Severe\}$, such that these values follow a predefined order. The *Noisy Max* and *Noisy Min* models generalize the Noisy OR and Noisy AND models, respectively, for multivalued ordered variables.

Next, we describe the Noisy OR model in detail; the other cases can be defined in a similar way.

**Noisy OR**

The Noisy OR model is applied when several variables or *causes* can produce an *effect* if any one of them is $True$, and as more of the *causes* are true, the probability of the effect increases. For instance, the effect could be a certain symptom or effect, $E$, and the causes are a number of possible diseases, $C_1, C_2, ..., C_n$, that can produce the symptom, such that if none of the diseases is present (all $False$) the symptom does not appear; and when any disease is present ($True$) the symptom is present with high probability and it increases as the number of $C_i = True$ increases. A graphical representation of a Noisy OR relation in a BN is depicted in Fig. 7.6.

Formally, the following two conditions must be satisfied for a Noisy OR canonical model to be applicable:

Responsibility: the effect is false if all the possible causes are false.

Independence of exceptions: if an effect is the manifestation of several causes, the mechanisms that inhibit the occurrence of the effect under one cause are independent of the mechanisms that inhibit it under the other causes.

**Table 7.1** Conditional probability table for a Noisy OR variable with three parents and parameters $q_1 = q_2 = q_3 = 0.1$

| $C_1$ | 0 | 0 | 0 | 0 | 1 | 1 | 1 | 1 |
|---|---|---|---|---|---|---|---|---|
| $C_2$ | 0 | 0 | 1 | [illegible] | 0 | 0 | 1 | 1 |
| $C_3$ | 0 | 1 | 0 | [illegible] | 0 | 1 | 0 | 1 |
| $P(E = 0)$ | 1 | 0.1 | 0.1 | 0.01 | 0.1 | 0.01 | 0.01 | 0.001 |
| $P(E = 1)$ | 0 | 0.9 | 0.9 | 0.99 | 0.9 | 0.99 | 0.99 | 0.999 |

The probability that the effect $E$ is inhibited (it does not occur) under cause $C_i$ is defined as:

$$q_i = P(E = False \mid C_i = True) \tag{7.3}$$

Given this definition and the previous conditions, the parameters in the CPT for a Noisy OR model can be obtained using the following expressions when all the $m$ causes are $True$:

$$P(E = False \mid C_1 = True, ...C_m = True) = \prod_{i=i}^{m} q_i \tag{7.4}$$

$$P(E = True \mid C_1 = True, ...C_m = True) = 1 - \prod_{i=i}^{m} q_i \tag{7.5}$$

In general, if $k$ of $m$ causes are $True$, then $P(E = False \mid C_1 = True, ...C_k = True) = \prod_{i=i}^{k} q_i$, so that if all the causes are $False$ then the effect is $False$ with a probability of one. Thus, only one parameter is required per parent variable to construct the CPT, the inhibition probability $q_i$. In this case the number of independent parameters $(q_1, q_2, ..., q_n)$ increases linearly with the number of parents, instead of exponentially.

As an example, consider a Noisy OR model with 3 causes, $C_1, C_2, C_3$, where the inhibition probability is the same for the three, $q_1 = q_2 = q_3 = 0.1$. Given these parameters we can obtain the CPT for the effect variable, as shown in Table 7.1

Canonical models can provide a considerable reduction in the number of parameters when a variable has *many* parents; and also some inference techniques take advantage of this compact representation.

#### 7.2.2.2 Other Representations

Canonical models apply in certain situations but do not provide a general solution for compact representations of CPTs. An alternative representation is based on the observation that in the probability tables for many domains, the same probability values tend to be repeated several times in the same table; for instance, it is common

**Table 7.2** Conditional probability table of $P(X \mid A, B, C, D, E, F, G)$, which has several repeated values. All variables are binary with values $T$ or $F$

| A | B | C | D | E | F | G | X |
|---|---|---|---|---|---|---|---|
| T | T/F | T/F | T/F | T/F | T/F | T/F | 0.9 |
| F | T | T/F | T | T/F | T | T | 0.9 |
| F | T | T/F | T | T/F | T | F | 0.0 |
| F | T | T/F | T | T/F | F | T/F | 0.0 |
| F | T | T | F | T | T/F | T | 0.9 |
| F | T | T | F | T | T/F | F | 0.0 |
| F | T | T | F | F | T/F | T/F | 0.0 |
| F | T | F | F | T/F | T/F | T/F | 0.0 |
| F | F | T | T/F | T | T/F | T | 0.9 |
| F | F | T | T/F | T | T/F | F | 0.0 |
| F | F | T | T/F | F | T/F | T/F | 0.0 |
| F | F | F | T/F | T/F | T/F | T/F | 0.0 |

to have many zero entries in a CPT. Thus, it is not necessary to represent these repeated values many times, is should be sufficient to represent each different value once.

A representation that takes advantage of this condition is a *decision tree* (DT), such that it could be used for representing a CPT in a compact way. In a DT, each internal node corresponds to a variable in the CPT, and the branches from a node correspond to the different values that a variable can take. The leaf nodes in the tree represent the different probability values. A trajectory from the root to a leaf, specifies a probability value for the corresponding variables—values in the trajectory. If a variable is omitted in a trajectory, it means that the CPT has the same probability for all values of this variable.

For example, Table 7.2 depicts the CPT $P(X \mid A, B, C, D, E, F, G)$, assuming all variables are binary $(F, T)$. Figure 7.7 shows a DT for the CPT in Table 7.2. In this example the savings in memory are not significant, however for *large* tables there could be a significant reduction in the memory space requirements.

A *decision diagram* (DD) extends a DT by considering a directed acyclic graph structure, such that it is not restricted to a tree. This avoids the need to duplicate repeated probability values in the leaf nodes, and in some cases provides an even more compact representation. An example of a decision diagram representation of the CPT of Table 7.2 is depicted in Fig. 7.8.

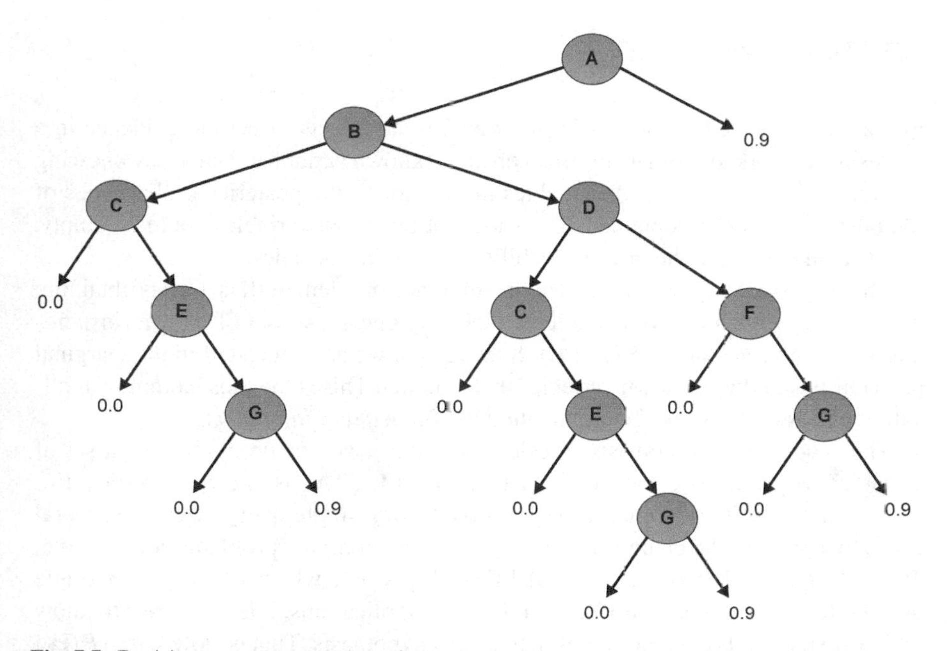


**Fig. 7.7** Decision tree representation of a CPT. The DT represents the CPT shown in Table 7.2. For each variable (node in the tree), the left arrow corresponds to the $F$ value and the right arrow to the $T$ value

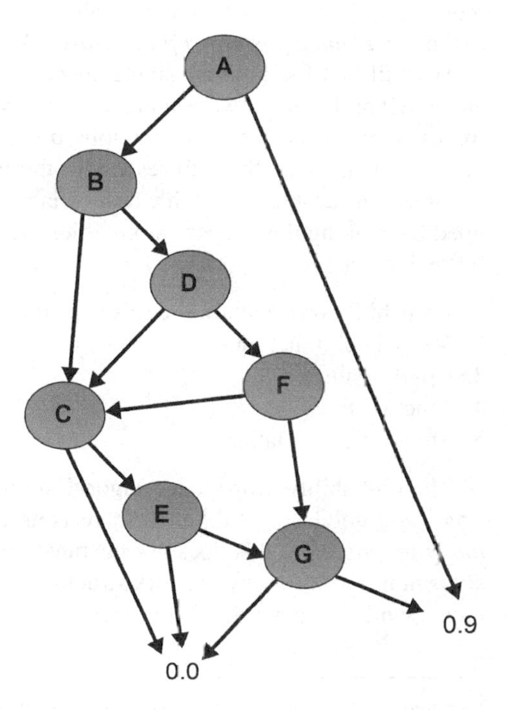


**Fig. 7.8** Decision diagram representation of a CPT. The DD represents the CPT shown in Table 7.2. As in Fig. 7.7, the left arrow corresponds to the $F$ value and the right arrow to the $T$ value

## 7.3 Inference

Probabilistic inference consists in *propagating* the effects of certain evidence in a Bayesian network to estimate its effect on the unknown variables. That is, by knowing the values for some subset of variables in the model, the posterior probabilities of the other variables are obtained. The subset of unknown variables could be empty, in this case we obtain the prior probabilities of all the variables.

There are basically two variants of the inference problem in BNs. One is obtaining the posterior probability of a single variable, $H$, given a subset of known (instantiated) variables, $\mathbf{E}$, that is, $P(H \mid \mathbf{E})$. Specifically, we are interested in the marginal probabilities of the unknown variables in the model. This is the most common application of BNs, and we will denominate it as *single query inference*.

The second variant consists in calculating the posterior probability of a set of variables, $\mathbf{H}$ given the evidence, $\mathbf{E}$, that is, $P(\mathbf{H} \mid \mathbf{E})$. This is known as *conjunctive query inference*. In principle, it can be solved using single query inference several times by applying the chain rule, making it a more complex problem. For example, $P(A, B \mid \mathbf{E})$ can be written as $P(A \mid \mathbf{E})P(B \mid A, \mathbf{E})$, which requires two single query inferences, and a multiplication. In some applications, it is of interest to know which are the most probable values in the set of hypothesis. That is, $ArgMax_{\mathbf{H}} P(\mathbf{H} \mid \mathbf{E})$. When $\mathbf{H}$ includes all non observed variables, it is known as the *most probable explanation* (MPE) or the *total abduction* problem. When we are interested in the most likely joint state of some (not all) of the unobserved variables, it corresponds to the *maximum a posteriori probability* (MAP) or *partial abduction* problem.

We will first focus on the single query inference problem, and later on the MPE and MAP problems. If we want to solve the inference problem using a direct (brute force) computation (i.e., from the joint distribution), the computational complexity increases exponentially with respect to the number of variables, and the problem becomes intractable even with few variables. Many algorithms have been developed for making this process more efficient, which can be roughly divided into the following classes:

1. Probability propagation (Pearl's algorithm [15]).
2. Variable elimination.
3. Conditioning.
4. Junction tree.
5. Stochastic simulation.

The probability propagation algorithm only applies to singly connected graphs (trees and polytrees[2]), although there is an extension for general networks called *loopy propagation*, this does not guarantee convergence. The other four classes of algorithms work for any network structure, the last one being an approximate technique, while the other three are exact.

---

[2] A polytree is a singly connected DAG in which some nodes have more than one parent; in a directed tree, each node has at most one parent.

In the worst case the inference problem is *NP-hard* for Bayesian networks [1]. However, there are efficient (polynomial) algorithms for certain types of structures (singly connected networks); while for other structures it depends on the connectivity of the graph. In many applications, the graphs are *sparse*, so in this case there are inference algorithms which are very efficient.

Next we will describe probability propagation for singly connected networks, and then the most common techniques used for multi connected BNs.

### 7.3.1 Singly Connected Networks: Belief Propagation

We now describe the tree propagation algorithm proposed by Pearl, which provides the basis for several of the most advanced and general techniques.

Given certain evidence, **E** (subset of instantiated variables), the posterior probability for a value $i$ of any variable $B$, can be obtained by applying the Bayes rule:

$$P(B_i|\mathbf{E}) = P(B_i)P(\mathbf{E}|B_i)/P(\mathbf{E}) \tag{7.6}$$

Given that the BN has a tree structure, any node divides the network into two independent subtrees. Thus, we can separate the evidence into (see Fig. 7.9):

**E-**: Evidence of the rooted tree in $B$.
**E+**: All other evidence.

Then:

$$P(B_i|\mathbf{E}) = P(B_i)P(\mathbf{E}-, \mathbf{E}+|B_i)/P(\mathbf{E}) \tag{7.7}$$

Given that **E**+ and **E**—are independent, by applying the Bayes rule again, we obtain:

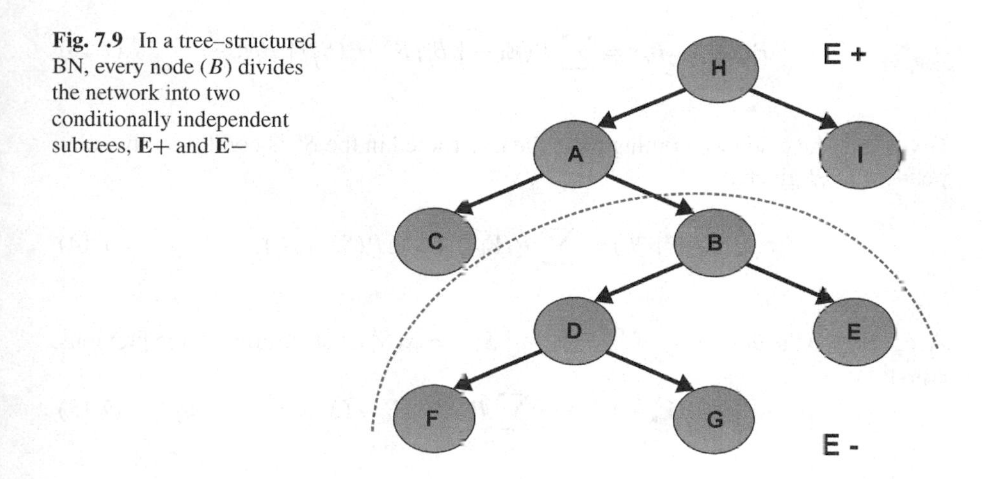


**Fig. 7.9** In a tree–structured BN, every node ($B$) divides the network into two conditionally independent subtrees, **E**+ and **E**−

$$P(B_i|\mathbf{E}) = \alpha P(B_i|\mathbf{E}+)P(\mathbf{E}-|B_i) \quad (7.8)$$

Where $\alpha$ is a normalization constant. If we define the following terms:

$$\lambda(B_i) = P(\mathbf{E}-|B_i) \quad (7.9)$$

$$\pi(B_i) = P(B_i|\mathbf{E}+) \quad (7.10)$$

Then Eq. 7.8 can be written as:

$$P(B_i|\mathbf{E}) = \alpha\pi(B_i)\lambda(B_i) \quad (7.11)$$

Equation 7.11 is the basis for a distributed propagation algorithm to obtain the posterior probability of all non-instantiated nodes. The computation of the posterior probability of any node $B$ is decomposed into two parts: (i) the evidence coming from the sons of $B$ in the tree ($\lambda$), and the evidence coming from the parent of $B$, ($\pi$). We can think of each node $B$ in the tree as a simple processor that stores its vectors $\pi(B)$ and $\lambda(B)$, and its conditional probability table, $P(B \mid A)$. The evidence is propagated via a message passing mechanism, in which each node sends the corresponding messages to its parent and sons in the tree.

Next we derive the equations for the messages. First the $\lambda$ messages that are propagated from the leaves to the root. Given that the sons of $B$ are conditionally independent given $B$:

$$\lambda(Bi) = P(\mathbf{E}-|Bi) = \prod_k P(\mathbf{E_k}- \mid B_i), \quad (7.12)$$

where $\mathbf{E_k}$—is the evidence coming from the tree rooted in the $S^k$ son of $B$. By applying the rule of total probability conditioning on $S^k$ we obtain:

$$P(\mathbf{E_k}- \mid B_i) = \sum_j P(\mathbf{E_k}- \mid B_i, S_j^K)P(S_j^K \mid B_i), \quad (7.13)$$

Given that the evidence coming from the tree rooted in the $S^k$ is conditionally independent of $B$ given $S^k$:

$$P(\mathbf{E_k}- \mid B_i) = \sum_j P(\mathbf{E_k}- \mid S_j^K)P(S_j^K \mid B_i), \quad (7.14)$$

According to the definition of $\lambda$, $P(\mathbf{E_k}- \mid S_j^K) = \lambda(S_j^k)$, substituting in the previous equation:

$$P(\mathbf{E_k}- \mid B_i) = \sum_j P(S_j^K \mid B_i)\lambda(S_j^k). \quad (7.15)$$

Now the $\pi$ messages propagated from the root to the leaves. We again apply the rule of total probability conditioning on $A$, the parent of $B$:

$$\pi(Bi) = P(B_i|\mathbf{E}+) = \sum_j P(B_i \mid \mathbf{E}+, A_j)P(A_j \mid \mathbf{E}+), \tag{7.16}$$

Given that $B$ is conditionally independent of the evidence coming from the rest of the tree except the subtree with root $B$ (the ascendants and other descendants of $A$) given $A$:

$$\pi(B_i) = \sum_j P(B_i \mid A_j)P(A_j \mid \mathbf{E}+), \tag{7.17}$$

$P(A_j \mid \mathbf{E}+)$ corresponds to the probability of $A_j$ given the evidence coming from all the tree except the subtree rooted on $B$, it can be written based on Eqs. 7.11 and 7.12 excluding the evidence coming $B$ and its descendants:

$$P(A_j \mid \mathbf{E}+) = \alpha\pi(A_j)\prod_{k\neq b} P(\mathbf{E}_k- \mid A_j) = \alpha\pi(A_j)\prod_{k\neq b}\lambda_k(A_j), \tag{7.18}$$

where $b$ indicates the variable $B$ (one of the children of $A$). Substituting in Eq. 7.17:

$$\pi(B_i) = \sum_j P(B_i \mid A_j)[\alpha\pi(A_j)\prod_{k\neq b}\lambda_k(A_j)]. \tag{7.19}$$

In summary, the message passing mechanism is the following.
Every node $B$ sends a message to its parent $A$:

$$\lambda_B(Ai) = \sum_j P(B_j \mid A_i)\lambda(B_j) \tag{7.20}$$

Each node can receive several $\lambda$ messages, which are combined via a term by term multiplication for the $\lambda$ messages received from each son. Therefore, the $\lambda$ for a node $A$ with $m$ sons is obtained as:

$$\lambda(Ai) = \prod_{j=1}^{m}\lambda_{Sj}(Ai) \tag{7.21}$$

Every node $B$ sends a message to each son $S_l$:

$$\pi_l(B_i) = \sum_j P(B_i \mid A_j)[\alpha\pi(A_j)\prod_{k\neq b}\lambda_k(A_j)]. \tag{7.22}$$

where $k$ refers to each one of the sons of $B$.

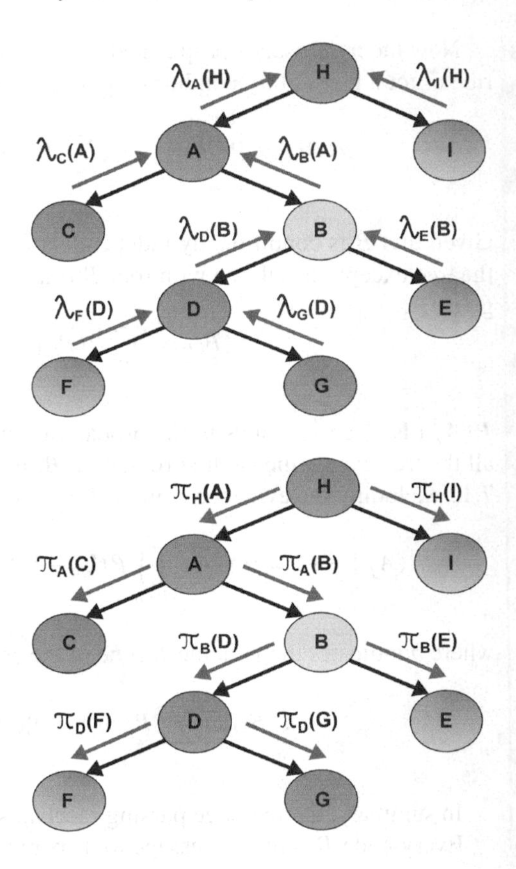


**Fig. 7.10** Bottom-up propagation. $\lambda$ messages are propagated from the leaf nodes to the root

**Fig. 7.11** Top-down propagation. $\pi$ messages are propagated from the root node to the leaves

The propagation algorithm starts by assigning the evidence to the known variables, and then propagating it through the message passing mechanism from the leaves until the root of the tree is reached for the $\lambda$ messages, and from the root until the leaves are reached for the $\pi$ messages. Figures 7.10 and 7.11 illustrate the propagation scheme. At the end of the propagation, each node has updated its $\lambda$ and $\pi$ vectors. The posterior probability of any variable $B$ is obtained by combining these vectors using Eq. 7.11 and normalizing.

For the root and leaf nodes we need to define some initial conditions:

Leaf nodes: If not known, $\lambda = [1, 1, ..., 1]$ (a uniform distribution). If known, $\lambda = [0, 0, ..., 1, ..., 0]$ (one for the assigned value and zero for all other values).

Root node: If not known, $\pi = P(A)$ (prior marginal probability vector). If known, $\pi = [0, 0, ..., 1, ..., 0]$ (one for the assigned value and zero for all other values).

We now illustrate the belief propagation algorithm with a simple example. Consider the BN in Fig. 7.12 with 4 binary variables (each with values $false$ and $true$), $C, E, F, D$, with the CPTs shown in the figure.

Consider that the only evidence is $F = false$. Then the initial conditions for the leaf nodes are: $\lambda_F = [1, 0]$ and $\lambda_D = [1, 1]$ (no evidence). Propagating to the parent

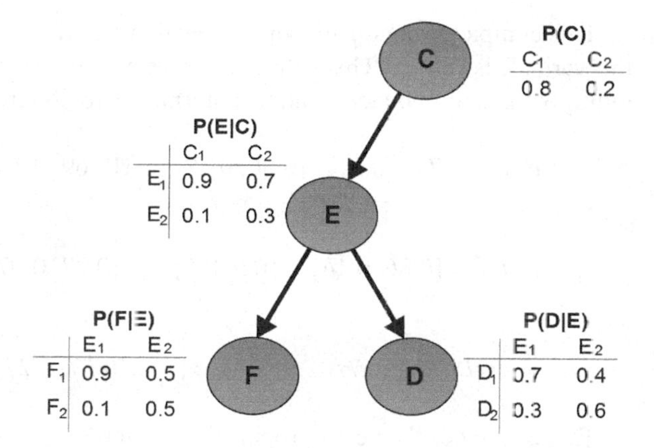


**Fig. 7.12** A simple BN used in the belief propagation example

node ($E$) is basically multiplying the $\lambda$ vectors by the corresponding CPTs:

$$\lambda_F(E) = [1, 0]\begin{bmatrix} 0.9, 0.5 \\ 0.1, 0.5 \end{bmatrix} = [0.9, 0.5]$$

$$\lambda_D(E) = [1, 1]\begin{bmatrix} 0.7, 0.4 \\ 0.3, 0.6 \end{bmatrix} = [1, 1]$$

Then, $\lambda(E)$ is obtained by combining the messages from its two sons:

$$\lambda(E) = [0.9, 0.5] \times [1, 1] = [0.9, 0.5]$$

And now it is propagated to its parent, $C$:

$$\lambda_E(C) = [0.9, 0.5]\begin{bmatrix} 0.9, 0.7 \\ 0.1, 0.3 \end{bmatrix} = [0.86, 0.78]$$

In this case $\lambda(C) = [0.86, 0.78]$, as $C$ has only one son. In this way, we complete the bottom-up propagation; we will now do it top–down.

Given that $C$ is not instantiated, $\pi(C) = [0.8, 0.2]$, we propagate to its son, $E$, which also corresponds to multiplying the $\pi$ vector by the corresponding CPT:

$$\pi(E) = [0.8, 0.2]\begin{bmatrix} 0.9, 0.7 \\ 0.1, 0.3 \end{bmatrix} = [0.86, 0.14]$$

We now propagate to its son $D$; however, given that $E$ has another son, $F$, we also need to consider the $\lambda$ message from this other son, thus:

$$\pi(D) = [0.86, 0.14] \times [0.9, 0.5]\begin{bmatrix} 0.7, 0.4 \\ 0.3, 0.6 \end{bmatrix} = [0.57, 0.27]$$

This completes the top–down propagation (we do not need to propagate to $F$ as this variable is known). Given the $\lambda$ and $\pi$ vectors for each unknown variable, we just multiply them term by term and then normalize to obtain the posterior probabilities:

$$P(C) = [0.8, 0.2] \times [0.86, 0.78] = \alpha[0.69, 0.16] = [0.815, 0.185]$$

$$P(E) = [0.86, 0.14] \times [0.9, 0.5] = \alpha[0.77, 0.07] = [0.917, 0.083]$$

$$P(D) = [0.57, 0.27] \times [1, 1] = \alpha[0.57, 0.27] = [0.68, 0.32]$$

This concludes the belief propagation example.

Probability propagation is a very efficient algorithm for tree structured BNs. The time complexity to obtain the posterior probability of all the variables in the tree is proportional to the *diameter* of the network (the number of arcs in the trajectory from the root to the most distant leaf).

The message passing mechanism can be directly extended to polytrees, as these are also singly connected networks. In this case, a node can have multiple parents, so the $\lambda$ messages should be sent from a node to all its parents. The time complexity is in the same order as for tree structures.

The propagation algorithm only applies to singly connected networks. Next we will present general algorithms that apply to any structure.

## 7.3.2 Multiple Connected Networks

There are several classes of algorithms for exact probabilistic inference on multi connected BNs. Next we review the main ones: (i) variable elimination, (ii) conditioning, (iii) junction tree.

### 7.3.2.1 Variable Elimination

The variable elimination technique is based on the idea of calculating the probability by marginalizing the joint distribution. However, in contrast to the naive approach, it takes advantage of the independence conditions of the BN and the associative and distributive properties of addition and multiplication to do the calculations more efficiently.

Assume a BN representing the joint probability distribution of $\mathbf{X} = \{X_1, X_2, ..., X_n\}$. We want to calculate the posterior probability of a certain variable or subset of variables, $X_H$, given a subset of evidence variables, $X_E$; the remaining variables are $X_R$, such that $\mathbf{X} = \{X_H \cup X_E \cup X_R\}$.

The posterior probability of $X_H$ given the evidence is:

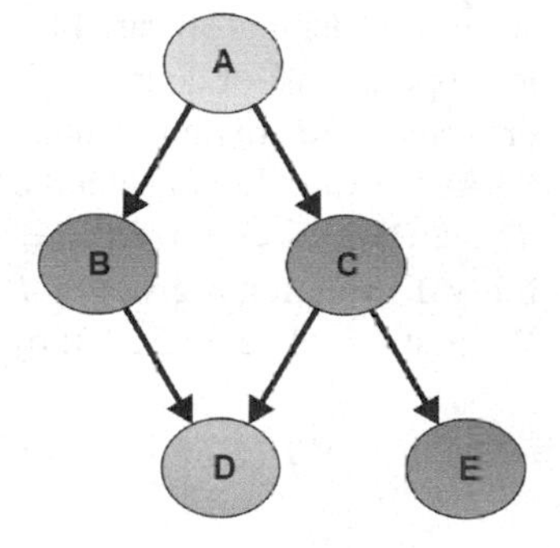


**Fig. 7.13** A Bayesian network used to illustrate the variable elimination algorithm

$$P(X_H \mid X_E) = P(X_H, X_E)/P(X_E) \tag{7.23}$$

We can obtain both terms via marginalization of the joint distribution:

$$P(X_H, X_E) = \sum_{X_R} P(\mathbf{X}) \tag{7.24}$$

and

$$P(X_E) = \sum_{X_H} P(X_H, X_E) \tag{7.25}$$

A particular case of interest is to obtain the marginal probability of the variables when there is no evidence; in this case $X_E = \emptyset$. Another calculation of interest is to obtain the probability of the evidence; this is given by the last equation.

The objective of the variable elimination technique is to perform these calculations efficiently. To achieve this, we can first represent the joint distribution as a product of local probabilities according to the network structure. Then, summations can be carried out only on the subset of terms which are a function of the variables being normalized. This approach takes advantage of the properties of summation and multiplication, resulting in the number of necessary operations being reduced. Next we will illustrate the method through an example.

Consider the BN in Fig. 7.13 where we want to obtain $P(A \mid D)$. In order to achieve this we need to obtain $P(A, D)$ and $P(D)$. To calculate the first term we must *eliminate* $B, C, E$ from the joint distribution, that is:

$$P(A, D) = \sum_B \sum_C \sum_E P(A)P(B \mid A)P(C \mid A)P(D \mid B, C)P(E \mid C) \tag{7.26}$$

By *distributing* the summations we can arrive to the following equivalent expression:

$$P(A, D) = P(A)\sum_B [P(B \mid A) \sum_C [P(C \mid A)P(D \mid B, C) \sum_E P(E \mid C)]] \tag{7.27}$$

If we consider that all variables are binary, this implies a reduction from 32 operations to 9 operations; of course, this reduction will be more significant for larger models or when there are more values per variable.

As an example, consider the BN in Fig. 7.12 and that we want to obtain $P(E \mid F = f_1) = P(E, F = f_1)/P(F = f_1)$. Given the structure of the BN, the joint probability distribution is given by $P(C, E, F, D) = P(C)P(E \mid C)P(F \mid E)P(D \mid E)$. We first calculate $P(E, F)$; by reordering the operations:

$$P(E, F) = P(F \mid E) \sum_D P(D \mid E) \sum_C P(C)P(E \mid C)$$

We must do this calculation for each value of $E$, given $F = f_1$:

$$P(e_1, f_1) = P(f_1 \mid e_1) \sum_D P(D \mid e_1) \sum_C P(C)P(e_1 \mid C)$$

$$P(e_1, f_1) = P(f_1 \mid e_1) \sum_D P(D \mid e_1)[0.9 \times 0.8 + 0.7 \times 0.2]$$

$$P(e_1, f_1) = P(f_1 \mid e_1) \sum_D P(D \mid e_1)[0.86]$$

$$P(e_1, f_1) = P(f_1 \mid e_1)[0.7 + 0.3][0.86]$$

$$P(e_1, f_1) = [0.9][1][0.86] = 0.774$$

In a similar way we obtain $P(e_2, f_1)$; and then from these values we can calculate $P(f_1) = \sum_E P(E, f_1)$. Finally, we calculate the posterior probability of $E$ given $f_1$: $P(e_1 \mid f_1) = P(e_1, f_1)/P(f_1)$ and $P(e_2 \mid f_1) = P(e_2, f_1)/P(f_1)$.

The critical aspect of the variable elimination algorithm is to select the appropriate order for eliminating each variable, as this has an important effect on the number of required operations. The different terms that are generated during the calculations are known as *factors* which are functions over a subset of variables, that map each instantiation of these variables to a non-negative number (these numbers are not necessarily probabilities). In general a factor can be represented as $f(X_1, X_2, ...X_m)$. For instance, in the previous example, one of the factors is $f(C, E) = P(C)P(E \mid C)$, which is a function of two variables.

The computational complexity in terms of space and time of the variable elimination algorithm is determined by the size of the factors; that is, the number of variables, $w$, on which the factor is defined. Basically, the complexity for eliminating (marginalize) any number of variables is exponential on the number of variables in the factor, $O(exp(w))$ [3]. Thus, the order in which the variables are eliminated,

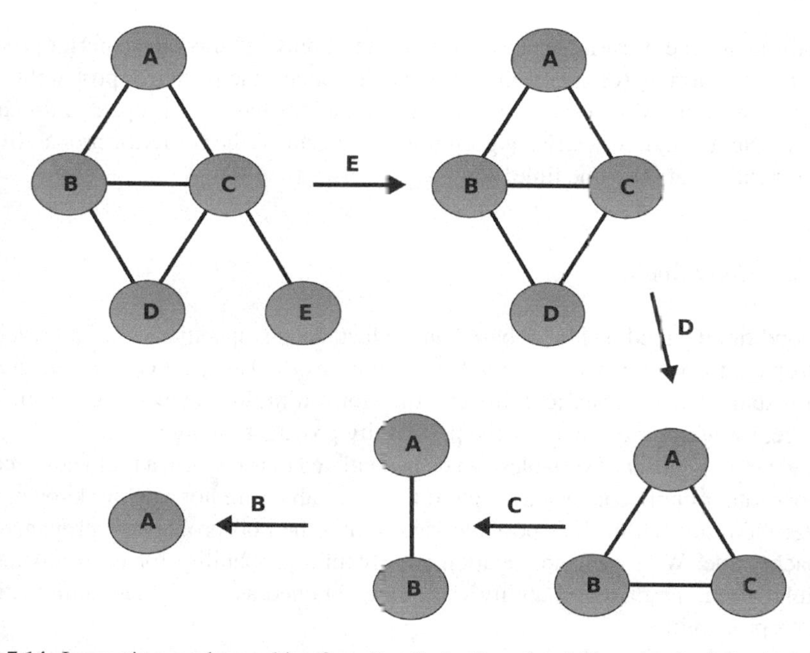


**Fig. 7.14** Interaction graphs resulting from the elimination of variables with the following elimination ordering: $E, D, C, B$ from the BN in Fig. 7.13

should be selected so that the largest factor is kept to a minimum. However finding the *best* order is in general a NP-Hard problem.

There are several heuristics that help to determine a *good* ordering for variable elimination. These heuristics can be explained based on the *interaction graph*—an undirected graph that is built during the process of variable elimination. The variables of each factor form a clique in the interaction graph. The initial interaction graph is obtained from the original BN structure by eliminating the direction of the arcs, and adding additional arcs between each pair of non-connected variables that have a common child. Then, each time a variable $X_j$ is eliminated, the interaction graph is modified by: (i) adding an arc between each pair of neighbors of $X_j$ that are not connected, (ii) deleting variable $X_j$ from the graph.

We illustrate the interaction graphs that result from the BN in Fig. 7.13 by the following elimination ordering: $E, D, C, B$,, depicted in Fig. 7.14.

Two popular heuristics for determining the elimination ordering, which can be obtained from the elimination graph, are the following:

Min-degree: eliminate the variable that leads to the smallest possible factor; which is equivalent to eliminating the variable with the smallest number of neighbors in the current elimination graph.

Min-fill: eliminate the variable that leads to adding the minimum number of edges to the interaction graph.

A disadvantage of variable elimination is that it only obtains the posterior probability of one variable (or subset of variables). To obtain the posterior probability of each non-instantiated variable in a BN, the calculations have to be repeated for each variable. Next, we describe two algorithms that calculate the posterior probabilities for all variables at the same time.

#### 7.3.2.2 Conditioning

The conditioning method [16] is based on the fact that an instantiated variable *blocks* the propagation of the evidence in a Bayesian network. Thus, we can *cut* the graph at an instantiated variable, and this can transform a multi connected graph into a polytree, for which we can apply the probability propagation algorithm.

In general, a subset of variables can be instantiated to transform a multi connected network into a singly connected graph. If these variables are not actually known, we can set them to each of their possible values, and then do probability propagation for each value. With each propagation we obtain a probability for each unknown variable. Then, the final probability values are obtained as a weighted combination of these probabilities.

First we will develop the conditioning algorithm assuming we only need to partition a single variable and then we will extend it for multiple variables. Formally, we want to obtain the probability of any variable, $B$, given the evidence $E$, conditioning on variable $A$. By the rule of total probability:

$$P(B \mid E) = \sum_i P(B \mid E, a_i) P(a_i \mid E) \tag{7.28}$$

Where:

$P(B \mid E, a_i)$ is the posterior probability of $B$ which is obtained by probability propagation for each possible value of $A$.

$P(a_i \mid E)$ is a *weight*.

By applying the Bayes rule we obtain the following equation to estimate the weights:

$$P(a_i \mid E) = \alpha P(a_i) P(E \mid a_i) \tag{7.29}$$

The first term, $P(a_i)$, can be obtained by propagating without evidence. The second term, $P(E \mid a_i)$, is calculated by propagation with $A = a_i$ to obtain the probability of the evidence variables. $\alpha$ is a normalizing constant.

For example, consider the BN in Fig. 7.13. This multi connected network can be transformed into a polytree by assuming $A$ is instantiated (see Fig. 7.15). If the evidence is $D, E$, then probabilities for the other variables, $A, B, C$ can be obtained via conditioning following these steps:

1. Obtain the prior probability of $A$ (in this case it is already given as it is a root node).

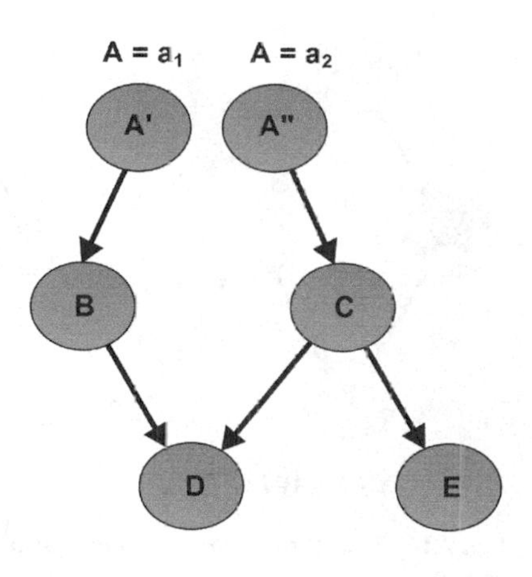


**Fig. 7.15** The Bayesian network in Fig. 7.13 is transformed into a singly connected network by instantiating $A$

2. Obtain the probability of the evidence nodes $D$, $E$ for each value of $A$ by propagation in the polytree.
3. Calculate the weights, $P(a_i \mid D, E)$, from (1) and (2) with the Bayes rule.
4. Estimate the probability of $B$ and $C$ for each value of $A$ given the evidence by probability propagation in the polytree.
5. Obtain the posterior probabilities for $B$ and $C$ from (3) and (4) by applying Eq. 7.28.

In general, to transform a multi connected BN to a polytree we need to instantiate $m$ variables. Thus, propagation must be performed for all the combinations of values (cross product) of the instantiated variables. If each variable has $k$ values, then the number of propagations is $k^m$. The procedure is basically the same as described above for one variable, but the complexity increases.

#### 7.3.2.3 Junction Tree Algorithm

The junction tree method is based on a transformation of the BN to a junction tree, where each node in this tree is a group or cluster of variables from the original network. Probabilistic inference is performed over this new representation.

The intuition behind the junction tree method is based on a transformation of a BN (which is a directed graph) to a Markov network (undirected graph); and then a clustering of the variables so that the resulting graph is singly connected. Consider a simple BN represented as a chain:
$A \rightarrow B \rightarrow C \rightarrow D$
In this case the clusters (cliques) are $AB$, $BC$, and $CD$; and the common variables between the neighbor clusters (separators) are $B$ and $C$. According to the structure of the BN, the joint probability is $P(A, B, C, D) = P(A)P(B \mid A)P(C \mid B)P(D \mid C)$. Which can be written as:

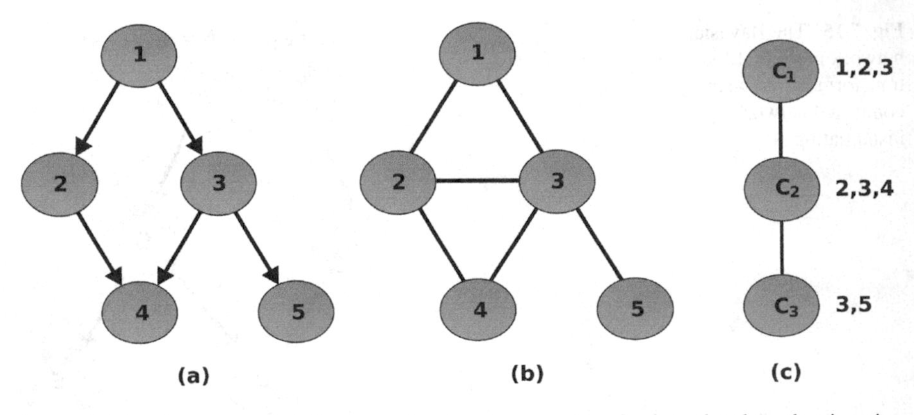


**Fig. 7.16** Transformation of a BN to a junction tree: **a** original net, **b** triangulated graph, **c** junction tree

$$P(A, B, C, D) = P(A)\frac{P(A, B)}{P(A)}\frac{P(B, C)}{P(B)}\frac{P(C, D)}{P(C)} \tag{7.30}$$

Which can be simplified to:

$$P(A, B, C, D) = \frac{P(A, B)P(B, C)P(C, D)}{P(B)P(C)} \tag{7.31}$$

That is basically the product of the probabilities of the clusters divided by the probabilities of the separators; which is the bases of the algorithm.

Next, we describe the junction tree algorithm. The algorithms consists of two phases: (i) transformation of the BN to a junction tree, (ii) probability propagation over the resulting singly connected network.

The transformation proceeds as follows (see Fig. 7.16):

1. Eliminate the directionality of the arcs.
2. Order the nodes in the graph (based on *maximum cardinality*).
3. Moralize the graph (add an arc between pairs of nodes with common children).
4. If necessary add additional arcs to make the graph *triangulated.*
5. Obtain the *cliques* of the graph (subsets of nodes that are fully connected and are not subsets of other fully connected sets).
6. Build a junction tree in which each node is a clique and its parent is any node that contains all common previous variables according to the ordering[3].

Ideally, we would like to find a triangulated graph with minimal maximal clique size to make the computations of marginals inside each clique more efficient. However, as we commented before in the case of variable elimination, this is an NP-Hard problem, and the same heuristics for determining the elimination order can be applied.

[3]Although a node could have multiple parents, only one is chosen when the junction tree is built.

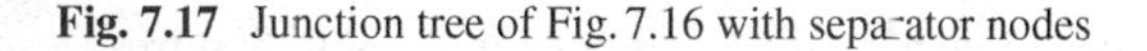


**Fig. 7.17** Junction tree of Fig. 7.16 with separator nodes

This transformation procedure guarantees that the resulting junction tree satisfies the *running intersection property*; that is that common variables with previous cliques are all in one clique. These common variables of neighbor cliques in the junction tree are called *separators*. Given the relevance of these separators, the junction tree is usually drawn including the separator nodes, depicted as rectangles. Figure 7.17 shows the junction tree of Fig. 7.16 including the separator nodes.

Once the junction tree is built, inference is based on probability propagation over the junction tree, in an analogous way as for tree–structured BNs. We must select a root node to begin. Each link between nodes and separators will be used twice during message passing, once in each direction. This is done by propagating messages up from each leaf to the root and then in reverse from the root to the leaves.

The junction tree algorithm can be divided in two stages: preprocessing and propagation. In the preprocessing phase the potentials of each clique are obtained following the next steps:

1. Determine the set of variables for each clique, $C_i$.
2. Determine the set of variables that are common with the previous (parent) clique, the separators, $S_i$.
3. Determine the variables that are in $C_i$ but not in $S_i$: $R_i = C_i - S_i$.
4. Calculate the potential of each clique, $clq_i$, as the product of the corresponding CPTs: $\psi(clq_i) = \prod_j P(X_j \mid Pa(X_j))$; where $X_j$ are the variables in $clq_i$.

For example, consider the BN in Fig. 7.16, with cliques: $clq_1 = \{1, 2, 3\}$, $clq_2 = \{2, 3, 4\}$, $clq_3 = \{3, 5\}$. Then the preprocessing phase is:

$C$: $C_1 = \{1, 2, 3\}$, $C_2 = \{2, 3, 4\}$, $C_3 = \{3, 5\}$.
$S$: $S_1 = \emptyset$, $S_2 = \{2, 3\}$, $S_3 = \{3\}$.
$R$: $R_1 = \{1, 2, 3\}$, $R_2 = \{4\}$, $R_3 = \{5\}$.
Potentials: $\psi(clq_1) = P(1)P(2 \mid 1)P(3 \mid 1)$, $\psi(clq_2) = P(4 \mid 3, 2)$, $\psi(clq_3) = P(5 \mid 3)$.

The propagation phase proceeds in a similar way to belief propagation for trees, by propagating $\lambda$ messages bottom-up and $\pi$ messages top-down.

**Bottom-Up Propagation**

1. Calculate the $\lambda$ message to send to the parent clique: $\lambda(C_i) = \sum_{R_i} \psi(C_i)$.

2. Update the potential of each clique with the $\lambda$ messages of its sons: $\psi(C_j)' = \lambda(C_i)\psi(C_j)$.
3. Repeat the previous two steps until reaching the root clique.
4. When reaching the root node obtain $P'(C_r) = \psi(C_r)'$.

**Top-Down Propagation**

1. Calculate the $\pi$ message to send to each child node $i$ by its parent $j$: $\pi(C_i) = \sum_{R_j} P'(C_j)$.
2. Update the potential of each clique when receiving the $\pi$ message of its parent: $P'(C_i) = \psi(C_i)' \frac{\pi(C_i)}{\lambda(C_i)}$.
3. Repeat the previous two steps until reaching the leaf nodes in the junction tree.

At the end of this propagation in both directions, each clique has the joint marginal probability of the variables that conform it. When there is evidence, the potentials for each clique are updated based on the evidence, and the same propagation procedure is followed.

After completing the bidirectional message passing phase, the messages at each clique correspond to its potential functions, which correspond to the joint probability distribution of the nodes in the clique. Because the nodes form a clique and thus are all connected, to find the probability densities of a subset of these nodes we need to marginalize the resulting potential function over the remaining variables. Thus, the marginal posterior probabilities of each variable are obtained from the clique potentials via marginalization: $P(X) = \sum_{C_i - X} \psi(C_i)$. The method guarantees that these probabilities will be the same not matter from which clique they are calculated. For instance, for the example in Fig. 7.16, we can obtain the probability of variable 2 from $C_1$ or $C_2$: $P(2) = \sum_{1,3} \psi(1, 2, 3) = \sum_{3,4} \psi(2, 3, 4)$

Continuing with the example of the BN in Fig. 7.16, we know illustrate the propagation without evidence.

First $C_3$ sends a $\lambda$ message to $C_2$: $\lambda(C_3) = \sum_5 \psi(3, 5)$. Next we update the potential of $C_2$: $\psi(C_2)' = \psi(2, 3, 4)\lambda(C_3)$. Then $C_2$ sends a $\lambda$ message to $C_1$: $\lambda(C_2) = \sum_4 \psi(C_2)'$. And then we update the potential of $C_1$: $\psi(C_1)' = \psi(1, 2, 3)\lambda(C_2)$; $P'(C_1) = \psi(C_1)'$. This completes the bottom-up propagation phase.

Now starts the top-down propagation. $C_1$ sends a $\pi$ message to $C_2$: $\pi(C_2) = \sum_1 P'(C_1)$. Next update the potential of $C_2$: $P'(C_2) = \psi(C_2)' \frac{\pi(C_2)}{\lambda(C_2)}$. Then $C_2$ sends a $\pi$ message to $C_3$: $\pi(C_3) = \sum_{2,4} P'(C_2)$. Finally, update the potential of $C_3$: $P'(C_3) = \psi(C_3)' \frac{\pi(C_3)}{\lambda(C_3)}$.

Let us verify that resulting potentials, $P'(C_i)$, are the joint marginal probability of the corresponding clique. Remember the original potentials of the BN: $\psi(clq_1) = P(1)P(2 \mid 1)P(3 \mid 1)$, $\psi(clq_2) = P(4 \mid 3, 2)$, $\psi(clq_3) = P(5 \mid 3)$.

Bottom-up phase:

$$\lambda(C_3) = \sum_5 P(5 \mid 3)$$

$$\psi(C_2)' = P(4 \mid 2, 3) \sum_5 P(5 \mid 3)$$

$$\lambda(C_2) = \sum_4 P(4 \mid 2, 3) \sum_5 P(5 \mid 3)$$

$$\psi(C_1)' = P(1)P(2 \mid 1)P(3 \mid 1) \sum_4 P(4 \mid 2, 3) \sum_5 P(5 \mid 3)$$

Which can be written as:

$$\psi(C_1)' = \sum_4 \sum_5 P(1)P(2 \mid 1)P(3 \mid 1)P(4 \mid 2, 3)P(5 \mid 3) = \sum_4 \sum_5 P(1, 2, 3, 4, 5)$$

That is the marginalization of 4 and 5 of the joint probability: $P'(C_1) = \psi(C_1)' = P(1, 2, 3)$. Which indeed is the joint of $C_1$.

Top-down phase:

$$\pi(C_2) = \sum_1 P(1, 2, 3)$$

$$P'(C_2) = \frac{P(4 \mid 2, 3) \sum_5 P(5 \mid 3) \sum_1 P(1, 2, 3)}{\sum_4 P(4 \mid 2, 3) \sum_5 P(5 \mid 3)} = \frac{P(4 \mid 2, 3) \sum_1 P(1, 2, 3)}{\sum_4 P(4 \mid 2, 3)}$$

Given that $\sum_1 P(1, 2, 3) = P(2, 3)$ and $\sum_4 P(4 \mid 2, 3) = 1$, then:

$$P'(C_2) = P(2, 3)P(4 \mid 2, 3) = P(2, 3, 4)$$

Which is the joint of $C_2$. And finally:

$$\pi(C_3) = \sum_{2,4} P(2, 3, 4)$$

$$P'(C_3) = P(5 \mid 3) \frac{\sum_{2,4} P(2, 3, 4)}{\sum_5 P(5 \mid 3)}$$

Given that $\sum_{2,4} P(2, 3, 4) = P(3)$ and $\sum_5 P(5 \mid 3) = 1$, then:

$$P'(C_3) = P(3)P(5 \mid 3) = P(3, 5)$$

That is the joint probability of $C_3$. Thus we have verified that the junction tree algorithm obtains the joint marginals of each cluster or clique, from which we can calculate the single variable's marginals via marginalization. In this case: $P(1) = \sum_{2,3} P'(C_1)$, $P(2) = \sum_{1,3} P'(C_1)$, $P(3) = \sum_{1,2} P'(C_1)$, $P(4) = \sum_{2,3} P'(C_2)$, and $P(5) = \sum_3 P'(C_3)$.

There are two main variations on the junction tree algorithm, which are known as the Hugin [8] and Shenoy-Shafer [18] architectures. The description above is based on the Hugin architecture. The main differences between them are in the information they store, and in the way they compute the messages. These differences have implications in their computational complexity. In general, the Shafer-Shenoy architecture will require less space but more time.

#### 7.3.2.4 Complexity Analysis

In the worst case, probabilistic inference for Bayesian networks is NP-Hard [1]. The time and space complexity is determined by what is known as the *tree-width*, and has to do with how close the structure of the network is to a tree. Thus, a tree-structured BN (maximum one parent per variable) has a tree-width of one. A polytree with at most $k$ parents per node has a tree-width of $k$. In general, the tree-width is determined by how *dense* the topology of the network is, and this affects: (i) the size of the largest factor in the variable elimination algorithm; (ii) the number of variables that need to be instantiated in the conditioning algorithm, (iii) the size of the largest clique in the junction tree algorithm.

In practice BNs tend to be sparse graphs, and in this case the exact inference techniques are very efficient even for models with hundreds of variables. In the case of complex networks, an alternative is to use approximate algorithms. These are described next.

### *7.3.3 Approximate Inference*

#### 7.3.3.1 Loopy Propagation

This is simply the application of the probability propagation algorithm for multi-connected networks. Although in this case the conditions for this algorithm are not satisfied, and it only provides an approximate solution for the inference problem, it is very efficient. Given that the BN is not singly connected, as the messages are propagated, these can *loop* through the network. Consequently, the propagation is repeated several times. The procedure is the following:

1. Initialize the $\lambda$ and $\pi$ values for all nodes to random values.
2. Repeat until convergence or a maximum number of iterations:
   a. Do probability propagation according to the algorithm for singly connected networks.
   b. Calculate the posterior probability for each variable.

The algorithm converges when the difference between the posterior probabilities for all variables of the current and previous iterations is below a certain threshold. It has been found empirically that for certain structures this algorithm converges to the true posterior probabilities; however, for other structures it does not converge [13].

An important application of loopy belief propagation is in "Turbo Codes"; which is a popular error detection and correction scheme used in data communications.

#### 7.3.3.2 Stochastic Simulation

Stochastic simulation algorithms consist in *simulating* the BN several times, where each simulation gives a sample value for all non-instantiated variables. These values are chosen randomly according to the conditional probability of each variable. This process is repeated $N$ times, and the posterior probability of each variable is approximated in terms of the frequency of each value in the sample space. This gives an estimate of the posterior probability which depends on the number of samples; however, the computational cost is not affected by the complexity of the network. Next we present two stochastic simulation algorithms for BNs: logic sampling and likelihood weighting.

**Logic Sampling**
Logic sampling is a basic stochastic simulation algorithm that generates samples according to the following procedure:

1. Generate sample values for the root nodes of the BN according to their prior probabilities. That is, a random value is generated for each root variable $X$, following a distribution according to $P(X)$.
2. Generate samples for the next *layer*, that is the sons of the already sampled nodes, according to their conditional probabilities, $P(Y \mid Pa(Y))$, where $Pa(Y)$ are the parents of $Y$.
3. Repeat (2) until all the leaf nodes are reached.

The previous procedure is repeated $N$ times to generate $N$ samples. The probability of each variable is estimated as the fraction of times (frequency) that a value occurs in the $N$ samples, that is, $P(X = x_i) \sim No(x_i)/N$; where $No(x_i)$ is the number of times that $X = x_i$ in all the samples.

The direct application of the previous procedure gives an estimate of the marginal probabilities of all the variables when there is no evidence. If there is evidence (some variables are instantiated), all samples that are not consistent with the evidence are discarded and the posterior probabilities are estimated from the remaining samples.

For example, consider the BN in Fig. 7.13, and 10 samples generated by logic sampling. Assuming all variables are binary, the 10 samples generated are shown in Table 7.3.

If there is no evidence, then given these samples, the marginal probabilities are estimated as follows:

- $P(A = T) = 4/10 = 0.4$
- $P(B = T) = 3/10 = 0.3$
- $P(C = T) = 5/10 = 0.5$
- $P(D = T) = 5/10 = 0.5$
- $P(E = T) = 3/10 = 0.3$

**Table 7.3** Samples generated using logic sampling for the BN in Fig. 7.13. All variables are binary with two possible values, $True = T$ or $False = F$

| Variables | A | B | C | D | E |
|---|---|---|---|---|---|
| $sample_1$ | T | F | F | F | T |
| $sample_2$ | F | T | T | F | F |
| $sample_3$ | T | F | F | T | F |
| $sample_4$ | F | F | T | F | T |
| $sample_5$ | T | F | T | T | F |
| $sample_6$ | F | F | F | F | T |
| $sample_7$ | F | T | T | T | F |
| $sample_8$ | F | F | F | F | F |
| $sample_9$ | F | F | F | T | F |
| $sample_{10}$ | T | T | T | T | F |

The remaining probabilities are just the complement, $P(X = F) = 1 - P(X = T)$.

In the case where there is evidence with $D = T$, we eliminate all the samples where $D = F$, and estimate the posterior probabilities from the remaining 5 samples:

- $P(A = T \mid D = T) = 3/5 = 0.6$
- $P(B = T \mid D = T) = 2/5 = 0.4$
- $P(C = T \mid D = T) = 3/5 = 0.6$
- $P(E = T \mid D = T) = 0/5 = 0.0$

A disadvantage of logic sampling when evidence exists is that many samples have to be discarded; this implies that a larger number of samples are required to have a *good* estimate. An alternative algorithm that does not waste samples is presented below.

**Likelihood Weighting**

Likelihood weighting generates samples in the same way as logic sampling, however when there is evidence the non-consistent samples are not discarded. Instead, each sample is given a weight according to the weight of the evidence for this sample. Given a sample $s$ and the evidence variables $\mathbf{E} = \{E_1, ..., E_m\}$, the weight of sample $s$ is estimated as:

$$W(\mathbf{E} \mid s) = P(E_1)P(E_2)...P(E_m) \tag{7.32}$$

where $P(E_i)$ is the probability of the evidence variable $E_i$ for that sample.

The posterior probability for each variable $X$ taking value $x_i$ is estimated by dividing the sum of the weights $W_i(X = x_i)$ for each sample where $X = x_i$ by the total weight for all the samples:

$$P(X = x_i) \sim \sum_i W_i(X = xi) / \sum_i W_i \tag{7.33}$$

### 7.3.4 Most Probable Explanation

The *most probable explanation* (MPE) or *abduction* problem consists in determining the most probable values for a subset of variables (explanation subset) in a BN given some evidence. There are two variants of this problem, *total abduction* and *partial abduction*. In the total abduction problem, the explanation subset is the set of all non-instantiated variables; while in partial abduction, the explanation subset is a proper subset of the non-instantiated variables. In general, the MPE is not the same as the union of the most probable value for each individual variable in the explanation subset.

Consider the set of variables $\mathbf{X} = \{X_E, X_R, X_H\}$, where $X_E$ is the subset of instantiated variables, $X_H$ the subset of hypothesis variables, and $X_R$ the rest of the variables; then we can formalize the MPE problems as follows:

Total abduction: $ArgMax_{X_H, X_R} P(X_H, X_R \mid X_E)$.
Partial abduction: $ArgMax_{X_H} P(X_H \mid X_E)$.

One way to solve the MPE problem is based on a modified version of the variable elimination algorithm. For the case of total abduction, we substitute the summations by maximizations:

$$max_{X_H, X_R} P(X_H, X_R \mid X_E)$$

For partial abduction, we sum over the variables that are not in the explanation subset and maximize over the explanation subset:

$$max_{X_H} \sum_{X_R} P(X_H, X_R \mid X_E)$$

The MPE problem is computationally more complex than the single query inference.

### 7.3.5 Continuous Variables

Up to now we have considered BNs with discrete multi-valued variables. When dealing with continuous variables, one option is to discretize them; however, this could result in a loss of information (few intervals) or in an unnecessary increase in computational requirements (many intervals). Another alternative is to operate directly on the continuous distributions. Probabilistic inference techniques have been developed for some distribution families, in particular for Gaussian variables. Next we describe the basic propagation algorithm for linear, Gaussian BNs [16].

The basic algorithm makes the following assumptions:

1. The structure of the network is a polytree.
2. All the sources of uncertainty are Gaussians and uncorrelated.

3. There is a linear relationship between each variable and its parents:

$$X = b_1U_1 + b_2U_2 + ... + b_nU_n + W_X$$

Where $U_i$ are parents of variable $X$, $b_i$ are constant coefficients and $W_X$ represents Gaussian noise with a zero mean.

The inference procedure is analogous to belief propagation in discrete BNs, but instead of propagating probabilities, it propagates means and standard deviations. In the case of Gaussian distributions, the marginal distributions of all the variables are also Gaussians. Thus, in general the posterior probability of a variable can be written as:

$$P(X \mid E) = N(\mu_X, \sigma_X)$$

Where $\mu_X$ and $\sigma_X$ are the mean and standard deviation of $X$ given the evidence $E$, respectively.

Next we describe how to calculate the mean and standard deviation via the propagation algorithm. Each variable sends to its parent variable $i$:

$$\mu_i^- = (1/b_i)\left[\mu_\lambda - \sum_{k\neq i} b_k\mu_k^+\right] \tag{7.34}$$

$$\sigma_i^- = (1/b_i^2)\left[\sigma_\lambda - \sum_{k\neq i} b_k^2\sigma_k^+\right] \tag{7.35}$$

Each variable sends to its child node $j$:

$$\mu_j^+ = \frac{\sum_{k\neq j}[\mu_k^-/\sigma_k + \mu_\pi/\sigma_\pi]}{\sum_{k\neq} 1/\sigma_k^- + \mu_\pi/\sigma_\pi} \tag{7.36}$$

$$\sigma_j^+ = \left[\sum_{k\neq j} 1/\sigma_k^- + 1/\sigma_\pi\right]^{-1} \tag{7.37}$$

Each variable integrates the messages it receives from its sons and parents via the following equations:

$$\mu_\pi = \sum_i b_i\mu_i^+ \tag{7.38}$$

$$\sigma_\pi = \sum_i b_i^2\sigma_i^+ \tag{7.39}$$

$$\mu_\lambda = \sigma_\lambda \sum_j \mu_j^- / \sigma_j^- \tag{7.40}$$

$$\sigma_\lambda = \left[ \sum_j 1/\sigma_j^- \right]^{-1} \tag{7.41}$$

Finally, each variable obtains its mean and standard deviation by combining the information from its parent and children nodes:

$$\mu_X = \frac{c_\pi \mu_\lambda + \sigma_\lambda \mu_\pi}{\sigma_\pi + \sigma_\lambda} \tag{7.42}$$

$$\sigma_X = \frac{\sigma_\pi \sigma_\lambda}{\sigma_\pi + \sigma_\lambda} \tag{7.43}$$

Propagation for other distributions is more difficult, as they do not have the same properties of the Gaussian; in particular, the product of Gaussians is also a Gaussian. An alternative for other types of distributions is to apply stochastic simulation techniques.

## 7.4 Applications

Bayesian networks have been applied in many domains, including medicine, industry, education, finance, biology, etc. To exemplify the application of BNs, in this chapter we will describe: (i) a technique for information validation, and (ii) a methodology for system reliability analysis. In the following chapters we will illustrate their application in other areas.

### *7.4.1 Information Validation*

Many systems use information to make decisions; if this information is erroneous it could lead to non-optimal decisions and in some cases decisions made based on erroneous data could be dangerous. Consider for example an intensive care unit of a hospital in which sensors monitor the status of an operated patient so that the body temperature is kept beneath certain levels. Given that the sensors are working constantly, there is potential for them to produce erroneous readings. If this happens two situations may arise:

- the temperature sensor indicates no changes in temperature even if it has increased to dangerous levels,

- the temperature sensor indicates a dangerous level even if it is normal.

The first situation may cause severe damage to the patient's health. The second situation may cause an emergency treatment of the patient that can also worsen his/her condition.

In many applications there are different sources of information, i.e. sensors, which are not independent; the information from one source gives us clues about the other sources. If we can represent these dependencies between the different sources, then we can use it to detect possible errors and avoid erroneous decisions. This section presents an information validation algorithm based on Bayesian networks [6]. The algorithm starts by building a model of the dependencies between sources of information (variables) represented as a Bayesian network. Subsequently, the validation is done in two phases. In the first phase, potential faults are detected by comparing the actual value with the one predicted from the related variables via propagation in the Bayesian network. In the second phase, the real faults are isolated by constructing an additional Bayesian network based on the Markov blanket property.

#### 7.4.1.1 Fault Detection

It is assumed that it is possible to build a probabilistic model relating all the variables in the application domain. Consider for example the network shown in Fig. 7.18 which represents the most basic function of a gas turbine.

Suppose it is required to validate the temperature measurements in the turbine. By reading the values of the rest of the sensors, and applying probability propagation, it is possible to calculate a posterior probability distribution of the temperature given all the evidence, i.e., $P(T \mid Mw, P, Fg, Pc, Pv, Ps)$. Assuming that all the variables are discrete or discretized if continuous, by propagation we obtain probability distributions for each value of $T$. If the real observed value coincides with a *valid*

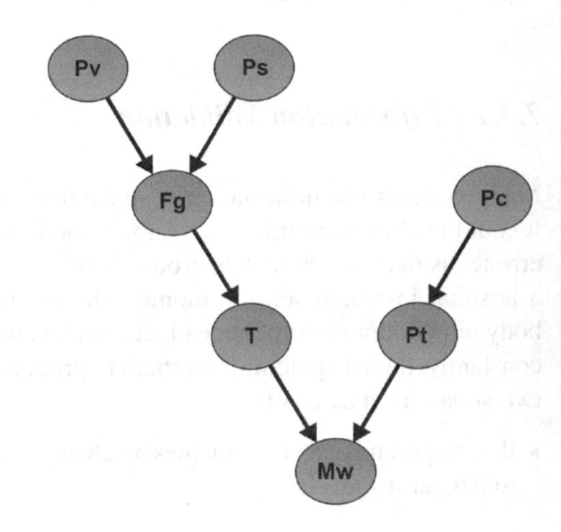


**Fig. 7.18** A basic probabilistic model of a gas turbine. The mega watts generated in a gas turbine (node $Mw$) depends on the temperature (node $T$) and pressure in the turbine (node $Pt$). Temperature depends on the flow of gas (node $Fg$) and this flow depends on the valve of gas position (node $Pv$) and the gas fuel pressure supply (node $Ps$). The pressure for the turbine depends on the pressure at the output of the compressor (node $Pc$)

value—that has a *high* probability, then the sensor is considered correct; otherwise it is considered faulty.

This procedure is repeated for all the sensors in the model. However, if a validation of a single sensor is made using a faulty sensor, then a faulty validation can be expected. In the example above, what happens if $T$ is validated using a faulty $Mw$ sensor? How do we know which of the sensors is faulty? Thus, by applying this validation procedure, we may only detect a faulty condition, but we are not able to identify which is the real faulty sensor. This is called an *apparent fault*. An isolation stage is needed.

#### 7.4.1.2 Fault Isolation

The isolation phase is based on the *Markov Blanket* (MB) property. For example, in the network of Fig. 7.18, the $MB(T) = \{Mw, Fg, Pt\}$, and the $MB(Pv) = \{Fg, Ps\}$. The set of nodes that constitute the MB of a sensor can be seen as a protection of the sensor against changes outside its MB. Additionally, we define the *Extended Markov Blanket* of a node $X$ ($EMB(X)$) as the set of sensors formed by the sensor itself plus its MB. For example, $EMB(T) = \{T, Mw, Fg, Pt\}$.

Using this property, if a fault exists in one of the sensors, it will be revealed in all of the sensors in its EMB. On the contrary, if a fault exists outside a sensor's EMB, it will not affect the estimation of that sensor. It can be said then, that the EMB of a sensor acts as its protection against others faults, and also protects others from its own failure. We utilize the EMB to create a *fault isolation* module that distinguishes the *real faults* from the apparent faults The full theory is developed in [7].

After a cycle of basic validations of all sensors is completed, a set $S$ of apparent faulty sensors is obtained. Thus, based on the comparison between $S$ and the EMB of all sensors, the theory establishes the following situations:

1. If $S = \phi$ there are no faults.
2. If $S$ is equal to the EMB of a sensor $X$, and there is no other EMB which is a subset of $S$, then there is a *single real fault* in $X$.
3. If $S$ is equal to the EMB of a sensor $X$, and there are one or more EMBs which are subsets of $S$, then there is a real fault in $X$, and possibly, real faults in the sensors whose EMBs are subsets of $S$. In this case, there are possibly *multiple indistinguishable* real faults.
4. If $S$ is equal to the union of several EMBs and the combination is unique, then there are *multiple distinguishable* real faults in all the sensors whose EMB are in $S$.
5. If none of the above cases is satisfied, then there are multiple faults but they can not be distinguished. All the variables whose EMBs are subsets of $S$ could have a real fault.

For example, considering the Bayesian network model in Fig. 7.18, some of the following situations may occur (among others):

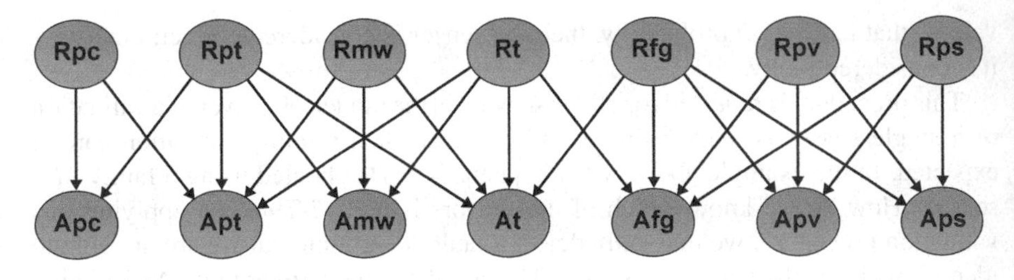


**Fig. 7.19** Isolation network of the example in Fig.7.18

- $S = \{T, Pt, Mw\}$, which corresponds to case 2, and confirms a single real fault in $Mw$,
- $S = \{T, Pc, Pt, Mw\}$, which corresponds to case 3, and as such, there is a real fault in $Pt$ and possibly in $Pc$ and $Mw$,
- $S = \{Pv, Ps, Fg\}$, which corresponds to case 4, and as such, there are real faults in $Pv$ and $Ps$.

The isolation of a real fault is carried out in the following manner. Based on the EMB property described above, there will be a real fault in sensor $X$ if an apparent fault is detected in its entire EMB. Also, we can say that an apparent fault will be revealed if a real fault exists in any sensor of its EMB. With these facts, we define the isolation network formed by two levels. The root nodes represent the real faults, where there is one per sensor or variable. The lower level is formed by one node representing the apparent fault for each variable. Notice that the arcs are defined by the EMB of each variable. Figure 7.19 shows the isolation network for the detection network of Fig.7.18. For instance, the apparent fault node corresponding to variable $Mw$ (node $A_{mw}$) is connected with nodes $R_{mw}$, $R_T$ and $R_{pt}$, which represent the real faults of the EMB nodes of $Mw$. At the same time, node $R_{mw}$ is connected with all the apparent faults that this real fault causes, i.e., to nodes $A_{mw}$, $A_T$, and $A_{pt}$.

The parameters of the isolation network are defined as following. The marginal probabilities of the real faults nodes (roots) are usually defined as 0.5, although if we have prior knowledge of the prior fault probability of a sensor, we can assign it this value. The conditional probabilities of the apparent faults nodes are set using the Noisy OR model, as its conditions are reasonably satisfied: (i) there are no apparent faults if there are no real faults, (ii) the mechanisms that inhibit an apparent fault given certain real fault are independent from the mechanisms of the other real faults.

The detection of the real faults is carried out by the isolation procedure described in Algorithm 7.1. This algorithm is applied for each sensor (variable) in the BN model. At the end the posterior probabilities of all the real fault nodes are updated; and those with a "high" probability could be the faulty ones.

**Algorithm 7.1** Function isolate

**Require:** A sensor $n$ and the state of sensor $n$.
1: Assign a value (instantiate) to the apparent fault node corresponding to $n$
2: Propagate probabilities and update the posterior probability of all *real fault* nodes
3: Update vector $P_f(sensors)$

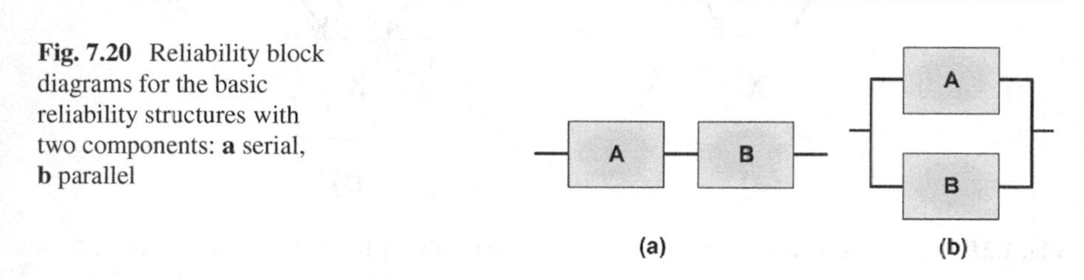


**Fig. 7.20** Reliability block diagrams for the basic reliability structures with two components: **a** serial, **b** parallel

## 7.4.2 *Reliability Analysis*

In the reliability analysis of a complex system, a common approach is to divide the system into smaller elements, units, subsystems, or components. The main assumption is that every entity has two states, success and failure. This subdivision generates a "block diagram" that is similar to the description of the system in operation. For each element, the *failure rate* is specified, and based on these, the reliability of the complete system is obtained.

Traditionally, fault trees are used for reliability analysis; however, this technique has its limitations, as it assumes independent events, thus it is difficult to model dependency between events or faults. Dependent events can be found in reliability analysis in the following cases: (i) common causes—condition or event which provokes multiple elemental failures; (ii) mutually exclusive primary events—the occurrence of one basic event precludes another; (iii) standby redundancies—when an operating component fails, a standby component is put into operation, and the redundant configuration continues to function; (iv) components supporting loads—failure of one component increases the load supported by the other components. Using Bayesian networks we can explicitly represent dependencies between failures, and in this way model complex systems that are difficult for traditional techniques [19].

### 7.4.2.1 Reliability Modeling with Bayesian Networks

Reliability analysis starts by representing the structure of the system in terms of a reliability block diagram. In this representation there are two basic structures: serial and parallel components (see Fig. 7.20). A serial structure implies that the two components should operate correctly for the system to function; or in other words, if one fails the entire system fails (this corresponds to an $AND$ gate in fault trees). In parallel structures, it is sufficient for one of the components to operate for the system to function ($OR$ gate in fault tress).

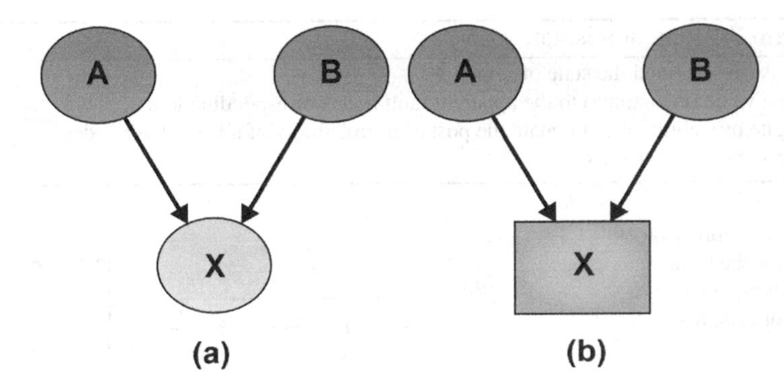


**Fig. 7.21** Bayesian network structure for the two basic reliability block diagrams from Fig. 7.20

**Table 7.4** Conditional probability table, $P(X \mid A, B)$, for two components with a serial structure. $A$ means that the component $A$ is operational and $\neg A$ that it has failed

| X | $A, B$ | $A, \neg B$ | $\neg A, B$ | $\neg A, \neg B$ |
|---|---|---|---|---|
| Success | 1 | 0 | 0 | 0 |
| Failure | 0 | 1 | 1 | 1 |

**Table 7.5** Conditional probability table, $P(X \mid A, B)$, for two components in parallel. $A$ means that the component $A$ is operational and $\neg A$ that it has failed

| X | $A, B$ | $A, \neg B$ | $\neg A, B$ | $\neg A, \neg B$ |
|---|---|---|---|---|
| Success | 1 | 1 | 1 | 0 |
| Failure | 0 | 0 | 0 | 1 |

We can represent the previous basic block diagrams with a Bayesian network as is depicted in Fig. 7.21. The structure is the same in both cases, the difference is the conditional probability matrix. The CPTs for both cases are depicted in Tables 7.4 and 7.5. In both cases, the prior probabilities of the basic components ($A$, $B$) will represent the failure rate. Thus, by applying probabilistic inference in the BN representation, we obtain the failure rate of the system, $X$.

The BN representation of the basic serial/parallel cases can be directly generalized to represent any block diagram that can be reduced to a set of serial and parallel combinations of components, which, in practice is the case for most systems. There are some structures that can not be decomposed to a serial/parallel combination, such as a *bridge*. However, it is also possible to model these cases using BNs [19].

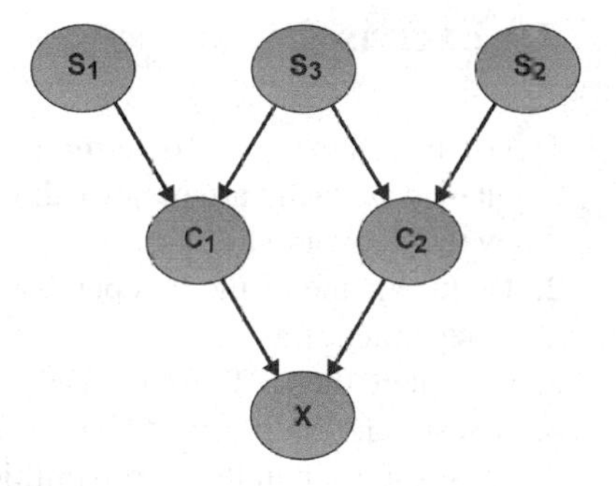


**Fig. 7.22** Bayesian network structure for the system with a common cause failures

#### 7.4.2.2 Modeling Dependent Failures

The main advantage of reliability modeling with BNs is that we can model dependent failures. We will illustrate this for the case of a system with common cause failures.

Suppose that a system has two components that are affected by three possible failure sources. Source $S_1$ affects component $C_1$, source $S_2$ affects component $C_2$, and source $S_3$ affects both components (common cause). For instance, the system could be a power plant with two subsystems; each subsystem has elements that can fail, but an earthquake can make both fail. A Bayesian network model for this example of dependent failures is depicted in Fig. 7.22. In this model, the CPT for all three non root nodes ($C_1$, $C_2$, $X$) is equivalent to that of a serial component combination. $X$ represents the failure rate of the system, which can be obtained by probability propagation given the failure rates for the three failure sources.

## 7.5 Additional Reading

An introduction to Bayesian networks is given in the classic book by Judea Pearl [16]. Other general books on BNs are [9, 14]. A more recent account with emphasis on modeling and inference is given by [3]; it includes a complexity analysis for the different inference techniques. Other books with more emphasis on applications are [10, 17]. An overview of canonical models is presented in [5]. Several variants of the conditioning algorithm have been proposed, including local conditioning [4] and recursive conditioning [2]. The junction tree algorithm was initially introduced by [11], and the two main architectures are described in [8, 18]. An analysis of loopy propagation can be seen in [13]. Inference for continuous Gaussian variables was introduced in [16], and a more general approach based on truncated exponentials is presented in [12].

## 7.6 Exercises

1. For the BN in Fig. 7.2 determine: (a) the contour of each variable, (b) the Markov blanket of each variable, (c) all the conditional independence relations implied by the network structure.
2. Deduce some of the independence relations in the previous problem using the independence axioms.
3. Complete the CPTs for the BN in Fig. 7.5 assuming all the variables are binary.
4. Investigate the Noisy AND model and obtain the CPT for a variable with 3 causes with inhibition probabilities equal to 0.05, 0.1 and 0.2, respectively.
5. Consider the belief propagation example in Sect. 7.3.1, obtain the posterior probabilities of all the variables via belief propagation considering that the only evidence is $C = true$.
6. Repeat the previous problem using the variable elimination procedure.
7. Estimate the posterior probabilities of the example in Sect. 7.3.1 under the same conditions as the previous two problems ($C = true$) using the logic sampling method for different numbers of samples (10, 20, ...) and compare the results using exact inference.
8. Given the BN structure of Fig. 7.13 and the following CPTs, considering all are binary variables:
   $P(A) = [0.6, 0.4]$, $P(B \mid A) = \frac{0.3,0.2}{0.7,0.8}$, $P(C \mid A) = \frac{0.6,0.1}{0.4,0.9}$, $P(D \mid B, C) = \frac{0.5,0.1,0.3,0.2}{0.5,0.9,0.7,0.8}$, $P(E \mid C) = \frac{0.7,0.7}{0.3,0.3}$.
   (a) Transform it to a junction tree, specify the structure and cliques. (b) Obtain the potentials for each junction. (c) Using the inference algorithm, obtain the marginal probability of each variable.
9. For the previous problem, recalculate the marginal probabilities given the evidence that $E$ has the value false (first value in the CPT).
10. Repeat the previous problem using the conditioning algorithm.
11. For the BN in Fig. 7.19: (a) moralize the graph, (b) triangulate the graph, (c) determine the cliques and obtain a junction tree, (d) obtain the sets $C$, $S$ and $R$ for each clique according to the junction tree algorithm.
12. Consider the BN for common cause failures in Fig. 7.22. Given the following probabilities of failure of the different sources: $S_1 = 0.05$, $S_2 = 0.03$ and $S_3 = 0.02$, calculate the probability of success and failure of the system, $X$.
13. *** Develop a program based on the Bayes ball procedure to illustrate D-Separation. Given a BN structure, the user selects two nodes and a separation subset. The program should find all the trajectories between the two nodes, and then determines if these are independent given the separation subset by applying the Bayes ball procedure, illustrating graphically if the ball goes through a trajectory or is blocked.
14. *** Develop a general program for the belief propagation algorithm for polytrees considering discrete variables. Develop a parallel version of the previous program, establishing how the processors are assigned for an efficient parallelization of the algorithm. Extend the previous programs for loopy belief propagation.

15. *** Develop a program that implements the information validation algorithm considering both phases. The input is BN that specifies the dependency structure of the variables and the associated CPTs. The program should (a) build automatically the isolation network; (b) Given values for all the variables, estimate the probabilities of apparent faults (c) Use the isolation network to estimate the probabilities of the real faults.

## References


1. Cooper, G.F.: The computational complexity of probabilistic inference using Bayesian networks. Artif. Intell. **42**, 393–405 (1990)
2. Darwiche, A.: Recursive conditioning. Artif. Intell. **126**, 5–41 (2001)
3. Darwiche, A.: Modeling and Reasoning with Bayesian Networks. Cambridge University Press, New York (2009)
4. Díez. J.F.: Local conditioning in Bayesian networks. Artif. Intell. **87**(1), 1–20 (1996)
5. Díez, F.J., Druzdzel, M.J.: Canonical probabilistic models for knowledge engineering. Technical Report CISIAD-06-01. Universidad Nacional de Educación a Distancia, Spain (2007)
6. Ibargüengoytia, P.H., Sucar L.E., Vadera, S.: A probabilistic model for sensor validation. In: Proceedings of the Twelfth Conference on Uncertainty in Artificial Intelligence UAI-96, pp. 332–339. Morgan Kaufmann Publishers Inc. (1996)
7. Ibargüengoytia, P.H., Vadera, S., Sucar, L.E.: A probabilistic model for information validation. British Comput. J. **49**(1), 113–126 (2006)
8. Jensen, F.V., Andersen, S.K.: Approximations in Bayesian belief universes for knowledge based systems. In: Proceedings of the Sixth Conference on Uncertainty in Artificial Intelligence UAI-90, pp. 162–169. Elsevier, New York (1990)
9. Jensen, F.V.: Bayesian Networks and Decision Graphs. Springer, New York (2001)
10. Korb, K.B., Nicholson, A.E.: Bayesian Artificial Intelligence, 2nd edn. CRC Press, Boca Raton (2010)
11. Lauritzen, S., Spiegelhalter, D.J.: Local computations with probabilities on graphical structures and their application to expert systems. J. R. Stat. Soc. Seri. B. **50**(2), 157–224 (1988)
12. Moral, S., Rumi, R., Salmerón, A.: Mixtures of truncated exponentials in hybrid Bayesian networks. Symb. Quant. Approach. Reas. Uncer. **2143**, 156–167 (2001)
13. Murphy, K.P., Weiss, Y., Jordan, M.: Loopy belief propagation for approximate inference: an empirical study. In: Proceedings of the Fifteenth Conference on Uncertainty in Artificial Intelligence, pp. 467–475. Morgan Kaufmann Publishers Inc. (1999)
14. Neapolitan, R.E.: Probabilistic Reasoning in Expert Systems. Wiley, New York (1990)
15. Pearl, J.: Fusion, propagation and structuring in belief networks. Artif. Intell. **29**, 241–288 (1986)
16. Pearl, J.: Probabilistic Reasoning in Intelligent Systems: Networks of Plausible Inference. Morgan Kaufmann, San Francisco (1988)
17. Pourret, O., Naim, P., Marcot, B. (eds.): Bayesian Belief Networks: A Practical Guide to Applications. Wiley, New Jersey (2008)
18. Shenoy, P., Shafer, G.: Axioms for probability and belief-function propagation. In: Uncertainty in Artificial Intelligence, vol. 4, pp. 169–198. Elsevier, New York (1990)
19. Torres-Toledano, J.G., Sucar, L.E.: Bayesian networks for reliability analysis of complex systems. In: Coelho, H. (ed.) IBERAMIA'98. Lecture Notes in Computer Science, vol. 1484, pp. 195–206. Springer, Berlin (1998)

# Chapter 8
# Bayesian Networks: Learning

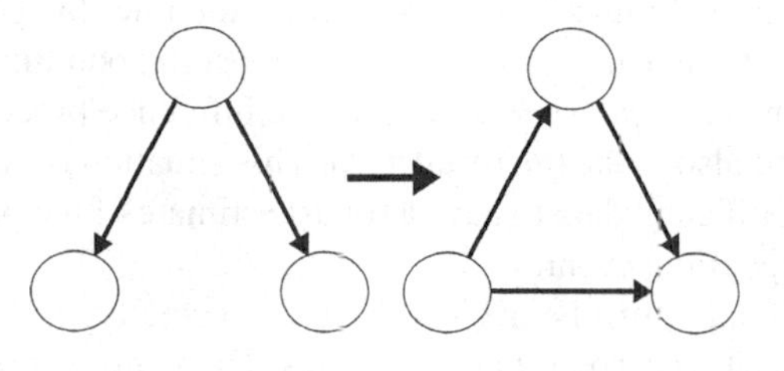

## 8.1 Introduction

Learning a Bayesian network includes two aspects: learning the structure and learning the parameters. When the structure is known, parameter learning consists in estimating the conditional probability tables (CPTs) from data. For structure learning there are two main types of methods: (i) global methods based on search and score, and (ii) local methods that use conditional independence tests. Next we describe both aspects, starting with parameter learning.

## 8.2 Parameter Learning

For this section we assume that the structure of the Bayesian network is known, and will focus on how to learn its parameters; later we will see how to learn the structure. We will consider that all the variables in the model are discrete; in the case of continuous variables, some discretization techniques will be described.



L. E. Sucar, *Probabilistic Graphical Models*, Advances in Computer Vision and Pattern Recognition, https://doi.org/10.1007/978-3-030-61943-5_8

If we have *sufficient* and complete data for all the variables, and we assume the topology of the BN is known, parameter learning is straight forward. The CPT for each variable can be estimated from the data based on the frequency of each value (or combination of values) obtaining a *maximum likelihood* (ML) estimator of the parameters. For example, to estimate the CPT of $B$ given it has two parents, $A, C$:

$$P(B_i \mid A_j, C_k) \sim N_{i,j,k}/N_{j,k} \tag{8.1}$$

where $N_{i,j,k}$ is the number of cases in the database in which $B = B_i$, $A = A_j$ and $C = C_k$, and $N_{j,k}$ is the total number of cases in which $A = A_j$ and $C = C_k$.

### 8.2.1 Smoothing

When we estimate probabilities from data, it can sometimes happen that a particular event never occurs in the data set. This leads to the corresponding probability value being zero, implying an *impossible* case; if in the inference process this probability is considered, it will also make the result zero. This situation occurs, in many cases, because there is insufficient data to have a robust estimate of the parameters, and not because it is an impossible event.

The previous situation can be avoided by using some type of *smoothing* for the probabilities, eliminating zero probability values. There are several smoothing techniques, one of the most common and simplest is *Laplacian* smoothing.

Laplacian smoothing consists in initializing the probabilities to a uniform distribution, and then updating these values based on the data. Consider a discrete variable, $X$, with $k$ possible values. Initially, each probability will be set to $P(x_i) = 1/k$. Then, consider a data set with $N$ samples, in which the value $x_i$ occurs $m$ times; the estimate of its probability will be the following:

$$P(x_i) = (1 + m)/(k + N) \tag{8.2}$$

### 8.2.2 Parameter Uncertainty

If there is not sufficient data, a situation common in practice, we have uncertainty in the parameters. This uncertainty can be modeled using a second order probability distribution, and could be propagated in the inference process so we have an estimate of the uncertainty in the resulting probabilities. For binary variables, the uncertainty in the parameters can be modeled using a Beta distribution:

$$\beta(a, b) = \frac{(a + b + 1)!}{a!b!} x^a (1 - x)^b \tag{8.3}$$

For multivalued variables, uncertainty in the parameters can be represented by an extension of the Beta known as the Dirichlet distribution.

For the binary case, the expected value of the Beta distribution is given by: $P(b_i) = a + 1/a + b + 2$, where $a$ and $b$ are the parameters of the Beta distribution. This representation could also be used when we have experts' estimates of the probabilities. The parameters of the Beta distribution can represent a measure of *confidence* in the expert's estimates, expressed by varying the term $a + b$ with the same probability value. For instance:

- Complete ignorance: $a = b = 0$.
- Low confidence: $a + b$ *small* (10).
- Medium confidence: $a + b$ *intermediate* (100).
- High confidence: $a + b$ large (1000).

This representation could be used to combine experts' estimations with data. For example, to approximate the probability value of a binary variable, $b_i$ we can use:

$$P(b_i) = k + c + 1/n + a + b + 2 \tag{8.4}$$

Where $a/a + b$ represents the expert's estimate, and $k/n$ is the probability obtained from the data ($k$ is the number of times $b_i$ occurs in $n$ samples).

For example, assume an expert gives an estimate of 0.7 for a certain parameter, and that the experimental data provides 40 positive cases among 100 samples. The parameter estimation for different confidences assigned to the expert will be the following:

Low confidence ($a + b = 10$): $P(b_i) = \frac{40+7+1}{100+10+2} = 0.43$
Medium confidence ($a + b = 100$): $P(b_i) = \frac{40+70+1}{100+100+2} = 0.55$
High confidence ($a + b = 1000$): $P(b_i) = \frac{40+700+1}{100+1000+2} = 0.67$

We observe that in the first case the estimate is dominated by the data, while in the third case the probability is closer to the expert's estimate; the second case provides a compromise.

### 8.2.3 *Missing Data*

Another common situation is to have incomplete data. There are two basic cases:

Missing values: In some registers there are missing values for one or more variables.

Hidden nodes: A variable or set of variables in the model for which there is no data at all.

For dealing with missing values, there are several alternatives:

1. Eliminate the registers with missing values.

2. Consider a special "unknown" value.
3. Substitute the missing value by the most common value (mode) of the variable.
4. Estimate the missing value based on the values of the other variables in the corresponding register.

The first and second alternatives are acceptable if there is sufficient data, otherwise we could be discarding useful information. The third alternative does not consider the other variables and as a result, it could bias the model. In general the best alternative is the fourth option. In this case, we first learn the parameters of the BN based on the complete registers, and then complete the data and re-estimate the parameters, applying the following process. For each register with missing values:

1. Instantiate all the known variables in the register.
2. Through probabilistic inference obtain the posterior probabilities of the missing variables.
3. Assign to each variable the value with highest posterior probability.
4. Add this completed register to the database and re-estimate the parameters.

An alternative to the previous process is that instead of assigning the value with the highest probability, we assign a *partial* case for each value of the variable proportional to the posterior probability.

For hidden nodes, the approach to estimate their parameters is based on the *Expectation–Maximization* (EM) technique.

#### 8.2.3.1 Hidden Nodes: EM

The EM algorithm is a statistical technique used for parameter estimation when there are non-observable variables. It consists of two phases which are repeated iteratively:

E step: the missing data values are estimated based on the current parameters.
M step: the parameters are updated based on the estimated data.

The algorithm starts by initializing the missing parameters with random values.

Given a database with one or more hidden nodes, $H_1, H_2, \ldots, H_k$, the EM algorithm to estimate their CPTs is the following:

1. Obtain the CPTs for all the *complete* variables (the values of the variable and all its parents are in the database) based on a ML estimator.
2. Initialize the unknown parameters with random values.
3. Considering the actual parameters, estimate the values of the hidden nodes based on the known variables via probabilistic inference.
4. Use the estimated values for the hidden nodes to complete/update the database.
5. Re-estimate the parameters for the hidden nodes with the updated data.
6. Repeat 3–5 until converge (no significant changes in the parameters).

The EM algorithm optimizes the unknown parameters and gives a *local maximum* (the final estimates depend on the initialization).

**Table 8.1** Data for the golf example with missing values for *Temperature* and a hidden variable, *Wind*

| Outlook | Temperature 1 | Humidity | Wind | Play |
|---|---|---|---|---|
| Sunny | xxx | High | – | N |
| Sunny | High | High | – | N |
| Overcast | High | High | – | P |
| Rainy | Medium | High | – | P |
| Rainy | Low | Normal | – | P |
| Rainy | Low | Normal | – | N |
| Overcast | Low | Normal | – | P |
| Sunny | Medium | High | – | N |
| Sunny | xxx | Normal | – | P |
| Rainy | Medium | Normal | – | P |
| Sunny | Medium | Normal | – | P |
| Overcast | Medium | High | – | P |
| Overcast | High | Normal | – | P |
| Rainy | Medium | High | – | N |

#### 8.2.3.2 Example

We now illustrate how to handle missing values and hidden variables using data from the *Golf* example (see Table 8.1). In this data set, there are some missing values for the variable *Temperature* (registers 1 and 9), and there is no information about *Wind*, which is a hidden node. We first illustrate how to fill-in the missing values for temperature and then how to manage the hidden node.

Assume that we learn a naive Bayes classifier (a NBC is a particular type of BN) based on the available data (12 complete registers without the wind variable), considering *Play* as the class variable and the other variables as attributes. Then, based on this model, we can estimate the probability of temperature for the registers in which it is missing, via probabilistic inference using the values of the other variables in the corresponding registers as evidence. Note that although the model is a NBC for *Play*, we can also apply probabilistic inference to estimate the value of an attribute given the class and the other attributes. That is:

Register 1: $P(Temperature \mid sunny, high, N)$
Register 9: $P(Temperature \mid sunny, normal, P)$

Then we can select as the value of temperature the one with highest posterior probability, and fill-in the missing values, as shown in Table 8.2.

For the case of the hidden node, *Wind*, we cannot obtain the corresponding CPT from the NBC, $P(Wind \mid Play)$, as there are no values for wind. However, we can apply the EM procedure, where we first pose initial random parameters for the CPT, which could be, for example, a uniform distribution:

**Table 8.2** Data for the golf example after completing the missing values for *Temperature* and one iteration of the EM procedure to estimate the values of *Wind*

| Outlook | Temperature 1 | Humidity | Wind | Play |
|---|---|---|---|---|
| Sunny | Medium | High | No | N |
| Sunny | High | High | No | N |
| Overcast | High | High | No | P |
| Rainy | Medium | High | No | P |
| Rainy | Low | Normal | Yes | P |
| Rainy | Low | Normal | Yes | N |
| Overcast | Low | Normal | Yes | P |
| Sunny | Medium | High | No | N |
| Sunny | Medium | Normal | No | P |
| Rainy | Medium | Normal | No | P |
| Sunny | Medium | Normal | Yes | P |
| Overcast | Medium | High | Yes | P |
| Overcast | High | Normal | Yes | P |
| Rainy | Medium | High | Yes | N |

$$P(Wind \mid Play) = \begin{matrix} 0.5 & 0.5 \\ 0.5 & 0.5 \end{matrix}$$

Given this CPT we have a complete, initial model for the NBC, and can estimate the probability of wind for each register based on the values of the other variables in the register. By selecting the highest probability value for each register, we can fill-in the table, as depicted in Table 8.2. Based on this new data table, we re-estimate the parameters, and obtain a new CPT:

$$P(Wind \mid Play) = \begin{matrix} 0.60 & 0.44 \\ 0.40 & 0.56 \end{matrix}$$

This completes one cycle of the EM algorithm; the process is then repeated until all parameters in the CPT have *almost* no change from the previous iteration. At this point, the EM procedure has converged, and we have an estimate of the missing parameters of the BN.

### 8.2.4 Discretization

Usually Bayesian networks consider discrete or nominal variables. Although there are some developments for continuous variables, these are restricted to certain distributions, in particular Gaussian variables and linear relations. An alternative to include continuous variables in BNs is to discretize them; that is, transform them to nominal variables. Discretization methods can be (i) unsupervised and (ii) supervised.

#### 8.2.4.1 Unsupervised Discretization

Unsupervised techniques do not consider the task for which the model is going to be used (e.g., classification), such that the intervals for each variable are determined independently. The two main types of unsupervised discretization approaches are: equal width and equal data.

Equal width consists in dividing the range of a variable, $[Xmin; Xmax]$, in $k$ equal bins; such that each bin has a size of $[Xmax - Xmin]/k$. The number of intervals, $k$, is usually set by the user.

Equal data divides the range of the variable in $k$ intervals, such that each interval includes the same number of data points from the training data. In other words, if there are $n$ data points, each interval will contain $n/k$ data points; this means that the intervals will not necessarily have the same width.

A way to determine how many intervals, specially useful for Bayesian classifiers, is Proportional k-Interval Discretization (PKID) [14]. This strategy seeks a trade-off between the bias and variance of the parameter estimates by adjusting the number and size of intervals to the number of training instances. Given a continuous variable with $N$ training instances, it is discretize to $\sqrt{N}$ intervals, with $\sqrt{N}$ instances in each interval. This method was compared experimentally with other discretization methods for implementing naive Bayesian classifiers; PKID achieves the lowest mean error [14].

#### 8.2.4.2 Supervised Discretization

Supervised discretization considers the task to be performed with the model, such that the variables are discretized to optimize this task, for instance classification accuracy. If we consider a BN for classification, i.e. a Bayesian classifier, then the supervised approach can be directly applied. Assuming continuous attribute variables, these are discretized according to the class values. This can be posed as an optimization problem.

Consider the attribute variable $X$ with range $[Xmin; Xmax]$ and a class variable $C$ with $m$ values $c_1, c_2, \ldots, c_m$. Given $n$ training samples, so that each one has a value for $C$ and $X$, the problem is to determine the *optimal* partition of $X$ such that the classifier precision is maximized. This a combinatorial problem that is computationally complex, and can be solved using a search process as follows:

1. Generate all potential divisions in $X$ which correspond to a value in $[Xmin; Xmax]$ where there is a change in the class value.
2. Based on the potential division points generate an initial set of $n$ intervals.

3. Test the classification accuracy of the Bayesian classifier (usually on a different set of data known as a validation set) according to the current discretization.
4. Modify the discretization by partitioning an interval or joining two intervals.
5. Repeat (3) and (4) until the accuracy of the classifier cannot be improved or some other termination criteria occurs.

Different search approaches can be used, including basic ones such as *hill-climbing* or more sophisticated methods like *genetic algorithms*.

The previous algorithm does not apply for the general case of a Bayesian network which can be used to predict different variables based on different evidence variables. In this case, there is a supervised method [8] that discretizes continuous attributes while it learns the structure of the BN. The method is based on the *Minimum Description Length* (MDL) principle –described in Sect. 8.3.3. For each continuous variable, the number of intervals is determined according to its neighbors in the network. The objective is to minimize the MDL (a compromise between the precision and complexity of the model), using a search and test approach analogous to the process for Bayesian classifiers. This is repeated iteratively for all continuous variables in the network.

## 8.3 Structure Learning

Structure learning consists in obtaining the topology of the BN from the data. This is a complex problem because: (i) the number of possible structures is *huge* even with a few variables (it is super-exponential on the number of variables; for example, for 10 variables the number of possible DAGs is in the order of $4 \times 10^{18}$), and (ii) a very large database is required to obtain good estimates of the statistical measures on which all methods depend.

For the particular case of a tree structure, there is a method that guarantees the *best* tree. For the general case several methods have been proposed, which can be divided into two main classes:

1. Global methods: these [3, 5] perform a heuristic search over the space of network structures, starting from some initial structure, and generating a variation of the structure at each step. The *best* structure is selected based on a score that measures how well the model represents the data. Common scores are BIC [3] and MDL [5].
2. Local methods: these are based on evaluating the independence relations between subsets of variables given the data, to sequentially obtain the structure of the network. The most well known variant of this approach is the PC algorithm [12].

Both classes of methods obtain similar results with *enough* data. Local methods tend to be more sensitive when there are few data samples, and global methods tend to be more computationally complex.

Next we review the tree learning algorithm developed by Chow and Liu [2] and its extension to polytrees. Then we present the techniques for learning a general structure.

### 8.3.1 Tree Learning

Chow and Liu [2] developed a method for approximating any multi-variable probability distribution as a product of second order distributions, which is the basis for learning tree-structured BNs. The joint probability of $n$ random variables can be approximated as:

$$P(X_1, X_2, \ldots, X_n) = \prod_{i=1}^{n} P(X_i \mid X_{j(i)}) \tag{8.5}$$

where $X_{j(i)}$ is the parent of $X_i$ in the tree.

The problem consists in obtaining the *best* tree, that is, the tree structure that best approximates the real distribution. A measure of how close the approximation is based on the information difference between the real distribution ($P$) and the tree approximation ($P^*$) is as follows:

$$DI(P, P^*) = \sum_{X} P(X) log(P(X)/P^*(X)) \tag{8.6}$$

Thus, now the problem consists in finding the tree that minimizes $DI$. However, evaluating this for all possible trees is very expensive. Chow and Liu proposed an alternative based on the mutual information between pairs of variables.

The mutual information between any pair of variables is defined as:

$$I(X_i, X_j) = \sum_{X_i, X_j} P(X_i, X_j) log(P(X_i, X_j)/P(X_i)P(X_j)) \tag{8.7}$$

Given a tree-structured BN with variables $X_1, X_2, \ldots, X_n$, we define its *weight*, $W$, as the sum of the mutual information of the arcs (pairs of variable) that constitute the tree:

$$W(X_1, X_2, \ldots X_n) = \sum_{i=1}^{n-1} I(X_i, X_j) \tag{8.8}$$

where $X_j$ is the parent of $X_i$ in the tree (a tree with $n$ nodes has $n - 1$ arcs).

It can be shown [2] that minimizing $DI$ is equivalent to maximizing $W$. Therefore, obtaining the optimal tree is equivalent to finding the *maximum weight spanning tree*, using the following algorithm:

**Table 8.3** Mutual information in descending order for the golf example

| No. | Var 1 | Var 2 | Mutual Info. |
|---|---|---|---|
| 1 | Temperature | Outlook | 0.2856 |
| 2 | Play | Outlook | 0.0743 |
| 3 | Play | Humidity | 0.0456 |
| 4 | Play | Wind | 0.0074 |
| 5 | Humidity | Outlook | 0.0060 |
| 6 | Wind | Temperature | 0.0052 |
| 7 | Wind | Outlook | 0.0017 |
| 8 | Play | Temperature | 0.0003 |
| 9 | Humidity | Temperature | 0 |
| 10 | Wind | Humidity | 0 |

1. Obtain the mutual information ($I$) between all pairs of variables (for $n$ variables, there are $n(n-1)/2$ pairs).
2. Order the mutual information values in descending order.
3. Select the pair with maximum $I$ and connect the two variables with an arc, this constitutes the initial tree
4. Add the pair with the next highest $I$ to the tree, while they do not make a cycle; otherwise skip it and continue with the following pair.
5. Repeat 4 until all the variables are in the tree ($n-1$ arcs).

This algorithm obtains the *skeleton* of the tree; that is, it does not provide the direction of the arcs in the BN. The directions of the links have to be obtained using external semantics or using higher order dependency tests (see Sect. 8.3.2).

To illustrate the tree learning method consider the classic *golf* example with 5 variables: *play, outlook, humidity, temperature, wind.* Given some data, we obtain the mutual information shown in Table 8.3.

In this case, we select the first 4 pairs (arcs) and obtain the tree in Fig. 8.1, where the directions were assigned arbitrarily.

### 8.3.2 Learning a Polytree

Rebane and Pearl [11] developed a method that can be used to direct the arcs in the skeleton, and in general, learn a *polytree* BN. The algorithm is based on independence tests for variable triplets, and in this way it can distinguish *convergent* substructures; once one or more substructures of this type are detected in the skeleton, it can direct additional arcs by applying the independence tests to neighboring nodes. However, there is no guarantee of obtaining the direction for all the arcs in the tree. This same idea is used in the PC algorithm for learning general structures.

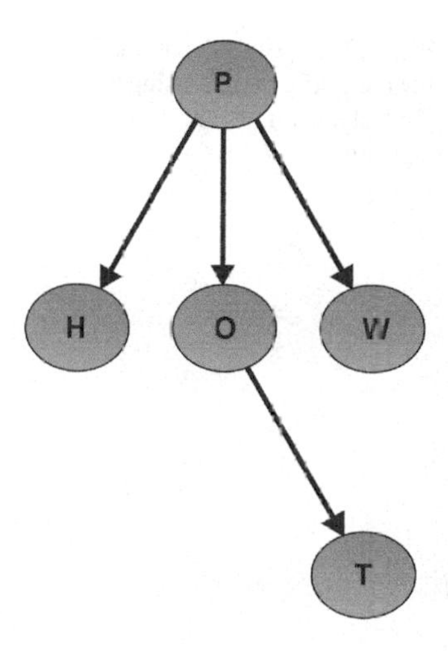


**Fig. 8.1** Tree structure obtained for the golf example ($P$ is play, $O$ is outlook, $H$ is humidity, $T$ is temperature, and $W$ is wind). Arc directions are set arbitrarily

The algorithm begins with the skeleton (undirected structure) obtained with the Chow and Liu algorithm. Subsequently, the direction of the arcs is learned using independence tests for variable triplets. Given three variables, there are three possibilities:

1. Sequential arcs: $X \rightarrow Y \rightarrow Z$.
2. Divergent arcs: $X \leftarrow Y \rightarrow Z$.
3. Convergent arcs: $X \rightarrow Y \leftarrow Z$.

The first two cases are indistinguishable under statistical independence testing; that is, they are equivalent. In both cases, $X$ and $Z$ are independent given $Y$. However the third case is different, since $X$ and $Z$ are NOT independent given $Y$. Consequently, this case can be used to determine the directions of the two arcs that connect these three variables; additionally, we can apply this knowledge to learn the directions of other arcs using independence tests. With this in mind, the following algorithm can be used for learning polytrees:

1. Obtain the skeleton using the Chow and Liu algorithm.
2. Iterate over the network until a convergent variable triplet is found. We will call the variable to which the arcs converge a *multi-parent node*.
3. Starting with a multi-parent node, determine the directions of other arcs using independence tests for variable triplets. Continue this procedure until it is no longer possible (causal base).
4. Repeat 2–3 until no other directions can be determined.
5. If any arcs are left undirected, use the external semantics to infer their directions.

To illustrate this algorithm, let us consider the golf example again, with the obtained skeleton (undirected structure). Suppose that the variable triplet $H, P, W$ falls in the convergent case. Then, the arcs will be directed such that $H$ points to $P$ and $W$ points to $P$. Subsequently, the dependence between $H$ and $W$ is measured with respect to $O$ given $P$. If $H$ and $W$ are independent from $O$ given $P$ then there

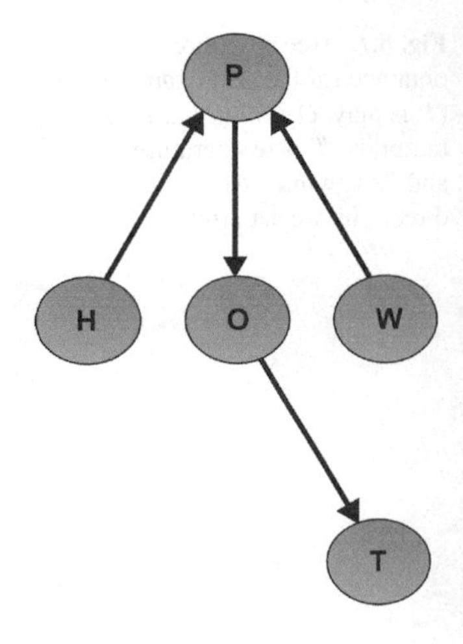


**Fig. 8.2** A polytree obtained for the golf example using the Rebane and Pearl algorithm

will be an arc that points from $P$ to $O$. Finally, the dependence relation between $P$ and $T$ given $O$ is tested, and if they are again found to be independent, then the arc points from $O$ to $T$. Figure 8.2 shows the resulting structure.

## 8.3.3 Search and Score Techniques

The previous methods are restricted to tree and polytree structures; in this and the following section we will cover general structure learning techniques, starting with global approaches.

Global methods search for the *best* structure based on a global metric. That is, different structures are generated and these are evaluated with respect to the data using some scoring method. There are different variants of these methods, all of which basically depend on two aspects: (i) a fitness measure between the structure and the data, and (ii) a method for searching for the best structure.

### 8.3.3.1 Scoring Functions

There are several possible fitness measures or scoring functions. Two desirable properties for scoring functions are [1]:

Decomposability: a scoring function is decomposable if the value assigned to each structure can be expressed as a sum (in the logarithmic space) of local values that depend only on each node and its parents. This is important for efficiency reasons during the search process; given this property when a local change is made to the structure, only a part of the score has to be re-evaluated.

Score Equivalence: a scoring function $S$ is score equivalent if it assigns the same value to all DAGs that are represented by the same essential graph. In this way, the result of evaluating an equivalence class will be the same regardless of the DAG that is selected from this class The structures of two BNs correspond to the same essential graph if they are equivalent in terms of the independence relations they represent. For instance, the following structures: $X \rightarrow Y \rightarrow Z$ and $X \leftarrow Y \rightarrow Z$ correspond to the same essential graph $(I(X, Y, Z))$.

Next we describe some common scoring functions, including: the maximum likelihood (ML), the Bayesian information criterion (BIC), the Bayesian score (BD), and the minimum description length (MDL) criterion.

The maximum likelihood score selects the structure that maximizes the probability of the data, $D$, given the structure, $G$:

$$G^* = ArgMax_G[P(D \mid \Theta_G, G_i)] \tag{8.9}$$

where $G_i$ is the candidate structure and $\Theta_G$ the corresponding vector of parameters (probability of each variable given its parents according to the structure).

The direct application of the ML score might result in a highly complex network, which usually implies overfitting the data (poor generalization) and also makes inference more complex. Therefore, a way to penalize complex models is required.

A commonly used scoring function that includes a penalty term is the Bayesian Information Criterion or BIC defined as:

$$BIC = logP(D \mid \Theta_G, G_i) - \frac{d}{2} logN \tag{8.10}$$

where $d$ is the number of parameters in the BN and $N$ the number of cases in the data. An advantage of this metric is that it does not require a prior probability specification and it is related to the MDL measure, compromising between the precision and complexity of the model. However, given the high penalty on the complexity of the model, it tends to choose structures that are too simple.

**Bayesian Scores**

An alternative metric is obtained by following a Bayesian approach, obtaining the posterior probability of the structure given the data with the Bayes rule:

$$P(G_i \mid D) = P(G_i)P(D \mid G_i)/P(D) \tag{8.11}$$

Given that P(D) is a constant that does not depend on the structure, it can be discarded from the metric to obtain de Bayesian or BD score:

$$BD = P(G_i)P(D \mid G_i) \tag{8.12}$$

$P(G_i)$ is the prior probability of the model. This can be specified by an expert or defined such that simpler structures are preferred; or just set to a uniform distribution.

The BDe score is a variation of the BD score which makes the following assumptions: (i) the parameters are independent and have a prior Dirichlet distribution, (ii) equivalent structures have the same score, (iii) the data samples are independent and identically distributed (iid). Under these assumptions the *virtual counts* required to compute the score can be estimated as:

$$N_{ijk} = P(X_i = k, Pa(X_i) = j \mid G_i, \Theta_G) \times N' \tag{8.13}$$

This is the estimated count of a certain *configuration*: $X_i = k$ given $Pa(X_i) = j$; $N'$ is the equivalent sample size.

By assuming that the hyper parameters of the priors are one, we can further simplify the calculation of the Bayesian score, and obtain what is known as the K2 metric.[1] This score is decomposable and it is calculated for each variable $X_i$ given its parents $Pa(X_i)$:

$$S_i = \prod_{j=1}^{q_i} \frac{(r_i - 1)!}{(N_{ij} + r_i - 1)!} \prod_{k=1}^{r_i} \alpha_{ijk}! \tag{8.14}$$

Where $r_i$ is the number of values of $X_i$, $q_i$ is the number of possible configurations for the parents of $X_i$, $\alpha_{ijk}$ is the number of cases in the database where $X_i = k$ and $Pa(X_i) = j$, and $N_{ij}$ is the number of cases in the database where $Pa(X_i) = j$.

This metric provides a practical alternative for evaluating a BN. Another common alternative that is based on the MDL principle is described next.

**MDL**

The MDL measure makes a compromise between accuracy and model complexity. Accuracy is estimated by measuring the mutual information between each variable and its parents (an extension of the tree learning algorithm). Model complexity is evaluated by counting the number of parameters. A constant, $\alpha$ within [0, 1], is used to balance the weight of each aspect, that is, accuracy against complexity. The fitness measure is given by the following equation:

$$MC = \alpha(W/Wmax) + (1 - \alpha)(1 - L/Lmax) \tag{8.15}$$

where $W$ represents the accuracy of the model, and $L$ the complexity. $Wmax$ and $Lmax$ represent the maximum accuracy and complexity, respectively. To determine the maximums, usually an upper bound is set on the number of parents each node is allowed to have. A value of $\alpha = 0.5$ gives equal importance to the model complexity and accuracy, while a value near 0 gives more importance to the complexity, and a value near 1 more importance to accuracy.

Complexity is given by the number of parameters required for representing the model, which can be measured with the following equation:

$$L = S[klog_2 n + d(S - 1)F] \tag{8.16}$$

[1] K2 is an algorithm for learning BNs described below.

where $n$ is the number of variables in the BN, $k$ is the average number of parents per variable, $S$ is the average number of values per variable, $F$ is the average number of values per parent variable, and $d$ the number of bits per parameter. For example, consider a BN with 16 variables, all variables are binary and have in average 3 parents, and each parameter is represented by 16 bits. Then:

$$L = 2 \times [3 \times log_2(16) + 16 \times (2 - 1) \times 2] = 2 \times [12 + 32] = 88$$

The accuracy can be estimated based on the 'weight' of each node; this is analogous to the weights in the methodology for learning trees. In this case, the weight of each node, $X_i$, is estimated based on its mutual information with its parents, $Pa(X_i)$:

$$w(X_i, Pa(X_i)) = \sum_{xi} P(X_i, Pa(X_i)) log[P(X_i, Pa(X_i))/P(X_i)P(Pa(X_i))] \tag{8.17}$$

and the weight (accuracy) total is given by the sum of the weights for each node:

$$W = \sum_{i} w(X_i, Pa(X_i)) \tag{8.18}$$

#### 8.3.3.2 Search Algorithms

Once a fitness measure for the structure has been established, we need to establish a method for choosing the 'best' structure among the possible options. Since the number of possible structures is exponential on the number of variables, it is impossible to evaluate every structure. To limit the number of structures that are evaluated, a heuristic search is carried out. Several different search methods can be applied. One common strategy is to use a *hill climbing* approach, where we begin with a simple tree structure that is improved until we obtain the 'best' structure. A basic greedy algorithm to search for the best structure is the following:

1. Generate an initial structure—tree
2. Calculate the fitness measure of the initial structure.
3. Add/invert an arc from the current structure.
4. Calculate the fitness measure of the new structure.
5. If the fitness improves, keep the change; if not, return to the previous structure.
6. Repeat 3–5 until no further improvements exist.

The previous algorithm is not guaranteed to find the optimum structure, since it is possible to reach only a local maximum. Figure 8.3 illustrates the search procedure for the golf example, starting with a tree structure that is improved until the final structure is obtained. Other search methods, such as genetic algorithms, simulated annealing, bidirectional searches, etc., can also be applied to obtain the best structure.

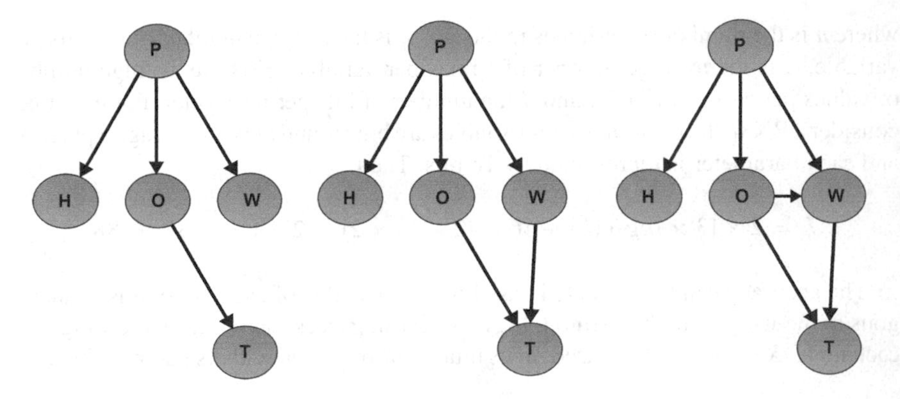


**Fig. 8.3** A few steps in the procedure for learning the structure of the "golf" example, beginning with a tree structure (left) until the final structure (right) is obtained

An alternative to reduce the number of potential structures to be evaluated, is to set an ordering on the variables, known as a *causal ordering*. Given this ordering, the arcs in the network are restricted to follow this order; that is, there could be NO arc from $V_j$ to $V_i$ if $j > i$ according to the ordering. The K2 algorithm [3] takes advantage of this, providing and efficient and popular method for learning BNs.

#### 8.3.3.3 The K2 Algorithm

Given a causal ordering for all the variables, learning the best structure is equivalent to selecting the best set of parents for each node independently. Initially, each variable has no parents. Then, the K2 algorithm incrementally adds parents to each node, as long as it increases the global score. When adding parents to any node does not increase the score, the search stops. Also, given a causal ordering it guarantees that there are not cycles in the graph.

Algorithm 8.1 provides a summary of the K2 procedure. The inputs to the algorithm are the set of $n$ variables with a causal ordering, $X_1, X_2, \ldots, X_n$, a data base $D$ containing $m$ cases, and, usually, a restriction on the maximum number of parents for each variable, $u$. The output is the set of parents, $Pa(X_i)$, for each variable, which defines the structure of the network. Starting from the first variable according to the ordering, the algorithm tests all possible parents of a variable that have not been added, and includes the one that makes the maximum increment in the score of the network. This is repeated until there is no additional parent that increases the score; for every node on the network.

**Algorithm 8.1** The K2 Algorithm

**Require:** Set of variables $X$ with a causal ordering, scoring function $S$, and maximum parents $u$
**Ensure:** Set of parents for each variable, $Pa(X_i)$

```
for i = 1 to n do
  oldScore = S(i, Pa(X_i))
  incrementScore = true
  Pa(X_i) = ∅
  while incrementScore and |Pa(X_i)| < u do
    let Z be the node in Predecessors(X_i) − Pa(X_i) that maximizes S
    newScore = S(i, Pa(X_i) ∪ Z)
    if newScore > oldScore then
      oldScore = newScore
      Pa(X_i) = Pa(X_i) ∪ Z
    else
      incrementScore = false
    end if
  end while
end for
return Pa(X_1), Pa(X_2)...Pa(X_n)
```

## 8.3.4 Independence Tests Techniques

The other class of structure learning techniques use a *local* approach instead of the *global* one used by the score and search techniques. The basic idea is to apply independence tests to sets of variables to recover the structure of the BN. An example of this type of techniques is the Chow and Liu algorithm for trees. Next we present a method for learning general structures, the PC algorithm.

### 8.3.4.1 The PC Algorithm

The PC algorithm [12] first recovers the skeleton (underlying undirected graph) of the BN, and then it determines the orientation of the edges.

To determine the skeleton, it starts from a fully connected undirected graph, and determines the conditional independence of each pair of variables given some subset of the other variables. For this it assumes that there is a procedure that can determine if two variables, $X$, $Y$, are independent given a subset of variables, $\mathbf{S}$, that is, $I(X, Y \mid \mathbf{S})$. An alternative for this procedure is the conditional cross entropy measure. If this measure is below a threshold value set according to a certain confidence level, the edge between the pair of variables is eliminated. These tests are iterated for all pairs of variables in the graph.

In the second phase the direction of the edges are set based on conditional independence tests between variable triplets. It proceeds by looking for substructures in the graph of the form $X - Z - Y$ such that there is no edge $X - Y$. If $X$, $Y$ are not independent given $Z$, it orients the edges creating a V-structure $X \rightarrow Z \leftarrow Y$. Once all the V-structures are found, it attempts to orient the other edges based on independence tests and avoiding cycles. Algorithm 8.2 summarizes the basic procedure.[2]

**Algorithm 8.2** The PC algorithm

**Require:** Set of variables **X**, Independence test $I$
**Ensure:** Directed Acyclic Graph $G$
1: Initialize a complete undirected graph $G'$
2: i = 0
3: **repeat**
4:   **for** $X \in \mathbf{X}$ **do**
5:     **for** $Y \in ADJ(X)$ **do**
6:       **for** $S \subseteq ADJ(X) - \{Y\}, \mid S \mid = i$ **do**
7:         **if** $I(X, Y \mid S)$ **then**
8:           Remove the edge $X - Y$ from $G'$
9:         **end if**
10:       **end for**
11:     **end for**
12:   **end for**
13:   i = i + 1
14: **until** $\mid ADJ(X) \mid \leq i, \forall X$
15: Orient edges in $G'$
16: Return $G$

If the set of independencies are faithful to a graph[3] and the independence tests are perfect, the algorithm produces a graph equivalent to the original one; that is, the BN structure that generated the data.

The independence test techniques rely on having *enough* data for obtaining good estimates from the independence tests. Search and score algorithms are more robust with respect to the size of the data set, however their performance is also affected by the size and quality of the available data. An alternative for when there is not *sufficient* data, is to combine expert knowledge and data.

## 8.4 Combining Expert Knowledge and Data

When domain expertise is available, this can be combined with learning algorithms to improve the model. In the case of parameter learning, we can combine data and expert estimates based on the Beta or Dirichlet distributions as described in Sect. 8.2.

For structure learning, there are two basic approaches to combine expert knowledge and data:

---

[2] $ADJ(X)$ is the set of nodes adjacent to $X$ in the graph.

[3] The *Faithfulness Condition* can be thought of as the assumption that conditional independence relations are due to the causal structure rather than to accidents in parameter values [12].

- Use expert knowledge as *restrictions* to reduce the search space for the learning algorithm.
- Start from a structure proposed by an expert and use data to validate and improve this structure.

There are several ways to use expert knowledge to aid the structure learning algorithm, such as:

1. Define an ordering for the variables (causal order), such that there could be an arc from $X_i$ to $X_j$ only if $X_j$ is after $X_i$ according to the specified ordering.
2. Define restrictions in terms of directed arcs that must exist between two variables, i.e. $X_i \rightarrow X_j$.
3. Define restrictions in terms of an arc between two variables that could be directed either way.
4. Define restrictions in terms of pairs of variables that are not directly related, that is, there must be no arc between $X_i$ and $X_j$.
5. Combinations of the previous restrictions.

Several variants of both types of techniques, search and score and independence tests, incorporate the previous restrictions.

In the case of the second approach, an example was presented in Chap. 4, with the structural improvement algorithm. This technique starts from a naive Bayes structure which is improved by eliminating, joining or inserting variables. This idea can be extended to general BN structures, in particular for tree-structured BNs.

## 8.5 Transfer Learning

Another alternative when there is not enough data for certain application is to *transfer* knowledge and/or data for related domains, what is known as *transfer learning*. Next we describe an approach based on transfer learning for Bayesian networks.

Luis et al. [7] proposed a transfer learning method that learns a BN, structure and parameters, for a target task, from data from this task and from other related auxiliary tasks. The structure learning method is based on the PC algorithm, combining the dependency measures obtained from data in the target task, with those obtained from data in the auxiliary tasks. The combination function takes into account the consistency between these measures.

For structure learning, the method modifies the independence tests of the PC algorithm by combining the independence measures in the target task with that of the *closest* auxiliary task. The similarity measure considers both, global and local similarity. The global similarity considers all pair-wise conditional independencies in the models, while the local similarity includes only those for specific variables. So the combined independence measure $I(X, \mathbf{S}, Y)$ is a linear weighted combination of the independence measures in the target and auxiliary tasks, considering the confidence and similarity measures. An alternative PC algorithm is defined, based on the combined independence measures.

The parameter learning technique uses an aggregation process, combining the parameters estimated from the target task, with those estimated from the auxiliary tasks. Based on linear combination techniques, they proposed two variants: (i) Distance-based linear pool, which takes into account the distance of the auxiliary parameters to the target parameters, and (ii) Local linear pool, which only includes auxiliary parameters that are close to the target one, weighted by the amount of data in each auxiliary task.

To combine the CPT for a variable $X$ from an auxiliary task, it is necessary that the variable has the same parents in the target task. If they do not have the same parents, the substructure in the auxiliary task is transformed to match the target one.

The Distance-based linear pool (DBPL) makes a weighted linear combination of the parameters estimated from data of the target task, with an average of the parameters of the auxiliary tasks:

$$P'_{target} = C_i \times P_{target} + (1 - C_i) \times P_{auxiliary} \quad (8.19)$$

where $C_i$ is a factor defined according to the similarity between the parameters estimated for that target task and those of the auxiliary tasks.

The Local linear pool (LoLP) takes into account only the most similar parameters from the auxiliary tasks, and weights them according to their confidence based on the amount of data. It is based also on Eq. 8.19, but only considering the parameters from the closest auxiliary tasks, those for which the difference with the parameters of the target task is below certain threshold.

For more details on the method see [7].

## 8.6 Applications

There are many domains in which learning Bayesian networks has been applied to get a better understanding of the domain or make predictions based on partial observations; for example medicine, finance, industry and the environment, among others. Next we present two application examples: modeling the air pollution in Mexico City and agricultural planning for coffee production.

### *8.6.1 Air Pollution Model for Mexico City*

Air quality in Mexico City is a major problem. There, air pollution is one of the highest in the world, with a high average of daily emissions of several primary pollutants, such as hydrocarbons, nitrogen oxides, carbon monoxide and others. The pollution is due primarily to transportation and industrial emissions. When the primary pollutants are exposed to sunshine, they undergo chemical reactions and yield a variety of secondary pollutants, ozone being the most important. Besides the health problems it may cause, ozone is considered an indicator of the air quality in urban areas.

The air quality is monitored in 25 stations within Mexico City, with five of these being the most complete. Nine variables are measured in each of the 5 main stations, including: wind direction and velocity, temperature, relative humidity, sulfur dioxide, carbon monoxide, nitrogen dioxide and ozone. These are measured every minute 24 hours a day, and are averaged every hour.

It is important to be able to forecast the pollution level several hours, or even a day in advance for several reasons, including:

1. To be able to take emergency measures if the pollution level is going to be above a certain threshold.
2. Helping industry make contingency plans in advance in order to minimize the cost of the emergency measures.
3. To estimate the pollution in an area where there are no measurements.
4. To take preventive actions in some places, such as schools, in order to reduce the health hazards produced by high pollution levels.

In Mexico City, the ozone level is used as a global indicator for the air quality within the different parts of the city. The concentrations of ozone are given in IMECA (Mexican air quality index). It is important to predict the daily ozone, or at least, predict it several hours in advance using the other variables measured at different stations.

It is useful to know the dependencies between the different variables that are measured, and specially their influence in the ozone concentration. This will provide a better understanding of the problem with several potential benefits:

- Determine which factors are more important for the ozone concentration in Mexico City.
- Simplify the estimation problem, by taking into account only the relevant information.
- Discover the most critical primary causes of pollution in Mexico City; these could help in future plans to reduce pollution.

We started by applying a learning algorithm to obtain an initial structure of the phenomena [13]. For this we considered 47 variables: 9 measurements for each of the 5 stations, plus the hour and month in which they were recorded. We used nearly 400 random samples, and applied the Chow and Liu algorithm to obtain the tree structure that best approximates the data distribution. This tree-structured Bayesian network is shown in Fig. 8.4.

We then considered the ozone in one station (Pedregal) as unknown, and we estimate it, one hour in advance, using the other measurements. Thus we make *ozone-Pedregal* the hypothesis variable and consider it as the root in the probabilistic tree, as shown in Fig. 8.4. From this initial structure we can get an idea of the relevance or

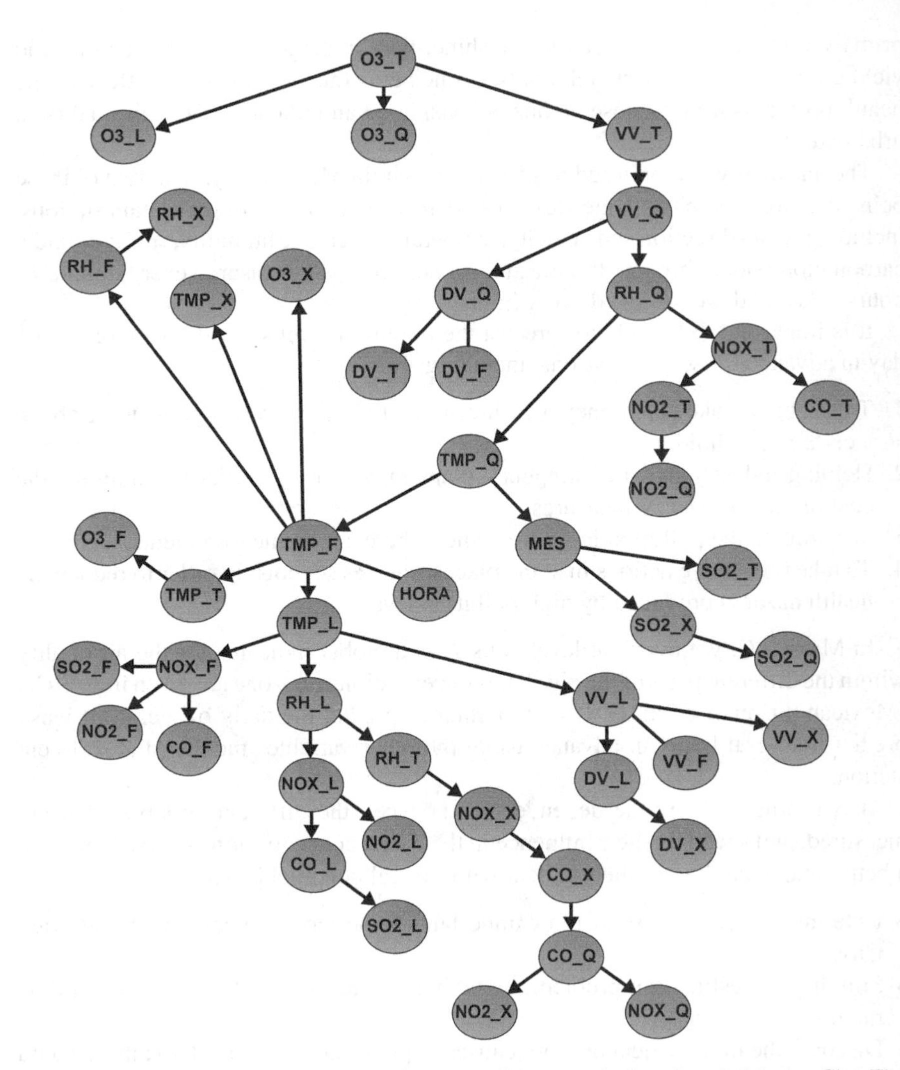


**Fig. 8.4** A Bayesian tree that represents the ozone phenomena in 5 stations in Mexico City. The nodes represent the 47 variables according to the following nomenclature. For the measured variables, each name is formed by two parts, "measurement-station", using the following abbreviations: the measurements, O3-ozone, SO2-sulfur dioxide, CO-carbon monoxide, NO2-nitrogen dioxide, NOX-nitrogen oxides, VV-wind velocity, DV-wind direction, TMP-temperature, RH-relative humidity; the monitoring stations, T-Pedregal, F-Tlalnepantla, Q-Merced, L-Xalostoc, X-Cerro de la Estrella. The other two variables correspond to the time when the measurements were taken, and they are: HORA-hour, MES-month

influence of the other variables for estimating *ozone-Pedregal*. The nodes "closest" to the root are the most important ones, and the "far-away" nodes are less important.

In this case we observe that there are 3 variables (ozone-Merced, ozone-Xalostoc, and wind velocity in Pedregal) that have the greatest influence in *ozone-Pedregal*. Furthermore, if the tree structure is a good approximation to the "real" structure, these 3 nodes make *ozone-Pedregal* independent from the rest of the variables (see Fig. 8.4). Thus, as a first test of this structure, we estimated *ozone-Pedregal* using only these 3 variables. We carried out two experiments: (1) estimate *ozone-Pedregal* using 100 random samples taken from the training data, and (2) estimate *ozone-Pedregal* with another 100 samples taken from separate data, not used for training. We observe that even with only three parameters, the estimations are quite good. For training data the average error (absolute difference between the real and the estimated ozone concentration) is 11 IMECA or 12%, and for non-training data it is 26 IMECA or 22%.

An interesting observation from the obtained structure is that *ozone-Pedregal* (located in the south of the city) is related basically to three variables, *ozone-Merced* and *ozone-Xalostoc* (located in the center and north of the city), and the wind velocity in Pedregal. Otherwise stated, pollution in the south depends basically on the pollution in the center and north of the city (where there is more transit and industry) and the wind velocity –which carries the pollution from the north to the south. This phenomenon was already known, but it was discovered automatically by learning a BN. Other, not so well known relations, can also be discovered and could be useful for making decisions to control pollution and take emergency measures.

### 8.6.2 Agricultural Planning Using Bayesian Networks

Climate data plays a key role in planning in the agricultural sector. However, data at the required spatial or temporal resolution is often lacking, or certain values are missing. An alternative is to learn a Bayesian network from the available data, and then use it to generate data for missing variables. This work is in the domain of coffee production in Central America and southern Mexico [6]. The variables considered are: precipitation, maximum and minimum air temperature, wind speed, and solar radiation; which are used to infer missing values for monthly relative humidity and relative humidity for the driest month.

The data was obtained from the surface reanalysis dataset, Climate Forecast System Reanalysis (CFSR), which includes daily values for the variables precipitation (mm), air temperature (°C, minimum and maximum at 2 m), wind speed (m/s, at 10 m), surface solar radiation (MJ/m$^2$) and relative humidity (%, at 2 m). The spatial resolution is 38 km $\times$ 38 km per pixel and data are available from 1979 to 2014

The BN model was built using the software package Netica (Version 6.04, Norsys Software Corp., Vancouver, BC, Canada). For each selected variable, nodes were created and discretized. Continuous variables were discretized in intervals of equal

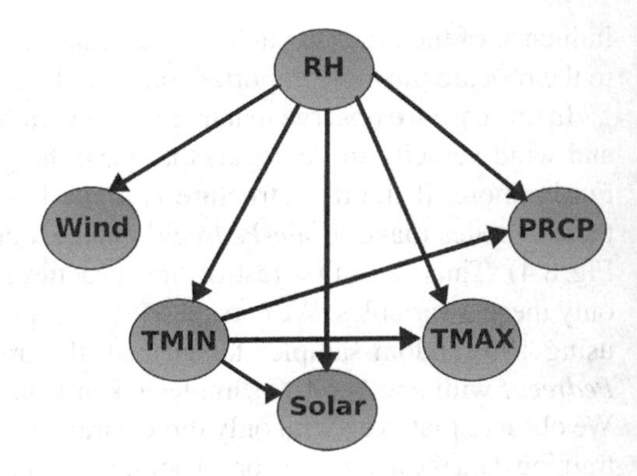


**Fig. 8.5** Bayesian network model to infer monthly relative humidity. Notation: RH: relative humidity, TMAX: maximum temperature, TMIN: minimum temperature, PRCP: total precipitation, Wind: wind speed, Solar: surface solar radiation

length; the number of intervals was selected for each variable according to the data distribution.

Once the variables were discretized, the graphical model was learned from 80% of the cases of the dataset (n = 180,530). The relative humidity node was set as the target variable, and the BN topology was restricted to a Tree Augmented Naive Bayes (TAN) classifier, considering *relative humidity* as the class. The resulting BN structure is depicted in Fig. 8.5. The parameters of this model were learned from the same dataset using Laplacian smoothing.

The model was validated in two ways. Firstly, they explored the capability of the model to infer relative humidity for all the months in a year, and the relative humidity of the driest month, based on all other variables. Secondly, they estimated relative humidity with the variables Solar and Wind missing, because these are hardly registered in the study region's weather stations. The rest of the data not used for training (45,190 cases) was used to evaluate the model under the two previous scenarios. In general, when comparing inferred values to the actual values, the metrics bias (less than the unit) and RMSE $\leq 4.1\%$ indicate a very close agreement. As expected, the best model performance was obtained when information on all other variables was available; however, even with missing variables (Solar and Wind), the results were very good (RMSE $\leq 5\%$).

## 8.7 Additional Reading

A general book on learning Bayesian networks is [9]; Heckerman [4] has comprehensive tutorial on learning BNs. The tree and polytree learning algorithms are described in [10]. A general introduction to learning BNs from a statistical perspective is given in [12]. An analysis of different scoring functions is presented in [1].

## 8.8 Exercises

1. The table below gives the original data for the golf example using numeric values for some of the variables. Discretize these variables using 3 intervals for each variable: (a) use equal width discretization, (b) use equal data.

| Outlook | Temperature | Humidity | Wind | Play |
|---|---|---|---|---|
| sunny | 19 | high | 5 | N |
| sunny | 25 | high | 3 | N |
| overcast | 26 | high | 3 | P |
| rainy | 17 | high | 6 | P |
| rainy | 11 | normal | 15 | P |
| rainy | 7 | normal | 17 | N |
| overcast | 8 | normal | 11 | P |
| sunny | 20 | high | 7 | N |
| sunny | 19 | normal | 1 | P |
| rainy | 22 | normal | 5 | P |
| sunny | 21 | normal | 20 | P |
| overcast | 22 | high | 18 | P |
| overcast | 28 | normal | 16 | P |
| rainy | 18 | high | 3 | N |

2. Using the discretized data from the previous problem, obtain the CPTs for the Bayesian network structure given in Fig. 8.2.
3. Discretize the continuous variables in the table of Problem 1 to maximize the accuracy of the prediction of *Play*, considering a naive Bayes classifier. Build an initial classifier with the discrete attributes, then apply the supervised discretization to *Temperature*, and then to *Wind*.
4. Obtain the CPTs for a naive Bayes classifier based on the data (only the complete registers) in Table 8.1, Sect. 8.2.3.2.
5. Estimate the missing values of *Temperature* for the example in Sect. 8.2.3.2. Re-estimate the CPTs including these values.
6. Continue the EM algorithm for the example in Sect. 8.2.3.2 until convergence. Show the final CPT, $P(Wind \mid Play)$, and the final data table.
7. Based on the data for the golf example in Table 8.2, learn the skeleton of a tree BN using Chow and Liu's algorithm.
8. Obtain the directions for the arcs of the skeleton from the previous problem by applying the polytree learning technique.
9. Based on the same dataset from the previous problem, Table 8.2, learn a BN using the PC algorithm. Use the conditional mutual information measure (seen in Chap. 2) for testing conditional independence.
10. Given the dataset of the golf example, Table 8.2, learn a BN using the K2 algorithm. Define a causal ordering for the variables, and use as an initial structure the tree obtained with Chow and Liu's algorithm. Compare the structure obtained with that of the previous problem.

11. Given the dataset in the following table, (a) learn a naive Bayesian classifier considering $C$ as the class, (b) learn a tree-structured BN and fix the directions of the arcs considering $C$ as the root, (c) compare the structures of both models.

| A1 | A2 | A3 | C |
|---|---|---|---|
| 0 | 0 | 0 | 0 |
| 0 | 1 | 1 | 1 |
| 0 | 1 | 0 | 1 |
| 0 | 0 | 1 | 1 |
| 0 | 0 | 0 | 0 |
| 0 | 1 | 1 | 0 |
| 1 | 1 | 0 | 1 |
| 0 | 0 | 0 | 1 |
| 0 | 1 | 0 | 0 |
| 0 | 1 | 1 | 0 |

12. For the dataset in the previous problem, use the Laplacian smoothing technique to obtain the conditional probability tables for both, the NBC and the tree BN. Compare the tables to the ones obtained without smoothing.
13. *** Develop a program that implements the polytree learning algorithm. Apply it to the golf data and compare with the results from Exercise 2.
14. *** Implement a program for learning a BN from data using a score and search technique based on the MDL scoring function, and another based on independence tests (PC algorithm). Apply both to different datasets and compare the results.
15. *** Develop a program for implementing the transfer learning approach for BNs. To evaluate it, select a BN as target task, and build several auxiliary tasks by modifying the original structure, adding, eliminating or inverting arcs. Generate *few* data samples for the target task (using logic sampling) and more data for the auxiliary tasks.

## References

1. De Campos, L.M.: A scoring function for learning Bayesian networks based on mutual information and conditional independence tests. J. Mach. Learn. Res. **7**, 2149–2187 (2006)
2. Chow, C.K., Liu, C.N.: Approximating discrete probability distributions with dependence trees. IEEE Trans. Inf. Theory **14**, 462–467 (1968)
3. Cooper, G.F., Herskovitz, E.: A Bayesian method for the induction of probabilistic networks from data. Mach. Learn. **9**(4), 309–348 (1992)
4. Heckerman, D.: A tutorial on learning with Bayesian networks. Innovations in Bayesian Networks, pp. 33–82. Springer, Netherlands (2008)
5. Lam, W., Bacchus, F.: Learning Bayesian belief networks: an approach based on the MDL principle. Comput. Intell. **10**, 269–293 (1994)
6. Lara-Estrada, L., Rasche, L., Sucar, L.E., Schneider, U.A.: Inferring missing climate data for agricultural planning using Bayesian networks. Land (2018)
7. Luis, R., Sucar, L.E., Morales, E.F.: Inductive transfer for learning Bayesian networks. Mach. Learn. **79**, 227–255 (2010)

8. Martínez, M., Sucar, L.E.: Learning an optimal Naive Bayes classifier. In: 18th International Conference on Pattern Recognition (ICPR), vol. 3, pp. 1236–1239 (2006)
9. Neapolitan, R.E.: Learning Bayesian Networks. Prentice Hall, New Jersey (2004)
10. Pearl, J.: Probabilistic Reasoning in Intelligent Systems: Networks of Plausible Inference. Morgan Kaufmann, San Francisco (1988)
11. Rebane, G., Pearl, J.: The recovery of causal poly-trees from statistical data. In: Kanal, L.N., Levitt, T.S., Lemmer, J.F. (eds.) Uncertainty in Artificial Intelligence, pp. 175–182 (1987)
12. Spirtes, P., Glymour, C., Scheines, R.: Causation, Prediction, and Search. Springer, Berlin (1993)
13. Sucar, L.E., Ruiz-Suarez, J.C.: Forecasting air pollution with causal probabilistic networks. In: Barnett, V., Turkman, K.F. (eds.) Statistics for the Environment 3: Statistical Aspects of Pollution, pp. 185–197. Wiley, Chichester (2007)
14. Yang, Y., Webb, G.I.: Proportional k-interval discretization for Naive-Bayes classifiers. In: 12th European Conference on Machine Learning (ECML), pp. 564–575 (2001)

# Chapter 9
# Dynamic and Temporal Bayesian Networks

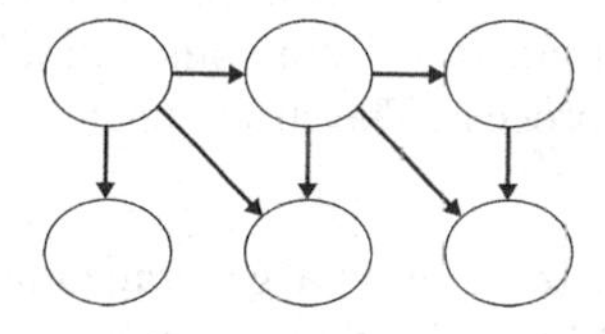

## 9.1 Introduction

Bayesian networks usually represent the *state* of certain phenomena at an instant in time. However, in many applications, we want to represent the temporal evolution of a certain process, that is, how the different variables evolve over time, known also as time series.

There are two basic types of Bayesian network models for dynamic processes: state based and event based. State based models represent the state of each variable at discrete time intervals, so that the networks consist of a series of *time slices*, where each time slice indicates the value of each variable at time $t$; these models are known as dynamic Bayesian networks. Event based models represent the changes in state of each state variable; each temporal variable will then correspond to the time in which a state change occurs. These types of models are known as event networks or temporal networks.

In this chapter we will review both, dynamic Bayesian networks and temporal networks, including representation, inference and learning.



L. E. Sucar, *Probabilistic Graphical Models*, Advances in Computer Vision and Pattern Recognition, https://doi.org/10.1007/978-3-030-61943-5_9

## 9.2 Dynamic Bayesian Networks

*Dynamic Bayesian networks* (DBNs) are an extension of Bayesian networks to model dynamic processes. A DBN consists of a series of *time slices* that represent the state of all the variables at a certain time, $t$; a kind of snapshot of the evolving temporal process. For each temporal slice, a dependency structure between the variables at that time is defined, called the *base network*. It is usually assumed that this structure is duplicated for all the temporal slices (except the first slice, which can be different). Additionally, there are edges between variables from different slices, with their directions following the direction of time, defining the *transition network*. Usually DBNs are restricted to have directed links between consecutive temporal slices, known as a first order Markov model; although, in general, this is not necessary. An example of a DBN with 3 variables and 4 time slices is depicted in Fig. 9.1.

Most of the DBNs considered in practice satisfy the following conditions:

- First order Markov model. The state variables at time $t$ depend only on the state variables at time $t-1$ (and other variables at time $t$).
- Stationary process. The structure and parameters of the model do not change over time.

DBNs can be seen as a generalization of Markov chains and hidden Markov models (HMMs). A Markov chain is the simplest DBN, in which there is only one variable, $X_t$, per time slice, directly influenced only by the variable in the previous time. In this case the joint distribution can be written as:

$$P(X_1, X_2, \ldots, X_T) = P(X_1)P(X_2 \mid X_1) \ldots P(X_T \mid X_{T-1}) \tag{9.1}$$

A hidden Markov model has two variables per time stage, one that is known as the *state variable*, $S$; and the other as the *observation variable*, $Y$. It is usually assumed that $S_t$ depends only on $S_{t-1}$ and $Y_t$ depends only on $S_t$. Thus, the joint probability can be factored as follows:

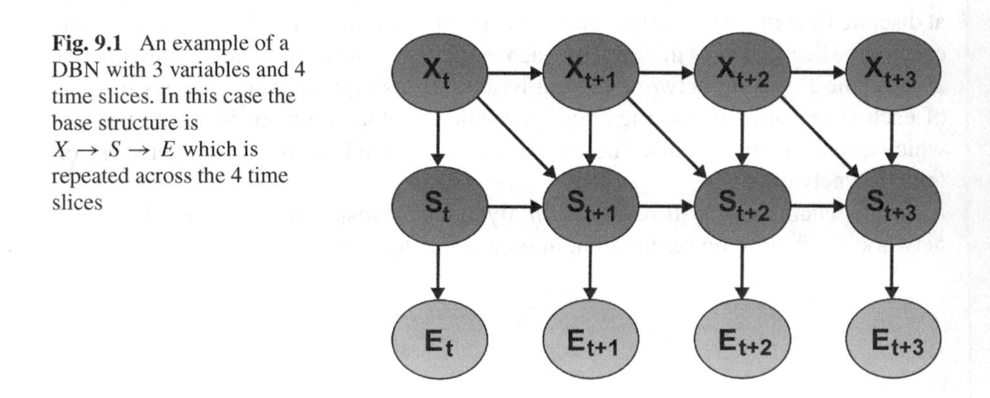


**Fig. 9.1** An example of a DBN with 3 variables and 4 time slices. In this case the base structure is $X \rightarrow S \rightarrow E$ which is repeated across the 4 time slices

$$P(\{S_{1:T}, Y_{1:T}\}) = P(S_1)P(Y_1 \mid S_1)\prod_{t=2}^{T} P(S_t \mid S_{t-1})P(Y_t \mid S_t) \quad (9.2)$$

Markov chains and HMMS are particular cases of DBNs, which in general can have $N$ variables per time step, with any base and transition structures. Another particular variant of DBNs are Kalman Filters, which also have one state and one observation variable, but both variables are continuous. The basic Kalman filter assumes Gaussian distributions and linear functions for the transition and observation models.

### 9.2.1 Inference

There are several classes of inferences that can be performed with DBNs. In the following, we briefly mention the main types of inference, where $\mathbf{X}$ are the unobserved (hidden) variables, and $\mathbf{Y}$ are the observed variables [9]:

- *Filtering*. Predict the next state based on past observations: $P(\mathbf{X}_{t+1} \mid \mathbf{Y}_{1:t})$.
- *Prediction*. Predict future states based on past observations: $P(\mathbf{X}_{t+n} \mid \mathbf{Y}_{1:t})$.
- *Smoothing*. Estimate the current state based on past and future observations (useful for learning): $P(\mathbf{X}_t \mid \mathbf{Y}_{1:T})$.
- *Decoding*. Find the most likely sequence of hidden variables given the observations: ArgMax($\mathbf{X}_{1:T}$) $P(\mathbf{X}_{1:T} \mid \mathbf{Y}_{1:T})$.

The inference methods for Bayesian networks (see Chap. 7) can be directly applied for DBNs. Efficient inference methods have been developed for particular types of models, such as HMMs [10] (see Chap. 5). However, for more complex models, inference may become computationally intractable. In these cases we can apply approximate methods based on sampling, such as Markov chain Monte Carlo [6]. A popular approximate method are *particle filters*, which approximate the state probability distribution (belief state) with a set of weighted *particles* or samples [9]. Next we review some sampling approaches for DBNs.

### 9.2.2 Sampling

The logic sampling method that we presented in Chap. 7 can also be applied to DBNs. For this, samples are generated for the first time slice, $t = 1$, according to the structure of the base network, and based on these samples, for the second time slice, and so on.

These types of sampling approaches are known as *ancestral sampling*, as they generate samples based on the ordering of the variables in a directed acyclic graph, as is the case of BNs and DBNs. They are adequate for when there is no evidence or for prediction; but for other types of inference they could become very inefficient when there is evidence. If we discard the samples that are no consistent with the evidence, then the probability of a sample being retained is in the order of $1/\prod_i \mid X_i \mid$, where $X_i$ are the evidence variables. For example, with 3 evidence variables with dimension 5, the probability will be one in 125; that is, less than 1% of the generated samples will be useful.

Given the limitations of ancestral sampling, alternative sampling techniques are more common, in particular for DBNs.

#### 9.2.2.1 Gibbs Sampling

Gibbs Sampling is an algorithm for obtaining a sequence of samples to approximate a multivariate probability distribution, when direct sampling is difficult. It can be used to approximate the joint distribution, the marginal distribution of one of the variables, or some subset of the variables. The idea is that it is easier to sample from the conditional distribution than to marginalize over the joint distribution. The algorithm is the following.

Given a set of variables $\mathbf{X} = (X_1, X_2, \ldots, X_n)$ with joint probability distribution $P(X_1, X_2, \ldots, X_n)$ we want to generate $k$ samples.

1. Set an initial random value for all the variables, $\mathbf{X^1}$.
2. Generate a next sample, $\mathbf{X^{i+1}}$ based on the values from the previous sample, $\mathbf{X^i}$, generating a sample for each variable conditioned on the other variables, according to the order of the variables: $X_1^{i+1} \mid (X_2^i, \ldots, X_n^i)$, $X_2^{i+1} \mid (X_1^i, X_3^i, \ldots, X_n^i)$, and so on. The samples are generated according to the conditional probability distributions: $P(X_i \mid X_1, \ldots, X_{i-1}, X_{i+1}, \ldots, X_n)$.
3. Repeat the previous step $k$ times

This is essentially a Markov chain Monte Carlo algorithm for approximating the joint distribution (that is, $\mathbf{X^{i+1}}$ depends only on $\mathbf{X^i}$); the sample distribution approximates the real one as $k$ increases if the Markov chain is irreducible. Given that the initial values are arbitrary, it is common to ignore some number of samples at the beginning, called *burn-in period.*

For the BNs and DBNs, the conditional distribution of each variable depends only on the Markov blanket of the variable. In the case of evidence, the evidence variables are set to their value and the algorithm proceeds in the same way, no need to discard samples.

As a simple example, consider two binary variables with the following conditional probability distributions:

$$P(X \mid Y) = \begin{array}{c|cc} & Y=0 & Y=1 \\ \hline X=0 & 0.80 & 0.40 \\ X=1 & 0.20 & 0.60 \end{array}$$

$$P(Y \mid X) = \begin{array}{c|cc} & X=0 & X=1 \\ \hline Y=0 & 0.85 & 0.50 \\ Y=1 & 0.15 & 0.50 \end{array}$$

Assume that the first sample is $(X^1 = 1, Y^1 = 1)$. Based on these values and according to the conditional distributions, a second sample is generated: $(X^2 = 1, Y^2 = 0)$. And so on. Lets say that after 20 samples we have: 13 times (1,1), 4 times (1,0), 2 times (0,1) and one time (0,0); then an estimate for the joint will be:

$$P(X, Y) = \begin{array}{c|cc} & X=0 & X=1 \\ \hline Y=0 & 0.05 & 0.20 \\ Y=1 & 0.10 & 0.65 \end{array}$$

#### 9.2.2.2 Importance Sampling

Monte Carlo sampling assumes that we can draw samples from the actual probability distribution; however this is not always possible. Importance sampling is a technique to estimate the expectation of a probability distribution, $p(X)$. However, it is very difficult to sample directly $p(X)$, so instead we sample another distribution, $q(X)$, and use these samples to estimate $p(X)$, via the following transformation:

$$E[f(x)] = \int f(X)p(X)dx = \int f(X)\frac{p(X)}{q(X)}q(X)dx \approx 1/n \sum_i f(X_i)\frac{p(X_i)}{q(X_i)} \tag{9.3}$$

So to estimate the expectation we can sample from another distribution, $q(X)$, known as the *proposal distribution*; $\frac{p(X)}{q(X)} = w(X)$ is called sampling ratio or sampling weight, which acts as a correction weight to offset as we are sampling from a different distribution. Then we can write the approximation as:

$$E[f(x)] \approx 1/n \sum_i f(X_i)w(X_i) \tag{9.4}$$

In principle $q(X)$ could be any distribution that satisfies $q(X) = 0 \Rightarrow p(X) = 0$. The estimation improves as the number of samples $n$ increases and also if both distributions are closer.

For DBNs, importance sampling is applied to temporal distributions, $p(x_1, x_2, \ldots x_t) = p(x_{1:t})$, what is known as *sequential importance sampling*. In this case the samples are paths from $q(x_{1:t})$. Given a set of sample paths, $x^k_{1:t-1}$, with their importance weights, $w^k_{t-1}$, we can find the new importance weights in $t$, $w^k_t$, by sampling from the transition distribution, $q(x_t \mid x_{1:t-1})$.

It can be shown that in this case the weights can be defined recursively as:

$$w^k_t = w^k_{t-1}\alpha^k_t \tag{9.5}$$

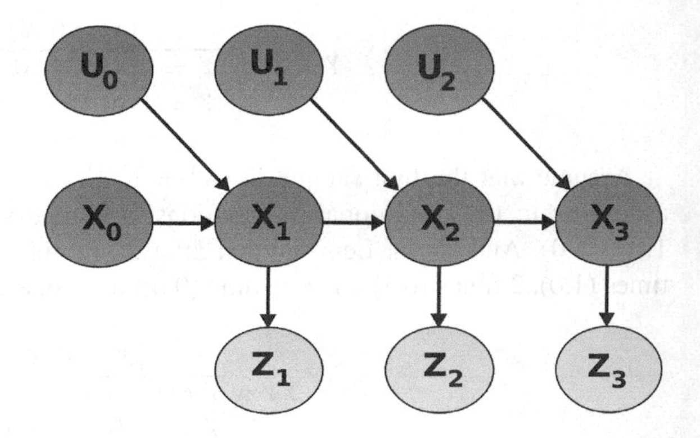


**Fig. 9.2** Graphical model of a particle filter. $X_t$ represent the state variables, $U_t$ the inputs and $Z_t$ the observations

where $\alpha_t^k = \frac{p(x_t^k|x_{1:t-1}^k)}{q(x_t^k|x_{1:t-1}^k)}$. Sequential importance sampling is also known as particle filtering.

#### 9.2.2.3 Particle Filters

Particle filters are a sampling technique for Bayesian filtering, that is predict the next state based on past observations and controls, $P(X_{t+1} \mid Z_{1:t}, U_{1:t})$; where $X$ is the state variable, $U$ the inputs and $Z$ the observations. An advantage is that it does not make any assumptions regarding the probability distribution of the state variable (for instance, Kalman filters assume Gaussian distributions). The graphical model of a particle filter is illustrated in Fig. 9.2.

The basic idea of particle filters is that the probability distribution can be represented as a set of samples (particles). The algorithm involves an initialization and then a cycle with three basic phases:

1. Initialization—set $P(X_0)$—initial distribution of particles (can be uniform).
2. Repeat for $t = 1$ to $t = T$:

   a. Prediction—for each particle, estimate the next state, $X_t = \int P(X_t \mid X_{t-1}, U_{t-1})P(X_{t-1})dx_{t-1}$.
   b. Update—estimate the weight of each particle as the likelihood of the observation given the particle hypothesis, $w(X_t) = P(Z_t \mid X_t)$; normalized so the sum of all the weights is 1.
   c. Resample—choose a new set of particles, such that the probability that a particle survives is proportional to its weight. Low probability particles are eliminated, and high probability particles are replicated.

A critical aspect of particle filters is the number of particles, $N$; in general the larger $N$ the better is the estimate of $P(X_t)$. Particle filters have many applications, including object tracking, robot localization, among others.

**Table 9.1** Learning dynamic Bayesian networks: 4 basic cases

| Structure | Observability | Method |
|---|---|---|
| Known | Full | Maximum likelihood estimation |
| Known | Partial | Expectation–maximization (EM) |
| Unknown | Full | Search (global) or tests (local) |
| Unknown | Partial | EM and global or local |

### 9.2.3 *Learning*

As with BNs, learning dynamic Bayesian networks involves two aspects: (i) learning the structure or graph topology, and (ii) learning the parameters or CPTs for each variable. Additionally, we can consider two cases in terms of the observability of the variables: (a) full observability, when there is data for all the variables, and (b) partial observability, when some variables are unobserved or hidden, or we have missing data. There are 4 basic cases for learning DBNs, see Table 9.1.

For all the cases we can apply extensions of the methods for parameter and structure learning for Bayesian networks that we reviewed in Chap. 8. We describe one of these extensions below, for the case of unknown structure and full observability.

Assuming that the DBN is stationary (time invariant), we can consider that the model is defined by two structures: (i) the base structure, and (ii) the transition structure. Thus, we can divide the learning of a DBN into two parts, first learn the base structure, and then, given the base structure, learn the transition structure, see Fig. 9.3.

For learning the base structure we can use all the available data for each variable, ignoring the temporal information. This is equivalent to learning a BN, so we can apply any of the methods used for learning BNs (see Chap. 8).

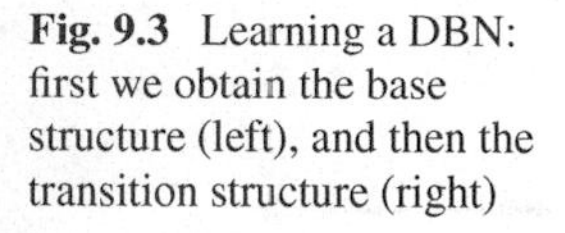
**Fig. 9.3** Learning a DBN: first we obtain the base structure (left), and then the transition structure (right)

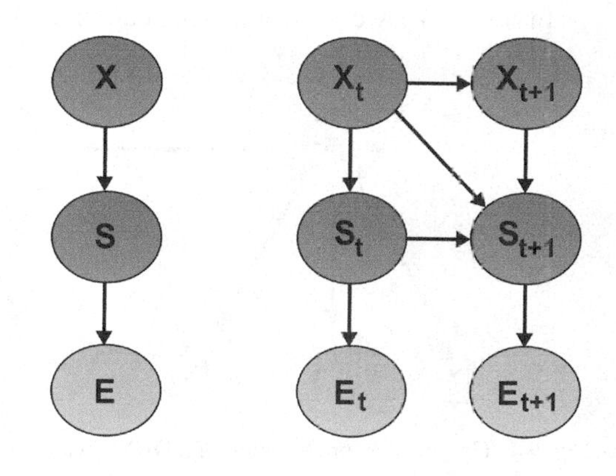

For learning the transition network we consider the temporal information, in particular the data for all variables in two consecutive time slices, $\mathbf{X_t}$ and $\mathbf{X_{t+1}}$. Considering the base structure, we can then learn the dependencies between the variables at time $t$ and $t+1$ (assuming a first–order Markov model), and restricting the direction of the edges from the past to the future.

Here we have described in a simple, general way the two phases for learning a DBN, however there are several variants of this idea that have been developed (see the additional readings section).

### 9.2.4 Dynamic Bayesian Network Classifiers

In this section we consider a particular type of DBN known as *dynamic Bayesian network classifier* (DBNC) [2]. Similarly to HMMs, a DBNC has a hidden state variable for each time instant, $S_t$; however, the observation variable is decomposed into $m$ attributes, $A_t^1, \ldots A_t^m$, which are assumed to be conditionally independent given $S_t$. Thus, the *base structure* for a DBNC has a star-like structure with the directed links from $S_t$ to each attribute $A_t^i$ (see Fig. 9.4).

The joint probability of a DBNC can be factored as follows:

$$P(\{S_{1:T}, \mathbf{A}_{1:T}\}) = P(S_1)\left[\prod_{m=1}^{M} P(A_1^m \mid S_1)\right]\left[\prod_{t=2}^{T} P(S_t \mid S_{t-1}) \prod_{m=1}^{M} P(A_t^m \mid S_t)\right] \tag{9.6}$$

where $\mathbf{A} = A^1, \ldots, A^m$.

The difference with the joint probability of a HMM is that instead of having $P(A_t \mid S_t)$, we have the product of each attribute given the state, $\prod_{m=1}^{M} P(A_t^m \mid S_t)$.

Parameter learning and classification with DBNCs are done in a similar manner as with HMMs, using a modified version of the Baum-Welch and Forward algorithms, respectively.

In Sect. 9.4.1 we present an application of DBNC for gesture recognition.

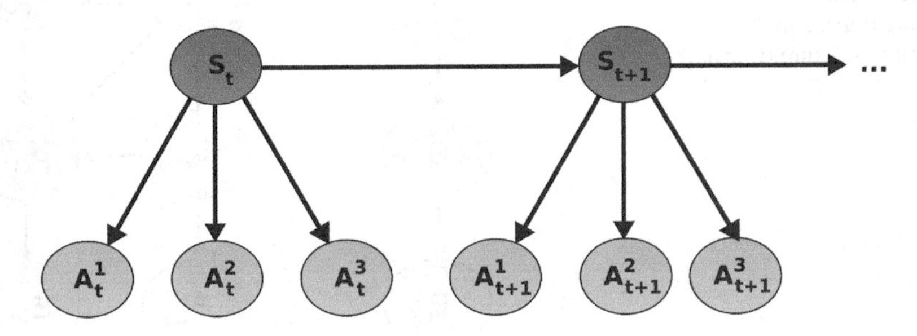


**Fig. 9.4** Graphical Representation of a DBNC with 3 attributes unrolled two times

## 9.3 Temporal Event Networks

Aside from DBNs, there are some alternative BN representations for describing temporal processes that have been developed. An example of a different type of temporal Bayesian networks are *temporal event networks* [1, 5].

Temporal event networks (TENs) are an alternative to DBNs for modeling dynamic processes. In a temporal event network, a node represents the *time* of occurrence of an event or state change of certain variable, in contrast to a node in a DBN that represents the state value of a variable at a certain time. For some problems, in which there are few state changes in the temporal range of interest, event networks provide a simpler and more efficient representation; however, for other applications such as monitoring or filtering, DBNs are more appropriate.

Several variants of TENs have been proposed, such as *time nets* and *temporal nodes Bayesian networks* (TNBNs). In the rest of this section we will focus on TNBNs.

### 9.3.1 Temporal Nodes Bayesian Networks

A Temporal Nodes Bayesian Network (TNBN) [1, 5] is composed of a set of Temporal Nodes (TNs). TNs are connected by edges, where each edge represents a causal-temporal relationship between TNs. There is at most one state change for each variable (TN) in the temporal range of interest. The value taken by the variable represents the interval in which the event occurs. Time is discretized in a finite number of intervals, allowing a different number and duration of intervals for each node (multiple granularity). Each interval defined for a child node represents the possible delays between the occurrence of one of its parent events (cause) and the corresponding child event (effect). Some Temporal Nodes do not have temporal intervals, these correspond to Instantaneous Nodes. Root nodes are instantaneous by definition [1]. Formally, a TNBN is defined as follows.

A TNBN is defined as a pair $B = (G, \Theta)$. G is a Directed Acyclic Graph, $G = (\mathbf{V}, \mathbf{E})$. G is composed of $\mathbf{V}$, a set of Temporal and Instantaneous Nodes; $\mathbf{E}$ a set of edges between Nodes. The $\Theta$ component corresponds to the set of parameters that quantify the network. $\Theta$ contains the values $\Theta_{v_i} = P(v_i | Pa(v_i))$ for each $v_i \in \mathbf{V}$; where $Pa(v_i)$ represents the set of parents of $v_i$ in $G$.

A Temporal Node, $v_i$, is defined by a set of states $\mathbf{S}$, each state is defined by an ordered pair $S = (\lambda, \tau)$, where $\lambda$ is the value of a random variable and $\tau = [a, b]$ is the interval associated, with an initial value a and a final value b. These values correspond to the time interval in which the state change occurs. In addition, each Temporal Node contains an extra default state $s_d =$ (“no change”, $[t_i, t_f]$), which has associated as interval the whole temporal range for that variable, $[t_i, t_f]$. If a Node has no intervals defined for any of its states, then it receives the name of Instantaneous Node.

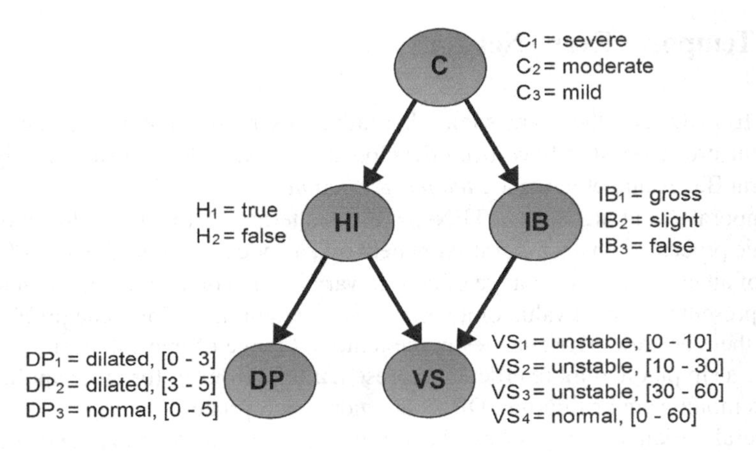


**Fig. 9.5** A TNBN for the accident example. C = Collision, HI = Head Injury, IB = Internal Bleeding, DP = Dilated Pupils, and VS = Vital Signs. C, HI, and IB are instantaneous nodes; while DP and VS are temporal nodes, depicted with their associated time intervals

The following is an example based on [1]:

**Example 9.1** Assume that at time $t = 0$, an automobile accident occurs, that is, a Collision. This kind of accident can be classified as *severe*, *moderate*, or *mild*. To simplify the model, we will consider only two immediate consequences for the person involved in the collision: Head Injury and Internal Bleeding. A Head Injury can bruise the brain, and chest injuries can lead to Internal Bleeding. These are all instantaneous events that may generate subsequent changes, for example the Head Injury event might generate Dilated Pupils and unstable Vital Signs. Suppose that we gathered information about accidents that occurred in a specific city. The information indicates that there is a strong causal relationship between the severity of the accident and the immediate effect of the patient's state. Additionally, a physician domain expert provided some important temporal information: If a head injury occurs, the brain will start to swell and if left unchecked the swelling will cause the pupils to dilate within 0–5 min. If internal bleeding begins, the blood volume will start to fall, which will tend to destabilize vital signs. The time required to destabilize vital signs will depend on the severity of the bleeding: if the bleeding is gross, it will take from 10 to 30 min; if the bleeding is slight it will take from 30 to 60 min. A head injury also tends to destabilize vital signs, taking from 0 to 10 min to make them unstable.

The graphical representation of a TNBN for the accident example is shown in Fig. 9.5. The model presents three instantaneous nodes: Collision, Head Injury and Internal Bleeding. These events will generate subsequent changes that are not immediate: Dilated Pupils and unstable Vital Signs, which depend on the severity of the accident, and therefore have temporal intervals associated to them. This TNBN contains only 5 nodes (in contrast to the 30 nodes that would be required for an equivalent DBN, considering 6 time slices with 5 variables each).

In TNBNs, each variable represents an event or state change. So, only one (or a few) instance(s) of each variable is required, assuming there is one (or a few) change(s) of a variable's state in the temporal range of interest. No copies of the model are needed, and no assumption about the Markovian nature of the process is made. TNBNs can deal with multiple granularity, because the number and the size of the intervals for each node can be different.

#### 9.3.1.1 Inference

A TNBN allows for reasoning about the probability of occurrence of certain events, for diagnosis (i.e., finding the most probable cause of a temporal event) or prediction (i.e., determining the probable future events that will occur given a certain event). For this, standard probability propagation techniques for BNs (see Chap. 7) can be applied. However, given that a TNBN represents relative times between events, the cases of prediction and diagnosis have to be differentiated for doing probabilistic inference:

*Prediction.* In the case where at least one of the root (instantaneous) nodes of the TNBN is part of the evidence, then the reference time for the model is fixed and probability propagation can be performed directly, obtaining the posterior probability of the subsequent events (a probability for each temporal interval of each non-instantiated temporal node). For example, considering the example in Fig. 9.5, if we know the severity of the *Collision* (e.g. moderate), then via probability propagation we obtain the posterior probability of the event *Dilated Pupils* occurring during each temporal interval and not occurring at all (False). The posterior probability of other temporal and instantaneous nodes is obtained in a similar manner. Note that the time intervals of the TNs will be with respect to the occurrence of the *Collision*, which will correspond to $t = 0$.

*Diagnosis.* In the case where none of the instantaneous nodes are known, and the evidence is given only for temporal nodes, then several *scenarios* need to be considered, as we do not know to which interval to assign the evidence given that there is no reference time. In this case, all the $n$ possible intervals for the TN have to be considered, performing inference $n$ times, one for each interval. The results for each scenario have to be maintained, until there is additional evidence, such as the occurrence of another event, that allows for discarding some scenarios. Considering the accident example of Fig. 9.5, assume that a paramedic arrives and finds that the person has his *Pupils Dilated.* As the time of the accident is unknown, then the evidence must be given to the first two temporal intervals for this variable, generating 2 scenarios (the last interval corresponds to the default value of no change). If later on, the time of the accident is determined, then the appropriate scenario is kept and the others are discarded.

#### 9.3.1.2 Learning

Learning a TNBN involves 3 aspects: (i) learning the temporal intervals for the temporal nodes, (ii) learning the structure of the model, (iii) learning the parameters of the model. As these three components are interrelated, an iterative procedure is required, that learns an initial estimate of one (or more) of these aspects, and then improves these initial estimates iteratively. Next we present an algorithm for learning the three components of a TNBN [7].

The algorithm assumes that the root nodes are instantaneous nodes and it obtains a finite number of non overlapping intervals for each temporal node. It uses the times (delays) between parent events and the current node as input for learning the intervals. With this top-down approach the algorithm is guaranteed to arrive at a local maximum in terms of predictive score.

A learning algorithm (known as *LIPS*) [7] for TNBNs is summarized as follows:

1. First, it performs an initial discretization of the temporal variables, for example using an Equal-Width discretization. With this process it obtains an initial approximation of the intervals for all the Temporal Nodes.
2. Then it performs standard BN structural learning. Specifically, the K2 learning algorithm [3] (see Chap. 8) is used to obtain an initial structure and corresponding parameters.
3. The interval learning algorithm refines the intervals for each temporal node (TN) by means of clustering. For this, it uses the information of the configurations of the parent nodes. To obtain a set of intervals a Gaussian mixture model is used as a clustering algorithm for the temporal data. Each cluster corresponds, in principle, to a temporal interval. The intervals are defined in terms of the mean and the standard deviation of the clusters. The algorithm obtains different sets of intervals that are merged and combined, this process generates different interval sets that will be evaluated in terms of the predictive accuracy. The algorithm applies two pruning techniques in order to remove sets of intervals that may not be useful and also to keep a low complexity for the TNBN. The best set of intervals (which may not be those obtained in the first step) for each TN is selected based on predictive accuracy. When a TN has as parents other Temporal Nodes, the configurations of the parent nodes are not initially known. In order to solve this problem, the intervals are sequentially selected in a top-down fashion according to the TNBN structure.
4. Finally, the parameters (CPTs) are updated according to the new set of intervals for each TN.

The algorithm then iterates between structure learning and interval learning.

We illustrate the process of obtaining the intervals for the TN *Dilated Pupils* (DP) of Fig. 9.5 (the intervals in this example are different from those shown in the figure). We can see that its parent node (Head Injury) has two configurations, *true* and *false*. Thus, the temporal data of Dilated Pupils is divided into two partitions, one for each configuration of the parent node. Then, for each partition the first approximation

**Table 9.2** Initial sets of intervals obtained for the node *Dilated Pupils*. There are three sets of intervals for each partition

| Partition | Intervals |
|---|---|
| Head Injury = true | [11 – 35] |
| | [11 – 27][32 – 53] |
| | [8 – 21][25 – 32][45 – 59] |
| Head Injury = false | [3 – 48] |
| | [0 – 19][39 – 62] |
| | [0 – 14][28 – 40][47 – 65] |

**Table 9.3** Collected data to learn the TNBN of Fig. 9.5. Top: original data showing the time of occurrence of the temporal events. Bottom: temporal data after the initial discretization. For Dilated Pupils and Vital Signs the temporal data represents the minutes after the collision occurred

| Collision | Head Injury | Internal Bleeding | Dilated Pupils | Vital Signs |
|---|---|---|---|---|
| Severe | True | Gross | 14 | 20 |
| Moderate | True | Gross | 25 | 25 |
| Mild | False | False | – | – |
| ... | ... | ... | ... | ... |

| Collision | Head Injury | Internal Bleeding | Dilated Pupils | Vital Signs |
|---|---|---|---|---|
| Severe | True | Gross | [10 – 20] | [15 – 30] |
| Moderate | True | Gross | [20 – 30] | [15 – 30] |
| Mild | False | False | – | – |
| ... | ... | ... | ... | ... |

of the interval learning step of the previous algorithm is applied. The expectation-maximization algorithm is applied to get Gaussian mixture models with parameters 1, 2 and 3 as the number of clusters. That gives six different sets of intervals, as shown in Table 9.2. Then each set of intervals is evaluated in terms of its prediction performance to measure its quality and the set of intervals with the best score is selected.

Now we present a complete example of the TNBN learning algorithm considering that we have data for the accident example illustrated in Fig. 9.5. First we assume we have data from other accidents that is similar to the one presented in the upper part of Table 9.3. The first three columns have nominal data, however, the last two columns have temporal data which represent the occurrence of those events after the collision. Those two columns would correspond to Temporal Nodes of the TNBN. We start by applying the equal width discretization on the numerical data, so it would yield results similar to the ones presented in the lower part of Table 9.3.

Using the discretized data we can apply a structure learning algorithm (like K2) using the partial ordering of the temporal events: {Collision}, {Head Injury, Internal Bleeding}, and {Dilated Pupils, Vital Signs}. Now we have an initial TNBN, however the obtained intervals are somewhat naive, so the interval learning step of the algorithm can be applied to improve the initial temporal intervals. This process will learn a TNBN similar to the one presented in Fig. 9.5.

## 9.4 Applications

We illustrate the application of dynamic BN models in two domains. First, dynamic Bayesian networks are used for dynamic gesture recognition. Then, temporal event networks are used for predicting HIV mutational pathways.

### *9.4.1 DBN: Gesture Recognition*

Dynamic Bayesian networks provide an alternative to HMMs for dynamic gesture recognition. They have the advantage of greater flexibility in terms of the structure of the models. Next we describe an application of dynamic Bayesian network classifiers for gesture recognition.

#### 9.4.1.1 Gesture Recognition with DBNCs

DBNC have been applied for recognizing different hand gestures oriented to command a mobile robot. A set of 9 different gestures were considered; a key frame for each type of gesture is illustrated in Fig. 9.6. Initially the hand of the person performing the gesture is detected and tracked using a camera and specialized vision software. A rectangle approximates the position of the hand in each image in the sequence, and from these rectangles a set of features is extracted which are the observations for the DBNCs.

The features include motion and posture information, including in total 7 attributes: (i) 3 features to describe motion, and (ii) 4 to describe posture. Motion features are $\Delta$area—or changes in the hand area—, $\Delta$x and $\Delta$y—or changes in hand position on the XY-plane of the image. The conjunction of these three attributes allow us to estimate hand motion in the Cartesian space, $XYZ$. Each one of these features takes only one of three possible values: $+$, $-$, 0, that indicate increment, decrement or no change, depending on the area and position of the hand in the previous image of the sequence.

Posture features named *form, right, above, and torso* describe hand orientation and spatial relations between the hand and other body parts, such as the face and torso. Hand orientation is represented by $form$. This feature is discretized into one of three values: $+$ if the hand is vertical, $-$ if the hand is horizontal, or 0 if the hand is leaning to the left or right over the $XY$ plane. $right$ indicates if the hand is to the right of the head, *above* if the hand is above the head, and $torso$ if the hand is in

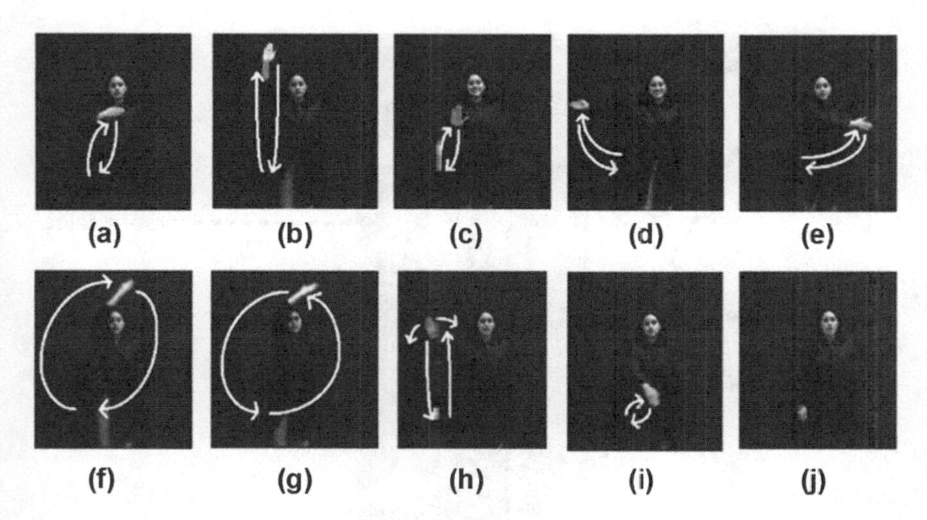

Fig. 9.6 Types of gestures considered in the experiments: **a** come, **b** attention, **c** stop, **d** right, **e** left, **f** turn left, **g** turn right, **h** waving-hand and **i** pointing; **j** initial and final position for each gesture

front of the torso. These three latter attributes take binary values, true or false, that represent if their corresponding condition is satisfied or not. An example of posture extraction in terms of these variables is depicted in Fig. 9.7.

As with HMMs, a DBNC is trained for each type of gesture; and for classification the probability of each model is evaluated and the one with the highest probability is selected as the recognized gesture.

#### 9.4.1.2 Experiments

Three experiments were done to compare classification and learning performances of DBNCs and HMMs. In the first experiment, gestures taken from the same person are used for recognition. In the second experiment, the generalization capabilities of the classifiers are evaluated by training and testing with gestures from different people. Experiment three considers gestures with variations on distance and rotation. Additionally, the DBNC and HMM models using only motion, and using motion and posture information were compared. The number of hidden states in each model was varied between 3 and 18.

For the first experiment, 50 executions of each type of gesture by one individual were recorded; 20 samples were used for training and 30 for testing. In the second experiment, the models learned for one person were evaluated with gestures executed by other 14 persons, 2 samples per person for each gesture class. For the experiment on distance variation, 15 samples were randomly extracted for each gesture performed at 2 m and 4 m, giving a test set of 30 samples per gesture. Similarly, for rotation variation, 15 random samples of each gesture at $+45°$ and $-45°$ were used.

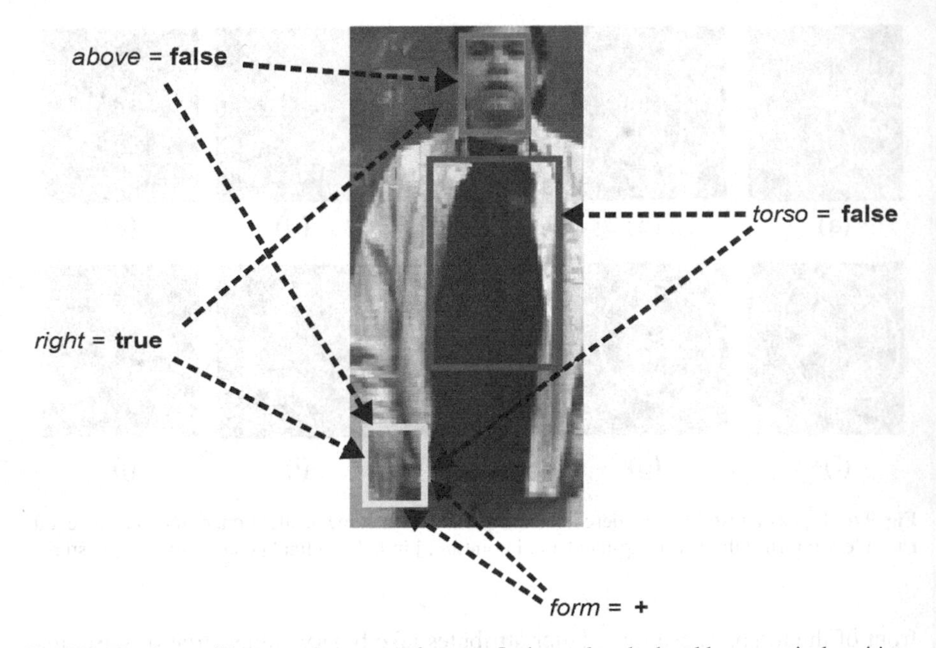


**Fig. 9.7** The image illustrates the posture features. It shows that the hand has a vertical position, below the head, to the right of the user and not over the user's torso. Thus, the attribute values are $above = false, right = true, torso = false$, and $form = +$

An important difference between the DBNCs and the HMMs is the number of parameters that are required for each model. The number of parameters to specify state observation distributions of HMMs with posture-motion features is 648 and with only motion data is 27. With DBNCs, parameters are 21 in the former case, and 12 in the latter case. This significant reduction in the number of parameters for the DBNC has an important impact in the training time, as well as in the classification accuracy when the number of training instances is low.

In the case of the experiments with the same person using motion and posture attributes, both HMMs and DBNCs obtain a very good performance, with recognition rates between 96 and 99%, depending on the number of states (best results with 12–15 states). DBNCs obtain slightly better recognition results, but with a significant reduction in training time, about ten times faster than HMMs.

For the experiments with multiple people, as expected the performance of both, HMMs and DBNCs decreases, to about 86% recognition with motion-posture attributes. If only motion attributes are used, the performance is in the order of 65%, about 20 points less than when incorporating posture! As with most classification problems, the selection of the appropriate set of attributes is critical.

In the case of the third experiments, variations in distance and orientation also have an impact in the recognition rates, with the second aspect having a greater effect. Varying the distance from the camera to the user between 2 and 4 m, reduces the recognition rate to about 90% (single user); if the orientation is varied in the interval between $+45°$ and $-45°$, the performance goes down to aprox. 70%. In both situations, the results with HMMs and DBNCs are similar in terms of performance.

Experimental results show the competitiveness in terms of recognition rates of DBNCs in comparison to standard HMMs in various issues in gesture recognition. Attribute factorization allows an important decrease on training time for a DBNC in comparison with equivalent HMMs, this allows online learning of gestures. Additionally, DBNCs require less training examples to achieve a similar performance as the corresponding HMMs, in particular when the number of attributes increases.

## 9.4.2 *TNBN: Predicting HIV Mutational Pathways*

In this section we explore the application of TNBNs to uncover the temporal relationships between drug resistance mutations of the HIV virus and antiretroviral drugs, unveiling possible mutational pathways and establishing their probabilistic-temporal sequence of appearance [8].

The human immunodeficiency virus (HIV) is one of the fastest evolving organisms on the planet. Its remarkable variation capability makes HIV able to escape from multiple evolutionary forces naturally or artificially acting on it, through the development and selection of adaptive mutations. This is the case of antiretroviral therapy (ART), a strong selective pressure acting on HIV that, under suboptimal conditions, readily selects for mutations that allow the virus to replicate even in the presence of highly potent antiretroviral drug combinations. In particular, we address the problem of finding mutation-mutation and drug-mutation associations in individuals receiving antiretroviral therapy. We have focused on protease inhibitors (PIs), a family of antiretroviral drugs widely used in modern antiretroviral therapy. The development of drug resistant viruses compromises HIV control, with a consequent further deterioration of the patient's immune system. Hence, there is interest in having a profound understanding of the dynamics of appearance of drug resistance mutations.

Historical data from HIV patients was used to learn a TNBN, and then this model was evaluated in terms of the discovered relationships according to a domain expert and its predictive power.

### 9.4.2.1 Data

Clinical data from 2373 patients with HIV subtype B was retrieved from the HIV Stanford Database (HIVDB) [11]. The isolates in the HIVDB were obtained from longitudinal treatment profiles reporting the evolution of mutations in individual sequences. For each patient, data consisted of an initial treatment (a combination of drugs) administered to the patient and a list of laboratory resistance tests at different times (in weeks). Each test included a list of the most frequent mutations in the viral population within the host at a specific time after the initiation of treatment. An

**Table 9.4** An example of the data. Patient $Pat_1$ with 3 temporal studies, and patient $Pat_2$ with two temporal studies

| Patient | Initial treatment | List of mutations | Time (weeks) |
|---|---|---|---|
| $Pat_1$ | LPV, FPV, RTV | L63P, L10I | 15 |
| | | V77I | 25 |
| | | I62V | 50 |
| $Pat_2$ | NFV, RTV, SQV | L10I | 25 |
| | | V77I | 45 |

example of the data is presented in Table 9.4. The number of studies available varied from 1 to 10 studies per patient history.

Antiretrovirals are usually classified according to the enzyme that they target. We focused on the viral protease, as this is the smallest of the viral enzymes in terms of number of aminoacids. Nine protease inhibitors were available at the time of this study, namely: Amprenavir (APV), Atazanavir (ATV), Darunavir (DRV), Lopinavir (LPV), Indinavir (IDV), Nelfinavir (NFV), Ritonavir (RTV), Tipranavir (TPV) and Saquinavir (SQV).

To test the ability of the model for predicting clinically relevant data, a subset of patients from the original dataset was selected, including individuals that received ART regimes including LPV, IDV, and SQV. Relevant major drug resistance mutations associated with the selected drugs were included. The mutations selected were: V32I, M46I, M46L, I47V, G48V, I54V, V82A, I84V, and L90M. Since we used a subset of drugs, the number of patients in the final dataset was reduced from the previous experiments to 300 patients.

#### 9.4.2.2 Model and Evaluation

A TNBN was learned from the reduced HIVDB with the learning algorithm described in Sect. 9.3. Two modifications to the original algorithm were made, one to measure the strength of the temporal-probabilistic relations, and another to vary the variable order given to the structure learning algorithm (K2), so the results are not biased by a particular predefined order.

In order to evaluate the models and to measure the statistical significance of edge strengths, non-parametric bootstrapping was used (obtaining several models). Two thresholds were defined for considering a relation as important. A strong relation was defined as one that appeared in at least 90% of the graphs, and a suggestive relation was defined as one that occurred with values between 70% and 90%. Since the approach for selecting the drugs and mutations is based on experts opinions, we used a more elaborate way to obtain the order for the K2 algorithm. For this experiment different orderings for the K2 algorithm were considered and the one with the highest predictive accuracy was selected.

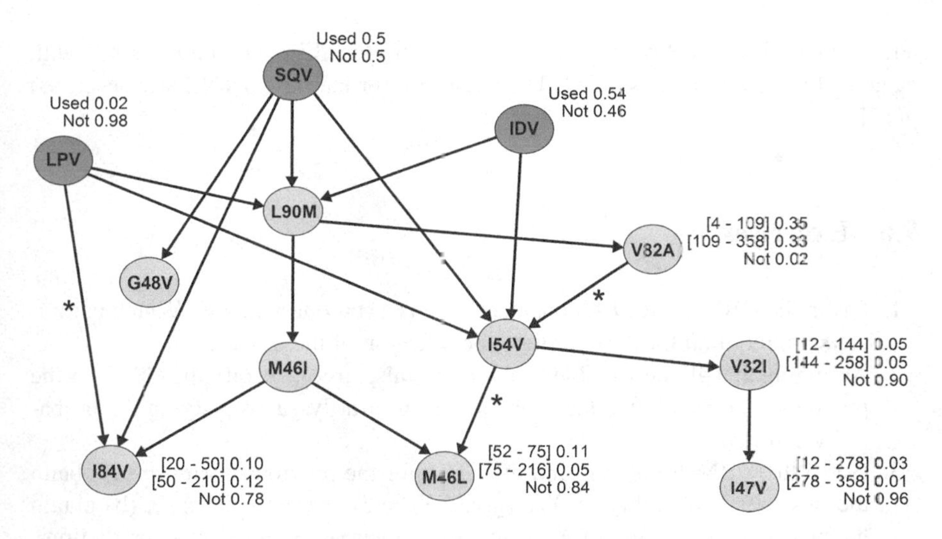


**Fig. 9.8** The learned TNBN model that depicts the temporal-probabilistic relations between a set of antiretroviral drugs (upper 3 nodes) and relevant HIV mutations (other nodes). An arc labeled with a * represents a strong relation. Only a few sets of intervals associated with their respective temporal nodes are shown

Figure 9.8 depicts the TNBN obtained. The green nodes represent the antiretroviral drugs, and the yellow nodes the mutations. For each mutation (temporal nodes), the associated temporal intervals with the probability of occurrence of the mutation in that interval are shown. Arcs that suggest a strong relation are marked with a $*$.

The model was able to predict clinically relevant associations between the chosen drugs and mutations. Indeed, a strong association between SQV, G48V, and I84V was readily predicted in the model, although no temporal associations were observed between the two mutations. All three drugs showed direct associations with L90M reflecting the fact that this mutation causes cross-resistance to many members of the PI family. Remarkably, the two possible mutational pathways for LPV resistance [8] were predicted:

- I54V $\rightarrow$ V32I $\rightarrow$ I47V
- L90M $\rightarrow$ M46IL $\rightarrow$ I84V

Whether the temporal order of mutations is relevant, still needs to be further evaluated. Also, the shared mutational pathway between IDV and LPV was observed, involving mutations L90M, M46IL, I54V, V82A and I84V.

## 9.5 Additional Reading

A comprehensive review of dynamic Bayesian networks is presented in [9]. References [4, 6] include techniques for learning DBNs. For more information on sampling techniques, see [6] for Markov chain Monte Carlo methods and [9] for particle filters. Temporal nodes Bayesian networks are introduced in [1], and extended with canonical temporal models in [5]. The algorithm for learning a TNBN is described in [7].

## 9.6 Exercises

1. Given the DBN in Fig. 9.1, consider that it is a stationary model. Which parameters are required for the complete specification of the model?
2. Assuming that all the variables are binary (false, true), specify the CPTs for the previous exercise (define the probabilities arbitrarily, just satisfying the probability axioms).
3. Given the DBN in Fig. 9.1, and the CPTs for the previous exercise: (a) obtain the posterior probability of $X_{t+1}$ given $X_t = S_t = true$ (filtering), (b) obtain the posterior probability of $X_{t+2}$ and $X_{t+3}$ given $X_t = S_t = true$ (prediction), (c) obtain the posterior probability of $X_{t+1}$ given $E_t = E_{t+1} = E_{t+2} = false$ (smoothing).
4. Repeat the previous problem using logic sampling with different number of samples (10, 20, ...).
5. Given the following data set for a dynamic process with four binary variables:

| Time | $X_1$ | $X_2$ | $X_3$ | $X_4$ |
|---|---|---|---|---|
| 1 | 0 | 0 | 0 | 0 |
| 2 | 0 | 1 | 1 | 1 |
| 3 | 0 | 1 | 0 | 1 |
| 4 | 0 | 0 | 1 | 1 |
| 5 | 0 | 0 | 0 | 0 |
| 6 | 0 | 1 | 1 | 0 |
| 7 | 1 | 1 | 0 | 1 |
| 8 | 0 | 0 | 0 | 1 |
| 9 | 0 | 1 | 0 | 0 |
| 10 | 0 | 1 | 1 | 0 |
| 11 | 0 | 1 | 0 | 0 |
| 12 | 0 | 0 | 0 | 1 |
| 13 | 0 | 1 | 0 | 1 |
| 14 | 0 | 0 | 1 | 0 |
| 15 | 0 | 1 | 0 | 0 |
| 16 | 0 | 0 | 1 | 0 |
| 17 | 0 | 1 | 0 | 0 |
| 18 | 0 | 1 | 0 | 1 |
| 19 | 1 | 1 | 1 | 0 |
| 20 | 0 | 1 | 1 | 1 |

Learn a DBN from this dataset including structure and parameters. Obtain first the base structure assuming has a tree structure, and then the transition network applying the PC algorithm.

6. Consider a robot that moves in a corridor (one dimension) from $X = 0$ to $X = 10$, with three possible actions: move left, move right and not move. Move left and right displace the robot half meter to the left or right, with Gaussian noise (zero mean, 0.1 m S.D.). In the corridor there are different marks every meter that the robot can detect with its camera and obtain its position with Gaussian noise (zero mean, 0.2 m S.D.). Specify a model for the localization ($X$) of the robot based on the actions and observations, including your assumptions.
7. Given the model of the previous problem, estimate the position of the robot, given that initially $X = 5$ and then it takes the following actions: (i) move right, (ii) move right, (iii) move left. Use particle filters with 10 samples, which are initially distributed uniformly in $X = [4.5, 5.5]$. Show the position of the particles after each cycle, and the estimated (mean) position of the robot.
8. Repeat the previous problem, given that the robot initially is *lost*, so the particles are distributed uniformly in $X = [0, 10]$.
9. Consider the TNBN in Fig. 9.5. Define the CPTs for all the variables according to the values/intervals shown in the figure. Specify the parameters according to your intuition (subjective estimates).
10. Considering the structure and parameters for the TNBN of the previous exercise, obtain the posterior probability of all the variables given the evidence $C = moderate$ using probabilistic inference (you can apply any of the inference techniques for BNs).
11. Repeat the previous problem considering as evidence $DP = dilated$. As the relative time of occurrence of this event is not known, consider the different possible scenarios. Which scenario will apply if we later find out that the pupils became dilated 4 time units after the accident?
12. Modify the inference (forward) and learning (Baum-Welch) algorithms for HMMs (see Chap. 5) so that they can be applied to dynamic Bayesian network classifiers.
13. *** Search for data sets in different dynamic domains to learn dynamic Bayesian models (see next exercises). For example, stock exchange data, weather data, medical longitudinal studies, etc. Determine, according to each application, which type of model, state based or event based, is more appropriate.
14. *** Develop a program that learns a DBN considering a two phase procedure: first learn the initial structure and parameters (for $t = 0$); then learn the transition structure and parameters (for $t = k + 1$ given $t = k$).
15. *** Develop a program that implements the $LIPS$ algorithm for learning TNBNs.

## References


1. Arroyo-Figueroa, G., Sucar, L.E.: A temporal Bayesian network for diagnosis and prediction. In: Proceedings of the 15th Conference on Uncertainty in Artificial Intelligence (UAI), pp. 13–20. Morgan-Kaufmann, San Mateo (1999)
2. Avilés-Arriaga, H.H., Sucar, L.E., Mendoza-Durán, C.E., Pineda-Cortés, L.A.: Comparison of dynamic naive Bayesian classifiers and hidden Markov models for gesture recognition. J. Appl. Res. Technol. **9**(1), 81–102 (2011)
3. Cooper, G.F., Herskovitz, E.: A Bayesian method for the induction of probabilistic networks from data. Mach. Learn. **9**(4), 309–348 (1992)
4. Friedman, N., Murphy, K., Russell, S.: Learning the structure of dynamic probabilistic networks. In: Proceedings of the 14th Conference on Uncertainty in Artificial (UAI), pp. 139–147. Morgan Kaufmann Publishers Inc. (1998)
5. Galán, S.F., Arroyo-Figueroa, G., Díez, F.J., Sucar, L.E.: Comparison of two types of event Bayesian networks: a case study. Appl. Artif. Intell. **21**(3), 185–209 (2007)
6. Ghahramani, Z.: Learning dynamic Bayesian networks. Lecture Notes in Computer Science, vol. 1387, pp. 168–197 (1998)
7. Hernández-Leal, P., González, J.A., Morales, E.F., Sucar, L.E.: Learning temporal nodes Bayesian networks. Int. J. Approx. Reason. **54**(8), 956–977 (2013)
8. Hernández-Leal, P., Rios-Flores, A., Ávila-Rios, S., Reyes-Terán, G., González, J.A., Fiedler-Cameras, L., Orihuela-Espina, F., Morales, E.F., Sucar, L.E.: Discovering HIV mutational pathways using temporal Bayesian networks. Artif. Intell. Med. **57**(3), 185–195 (2013)
9. Murphy, K.: Dynamic Bayesian networks: representation, inference and learning. Dissertation, University of California, Berkeley (2002)
10. Rabiner, R., Juang, B.: Fundamentals of Speech Recognition. Prentice Hall, New Jersey (1993)
11. Shafer, R.: Rationale and uses of a public HIV drug-resistance database. J. Infect. Dis. **194**(Supplement 1), S51–S58 (2006)

# Part III
# Decision Models

Probabilistic graphical models that include decisions are presented in this part. These models involve, besides random variables, decision variables and utilities; and their aim is to help decision-makers to take the *best* decisions under uncertainty. The first chapter is dedicated to models that have one or few decisions, including decision trees and influence diagrams. The other two chapters focus on sequential decision problems in which many decisions have to be taken over time: Markov decision processes and partially observable Markov decision processes.

# Chapter 10
# Decision Graphs

## 10.1 Introduction

The models that were covered in Part II have only random variables, so they can be used for estimating the posterior probability of a set of variables given some evidence; for example for classification, diagnosis or prediction. They can also provide the most probable combination of values for a subset of variables (most probable explanation) or the global probability of the model given some observations. However, they can not be directly used to make decisions

In this, and the following two chapters, we will present *decision models*, whose aim is to help the decision maker to choose the *best* decisions under uncertainty. We will consider that the best decisions are those that maximize the expected utility of an agent, given its current knowledge (evidence) and its objectives, under a decision-theoretic framework. These types of agents are known as *rational agents*.

After a brief introduction to decision theory, in this chapter we will describe two types of modeling techniques for problems with one or few decisions: decision trees and influence diagrams. As in the case of probabilistic models, these techniques take advantage of the dependency structure of the problem to have a more compact representation and a more efficient evaluation.



L. E. Sucar, *Probabilistic Graphical Models*, Advances in Computer Vision and Pattern Recognition, https://doi.org/10.1007/978-3-030-61943-5_10

## 10.2 Decision Theory

Decision Theory provides a normative framework for decision making under uncertainty. It is based on the concept of *rationality*, that is, an agent should try to maximize its utility or minimize its costs. This assumes that there is some way to assign utilities (usually a number, that can correspond to monetary value or any other scale) to the result of each alternative action, such that the best decision is the one that has the highest utility. In general, an agent is not sure about the results of each of its possible decisions, so it needs to take this into account when it estimates the value of each alternative. In decision theory we consider the *expected utility*, which makes an average of all the possible results of a decision, weighted by their probability. Thus, in a nutshell, a rational agent must select the decision that maximizes its expected utility.

Decision theory was initially developed in economics and operations research [14], but in recent years has attracted the attention of artificial intelligent (AI) researchers interested in understanding and building intelligent agents. These intelligent agents, such as robots, financial advisers, intelligent tutors, etc., must deal with similar problems such as those encountered in economics and operations research, but with two main differences. One difference has to do with the size of the problems, which in artificial intelligence tend to be much larger, in general, than in traditional applications in economics. The other main difference has to do with knowledge about the problem domain. In many AI applications a model is not known in advance, and could be difficult to obtain.

### 10.2.1 Fundamentals

The principles of decision theory were initially developed in the classic text by Von Neuman and Morgensten, *Theory of Games and Economic Behavior* [14]. They established a set of intuitive constraints that should guide the preferences of a rational agent, which are known as the axioms of utility theory. Before we list these axioms, we need to establish some notation. In a decision scenario there are four elements:

*Alternatives.* Are the choices that the agent has and are under his control. Each decision has at least two alternatives (e.g. to do or not do some action).

*Events.* Are produced by the environment or by other agents; they are outside of the agent's control. Each random event has at least two possible results, and although we do not know in advance which result will occur, we can assign a probability to each one.

*Outcomes.* Are the results of the combination of the agents decisions and the random events. Each possible outcome has a preference (utility) for the agent.

*Preferences.* These are established according to the agent's goals and objectives and are assigned by the agent to each possible outcome. They establish a value for the agent for each possible result of its decisions.

In utility theory, the different scenarios are called *lotteries*. In a lottery each possible outcome or *state*, $A$, has a certain probability, $p$, and an associated preference to the agent which is quantified by a real number, $U$. For instance, a lottery $L$ with two possible outcomes, $A$ with probability $p$, and $B$ with probability $1 - p$, will be denoted as:

$$L = [p, A; 1 - p, B]$$

If an agent prefers $A$ rather than $B$ it is written as $A \succ B$, and if it is indifferent between both outcomes it is denoted as $A \sim B$. In general a lottery can have any number of outcomes; an outcome can be an atomic state or another lottery.

Based on these concepts, we can define utility theory in an analogous way as probability theory, by establishing a set of reasonable constraints on the preferences for a rational agent; these are the **axioms of utility theory**:

*Order:* Given two states, an agent prefers one or the other or it is indifferent between them.

*Transitivity:* If an agent prefers outcome $A$ to $B$ and prefers $B$ to $C$, then it must prefer $A$ to $C$.

*Continuity:* If $A \succ B \succ C$, then there is some probability $p$ such that the agent is indifferent between getting $B$ with probability one, or the lottery $L = [p, A; 1 - p, C]$.

*Substitutability:* If an agent is indifferent between two lotteries $A$ and $B$, then the agent is indifferent between two more complex lotteries that are the same except that $B$ is substituted for $A$ in one of them.

*Monotonicity:* There are two lotteries that have the same outcomes, $A$ and $B$. If the agent prefers $A$, then it must prefer the lottery in which $A$ has higher probability.

*Decomposability:* Compound lotteries can be decomposed into simple ones using the rules of probability.

Then, the definition of a utility function follows from the axioms of utility.

**Utility Principle**: If an agent's preferences follow the axioms of utility, then there is a real-valued utility function $U$ such that:

1. $U(A) \succ U(B)$ if and only if the agent prefers $A$ over $B$,
2. $U(A) = U(B)$ if and only if the agent is indifferent between $A$ and $B$.

**Maximum Expected Utility Principle**: The utility of a lottery is the sum of the utilities of each outcome multiplied by its probability:

$$U[P_1, S_1; P_2, S_2; P_3, S_3; ...] = \sum_j P_j U_j$$

Based on this concept of a utility function, we can now define the expected utility (EU) of a certain decision $D$ taken by an agent, considering that there are $N$ possible results of this decision, each with probability $P$:

$$EU(D) = \sum_{j=1}^{N} P(result_j(D))U(result_j(D))$$

The principle of **Maximum Expected Utility** states that a rational agent should choose an action that maximizes its expected utility.

#### 10.2.1.1 Utility of Money

In many cases it seems natural to measure utility in monetary terms; the more money we make based on our decisions, the better. Thus, we can think of applying the maximum expected utility principle measuring utility in terms of its monetary value. But this is not as straight forward as it seems.

Suppose that you are participating in a game, such as those typical TV shows, and that you have already won one million dollars. The host of the game asks you if you want to keep what you have already won and finish your participation in the game, or continue to the next stage and gain \$3,000,000. Instead of asking some difficult question, the host will just flip a coin and if it lands on *heads* you will get three million, but if it lands on *tails* you will loose all the money you have already won. You have to make a decision with two options: (D1) keep the money you have already won, (D2) go to the next stage, with a possibility of winning three million (or loosing everything). What will your decision be?

Let's see what the principle of maximum expected utility will advise us if we measure utility in dollars (known as *Expected Monetary Value* or EMV). We calculate the EMV for both options:

D1: $\text{EMV(D1)} = 1 \times \$1,000,000 = \$1,000,000$
D2: $\text{EMV(D2)} = 0.5 \times 0 + 0.5 \times \$3,000,000 = \$1,500,000$

Thus, it seems that if we want to maximize the expected utility in terms of dollars we must take the bet. However, most of us would probably select to keep the one million that we already have and not take the risk of loosing it! What is the reasoning behind this? Are we not being *rational*?

The relation between utility and monetary value is not linear for most people; instead they have a logarithmic relation which denotes *risk aversion* (see Fig. 10.1). It is approximately linear for low values of money (for instance, if we have \$10 instead of \$1,000,000 on the line, we will probably go for the bet); but once we have a large amount of money (the amount will depend on each individual), the increase in utility given more money is not longer linear.

The utility–monetary value relation varies from person to person (and organizations) depending on their perception of risk; there are three basic types: risk aversion, risk neutral and risk seeking; these are depicted in Fig. 10.1. Thus, we have to transform the monetary value to utility according to this relation, before we calculate the expected utility.

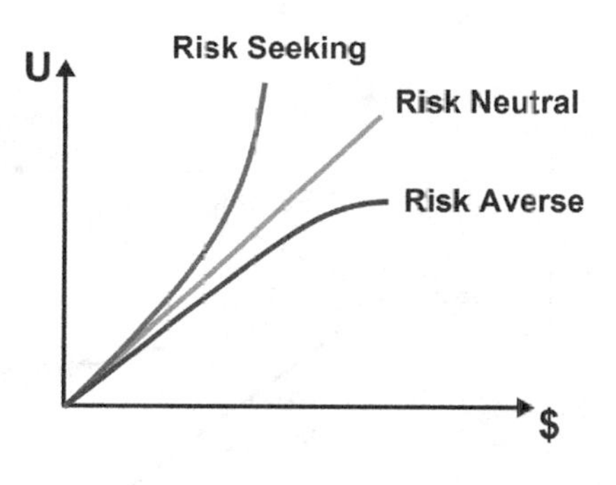


**Fig. 10.1** The graphs show typical relations between utility (U) and monetary value ($). Top: risk seeking, middle: neutral, bottom: risk averse

## 10.3 Decision Trees

Although it seems straight-forward to apply the principle of maximum expected utility to determine the best decision, as the decision problems become more complex, involving several decisions, events and possible outcomes, it is not as easy as it seems and a systematic approach is required to model and solve complex decision problems. One of the earliest modeling tools developed for solving decision problems are **decision trees** [2].

A decision tree is a graphical representation of a decision problem, which has three types of elements or nodes that represent the three basic components of a decision problem: decisions, uncertain events and results.

A *decision node* is depicted as a rectangle which has several *branches*, each branch represents each of the possible alternatives present at this decision node. At the end of each branch there could be another decision node, an event or a result.

An *event node* is depicted as a circle, and also has several branches, each branch represents one of the possible outcomes of this uncertain event. These outcomes correspond to all the possible results of this event, that is, they should be mutually exclusive and exhaustive. A probability value is assigned to each branch, such that the sum of the probabilities for all the branches is equal to one. At the end of each branch there could be another event node, a decision node or a result.

The *results* are annotated with the utility they express for the agent, and are usually at the end of each branch of the tree (the leaves).

Decision trees are usually drawn from left to right, with the root of the tree (a decision node) at the extreme left, and the leaves of the tree to the right. An example of a hypothetical decision problem (based on an example in [1]) is shown in Fig. 10.2. It represents an investment decision with 3 alternatives: (i) Stocks, (ii) Gold, and (iii) No investment. Assuming that the investment is for one year, if we invest in stock, depending on how the stock market behaves (uncertain event), we could gain $1000 or loose $300, both with equal probability. If we invest in Gold, we have to make another decision, to have insurance or not. If we get insurance, then we are sure to gain $200; otherwise we win or loose depending on if the price of gold is up, stable or down; this is represented as another event. Each possible outcome has a certain value and probability assigned, as shown in Fig. 10.2. What should the investor decide?

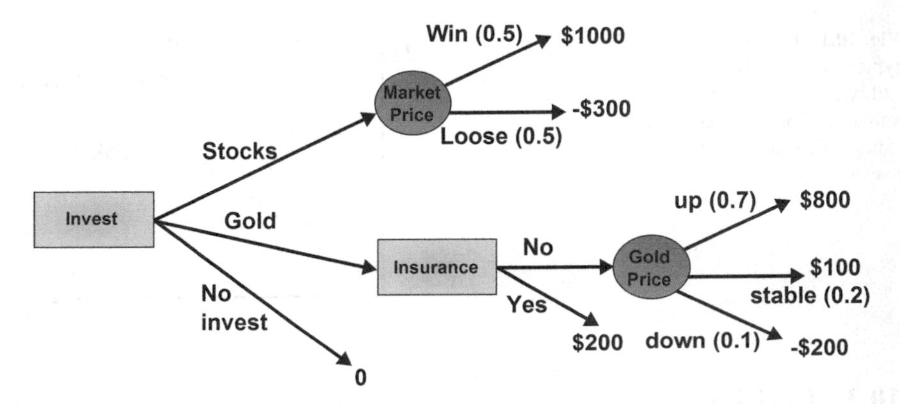


**Fig. 10.2** An example of a decision tree (see text for details)

To determine the best decision for each decision point, according to the maximum expected utility principle, we need to *evaluate* the decision tree. The evaluation of a decision tree consists in determining the values of both types of nodes, decision and event nodes. It is done from right to left, starting from any node that has only results for all its branches:

- The value of a decision node $D$ is the maximum value of all the branches that emanate from it:
$$V(D) = max_j U(result_j(D))$$
.
- The value of an event node $E$ is the expected value of all the branches that emanate from it, obtained as the weighted sum of the result values multiplied by their probabilities:
$$V(E) = \sum_j P(result_j(E))U(result_j(E))$$
.

Following this procedure we can evaluate the decision tree of Fig. 10.2:

Event 1 - Market Price: $V(E_1) = 1000 \times 0.5 - 300 \times 0.5 = 350$.
Event 2 - Gold Price: $V(E_2) = 800 \times 0.7 + 100 \times 0.2 - 200 \times 0.1 = 560$.
Decision 2 - Insurance: $V(D_2) = max(200, 560) = 560$ – No insurance.
Decision 1 - Investment: $V(D_1) = max(150, 560, 0) = 560$ – Invest in Gold.

Thus, in this case the best decisions are to invest in Gold without insurance.

Decision trees are a tool for modeling and solving sequential decision problems, as decisions have to be represented in sequence as in the previous example. However, the size of the tree (number of branches) grows exponentially with the number of decision and event nodes, so this representation is practical only for *small* problems. An alternative modeling tool is the *Influence Diagram* [5, 11], which provides a compact representation of a decision problem.

## 10.4 Influence Diagrams

Influence Diagrams (IDs) are a tool for solving decision problems that were introduced by Howard and Matheson [5] as an alternative to decision trees to simplify modeling and analysis. From another perspective, we can view IDs as an extension of Bayesian networks that incorporates decision and utility nodes. In the following sections we present an introduction to IDs, including their representation and basic inference techniques.

### *10.4.1 Modeling*

An influence diagram is a directed acyclic graph, $G$, that contains nodes that represent random, decision, and utility variables:

Random nodes ($X$): represent random variables as in BNs, with an associated CPT. These are represented as ovals.

Decision nodes ($D$): represent decisions to be made. The arcs pointing towards a decision node are *informational*; that is, it means that the random or decision node at the origin of the arc must be known before the decision is made. Decision nodes are represented as rectangles.

Utility nodes ($U$): represent the costs or utilities associated to the model. Associated to each utility node there is a function that maps each permutation of its parents to a utility value. Utility nodes are represented as diamonds. Utility nodes can be divided into *ordinary* utility nodes, whose parents are random and/or decision nodes; and *super-value* utility nodes, whose parents are ordinary utility nodes. Usually the super-value utility node is the (weighted) sum of the ordinary utility nodes.

There are three types of arcs in an ID:

Probabilistic: they indicate probabilistic dependencies, pointing towards random nodes.

Informational: they indicate information availability, pointing towards decision nodes. That is, $X \rightarrow D$ indicates that value of $X$ is known before the decision $D$ is taken.

Functional: they indicate functional dependency, pointing towards utility nodes.

An example of an ID is depicted in Fig. 10.3, which gives a simplified model for the problem of determining the location of a new airport, considering that the probability of accidents, the noise level and the estimated construction costs are the factors that directly affect the *utility*.

In an ID there must be a directed path in the underlying directed graph that includes all the decision nodes, indicating the order in which the decisions are made. This order induces a partition on the random variables in the ID, such that if there are

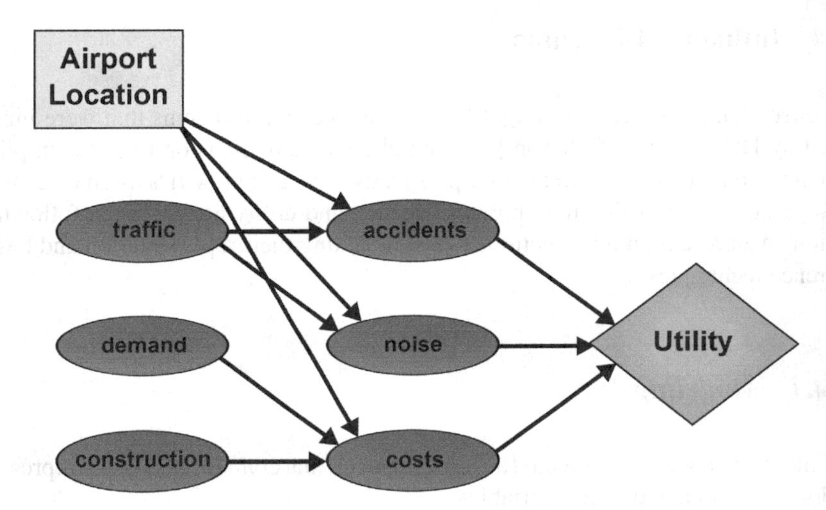


**Fig. 10.3** A simplified model of the airport location problem represented as a simple ID. The decision node represents the different options for the location of a new airport, and the utility node represents the utility (or cost) which depends on several factors, which in turn depend on other random variables

$n$ decision variables, the random variables are partitioned into $n+1$ subsets. Each subset, $\mathbf{R_i}$, contains all the random variables that are known before decision $D_i$ and unknown for previous decisions. Some of the algorithms for evaluating influence diagrams take advantage of these properties to make the evaluation more efficient.

IDs are used to aid a decision maker in finding the decisions that maximize its expected utility. That is, the goal in decision analysis is to find an *optimal policy*, $\pi = \{d_1, d_2, ..., d_n\}$, which selects the best decisions for each decision node to maximize the expected utility, $E_\pi(U)$. If there are several utility nodes, in general we consider that we have additive utility, so we will maximize the sum of these individual utilities:

$$E_\pi(U) = \sum_{u_i \in U} E_\pi(u_i) \tag{10.1}$$

The evaluation of influence diagrams to find the optimal decisiones is covered in the next section. Regarding learning influence diagrams, the parameters and dependency relations of the random nodes can be obtained using the techniques for Bayesian networks. How to define the utilities is an area known as *utility elicitation* [4] which is outside the scope of this book.

### *10.4.2 Evaluation*

Evaluating an influence diagram is finding the sequence of best decisions or optimal policy. First we will see how we can solve a simple influence diagram with only one decision; then we will cover general techniques for solving IDs.

We define a *simple* influence diagram as one that has a single decision node and a single utility node. For this case we can simple apply BN inference techniques to obtain the optimal policy following this algorithm:

1. For all $d_i \in D$:

   a. Set $D = d_i$.
   b. Instantiate all the known random variables.
   c. Propagate the probabilities as in a BN.
   d. Obtain the expected value of the utility node, $U$.

2. Select the decision, $d_k$, that maximizes $U$.

For more complex decision problems in which there are several decision nodes, the previous algorithm becomes impractical. In general, there are three main types of approaches for solving IDs:

- Transform the ID to a decision tree and apply standard solution techniques for decision trees.
- Solve the ID directly by variable elimination, applying a series of transformations to the graph.
- Transform the ID to a Bayesian network and use BN inference techniques.

Next we describe the three alternatives.

#### 10.4.2.1 Transformation to a Decision Tree

One way to solve an influence diagram is to convert it into a decision tree [5] and then solve the decision tree using the approach described before. To be able to do this, the influence diagram must be a *decision tree network*, where all the ancestors of any decision node are parents of that node. If the ID does not satisfy the previous assumption, it can be transformed into a decision tree network through a sequence of *arc reversals*. It is also assumed that the ID is a directed acyclic graph and that the decision nodes are ordered.

To invert an arc between nodes $X$ and $Y$ it is required that there is no other trajectory between these nodes. Then the arc $X \rightarrow Y$ is inverted and each node inherits the parents of the other node. The modified CPTs are obtained from the previous CPTs applying Bayes rule.

For example, consider that in the original ID, there is an arc $X \rightarrow Y, X$ has a parent $A$, and $Y$ has another parent $B$, then: $Pa(X) = A$ and $Pa(Y) = (X, B)$. The CPTs in the original structure are $P(X \mid A)$ and $P(Y \mid X, B)$. After the arc reversal, $X \leftarrow Y, X$ and $Y$ inherent each others parents, so now $Pa(X) = (A, B, Y)$ and $Pa(Y) = (A, B)$. Then we need to obtain the new CPTs: $P(X \mid A, B, Y)$ and $P(Y \mid A, B)$. This is done following the next steps:

- Obtain the joint distribution:

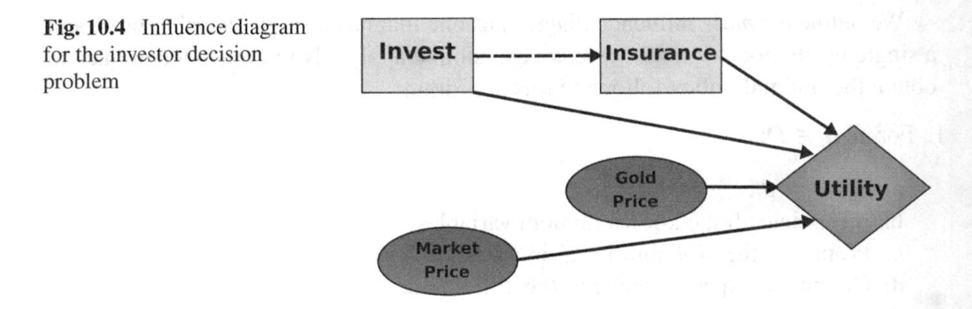


**Fig. 10.4** Influence diagram for the investor decision problem

$$P(X, Y \mid A, B) = P(X \mid A)P(Y \mid X, B)$$

- Calculate $P(Y \mid A, B)$ via marginalization:

$$P(Y \mid A, B) = \sum_{X} P(X, Y \mid A, B)$$

- Calculate $P(X \mid A, B, Y)$ via de definition of conditional probability:

$$P(X \mid A, B, Y) = \frac{P(X, Y \mid A, B)}{P(Y \mid A, B)}$$

Once the ID is a decision tree network, a decision tree is constructed from the decision tree network as follows. First, define a total order $<$ over the set of random nodes, decision nodes and utility nodes, $X \cup D \cup U$, satisfying the following conditions:

1. $X < Y$ if $Y$ is a utility node.
2. $X < Y$ if there is a directed path from $X$ to $Y$.
3. $X < Y$ if $X$ is a decision node and there is no directed path from $Y$ to $X$.

Then construct a decision tree by considering the variables one by one in the order. Each layer in the decision tree corresponds to a variable.

For example, consider the investor decision problem represented as an influence diagram in Fig. 10.4. In this case it is already a decision tree network. From the graph, we can deduce the following order of the variables in the investor ID:

$$Invest < MarketPrice < Insurance < GoldPrice < Utility$$

Based on this order, expanding each variable (drawing an arc for each value) and joining the corresponding variables, we obtain the decision tree of Fig. 10.2.

The major problem with this approach is that the resultant decision tree tends to be large. The depth of the decision tree obtained from an ID is equal to the number of variables in the influence diagram. Thus, the size of the decision tree is exponential in the number of variables in the influence diagram.

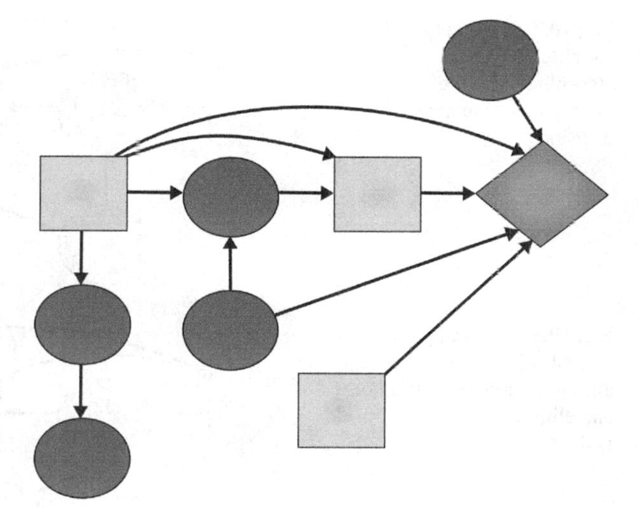

**Fig. 10.5** An example of the variable elimination algorithm: initial influence diagram

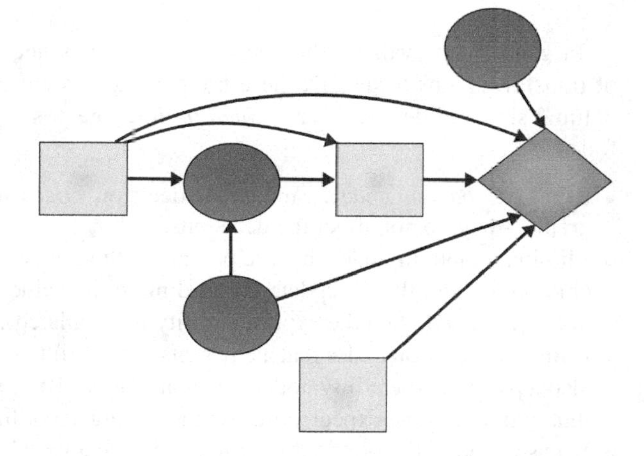

**Fig. 10.6** An example of the variable elimination algorithm: after eliminating two barren nodes

#### 10.4.2.2 Variable Elimination

The variable elimination algorithm [11] is based on evaluating the decision nodes one by one according to a certain order. Decision nodes that have been evaluated can be eliminated from the model, and this process continues until all the decision nodes have been evaluated. To apply this technique the influence diagram must be *regular*; that is, it satisfies the following conditions:

1. The structure of the ID is a directed acyclic graph.
2. The utility nodes do not have successors.
3. There is a directed path in the underlying directed graph that includes all the decision nodes, indicating the order in which the decisions are made.

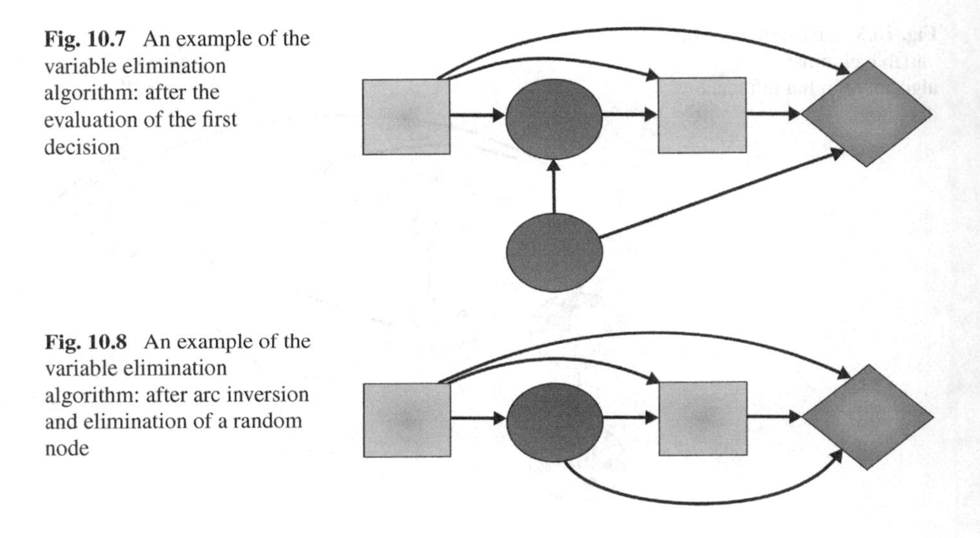

**Fig. 10.7** An example of the variable elimination algorithm: after the evaluation of the first decision

**Fig. 10.8** An example of the variable elimination algorithm: after arc inversion and elimination of a random node

In general, to evaluate the decision nodes, it is necessary to perform a series of transformations to the ID; these transformations are guaranteed to preserve the optimal series of decisions or *optimal policy*. The possible transformations are the following:

- Eliminate *barren* nodes, random or decision nodes that are leaf nodes in the graph—they do not affect the decisions.
- Eliminate random nodes that are parents of the utility node and do not have other children—the utility is updated according to the value of the node (if the node is not instantiated, then the expected utility is calculated).
- Eliminate decision nodes that are parents of the utility node where their parents are also a parent to the utility node—evaluate the decision node and take the decision that maximizes the expected utility; modifying accordingly the utility function.
- In case none of the previous operations can be applied, invert an arc between two random variables (see Sect. 10.4.2.1).

We illustrate graphically the variable elimination algorithm with an example. Assume that we initially have the ID depicted in Fig. 10.5. We can eliminate the barren random node at the bottom left; when this node is eliminated, its parent also becomes a barren node, which is also eliminated and we obtain the ID shown in Fig. 10.6.

Next we eliminate the random node on the top, parent of the utility node *absorbing* it in the utility, and we then evaluate the first decision, that is, the decision node at the bottom, resulting in the ID seen in Fig. 10.7.

Then we invert the arc between the two remaining random nodes, so we can eliminate the bottom random node, obtaining the model shown in Fig. 10.8. From this graph we can evaluate the decision node on the right, then eliminate the remaining random node, and finally evaluate the decision node to the left.

**Table 10.1** Probability table (marginals) for market price

| Value | Win | Loose |
|---|---|---|
| Probability | 0.5 | 0.5 |

**Table 10.2** Probability table (marginals) for gold price

| Value | Up | Stable | Down |
|---|---|---|---|
| Probability | 0.7 | 0.2 | 0.1 |

**Table 10.3** Utility table (in dollars) for the investor decision problem. An "*" indicates any value

| Invest | Insurance | Market price | Gold price | Value |
|---|---|---|---|---|
| No | * | * | * | 0 |
| Gold | Yes | * | * | 200 |
| Stocks | * | Win | * | 1000 |
| Stocks | * | Loose | * | −300 |
| Gold | No | * | Up | 800 |
| Gold | No | * | Stable | 100 |
| Gold | No | * | Down | −200 |

**Example**

We now present a more detailed, quantitative example of the elimination procedure considering the investor decision problem, Fig. 10.4. Before we start, we need to specify the parameters for this model, that is the CPTs for Market Price and Gold Price, and the Utility table, which are depicted in Tables 10.1, 10.2, 10.3, respectively.

We now apply the variable elimination process to this example:

1. Eliminate the random variable "Gold Price", updating the utility table:

| Invest | Insurance | Market Price | Value |
|---|---|---|---|
| No | * | * | 0 |
| Gold | Yes | * | 200 |
| Stocks | * | Win | 1000 |
| Stocks | * | Loose | -300 |
| Gold | No | * | 560 |

2. Eliminate the random variable "Market Price" and update the utility table:

| Invest | Insurance | Value |
|---|---|---|
| No | * | 0 |
| Gold | Yes | 200 |
| Stocks | * | 350 |
| Gold | No | 560 |

3. Evaluate the decision "Insurance" by taking the value (decision) of maximum utility. From the previous table, $U(Insurance = Yes) = 200$ and $U(Insurance = No) = 560$, so the optimal decision is $Insurance = No$. We also update the table (keeping the best option).

| Invest | Value |
|---|---|
| No | 0 |
| Stocks | 350 |
| Gold | 560 |

4. Finally we evaluate the decision "Invest". From the last table it is obvious that the optimal choice is $Invest = Gold$.

The resulting decisions are, of course, the same we obtained when we evaluated the corresponding decision tree in Sect. 11.4.

Next we describe an alternative evaluation technique for IDs that takes advantage of the efficient inference algorithms that have been developed for BNs.

#### 10.4.2.3 Transformation to a BN

The idea to reduce an ID to a BN was originally proposed by Cooper [3]. To transform an ID to a BN, the basic idea is to transform decision and utility nodes to random nodes, with an associated probability distribution. A decision node is converted to a discrete random variable by considering each decision, $d_i$, as a value for this variable, and using a uniform distribution as a CPT (a decision node has no parents as all incoming arcs are informational). A utility node is transformed to a binary random variable by *normalizing* the utility function so it is in the range from 0 to 1, that is:

$$P(u_i = 1 \mid Pa(u_i)) = val(Pa(u_i))/maximum(val(Pa(u_i))) \tag{10.2}$$

where $Pa(u_i)$ are the parents of the utility node in the ID, and *val* is the value assigned to each combination of values of the parent nodes. As it is a binary variable, the probability of $u_i = 0$ is just the complement: $P(u_i = 0 \mid Pa(u_i)) = 1 - P(u_i = 1 \mid Pa(u_i))$.

After the previous transformation, and considering a single utility node, the problem of finding the optimal policy is reduced to finding the values of the decision nodes that maximize the probability of the utility node: $P(u = 1 \mid D, R)$, where $D$ is the set of decision nodes, and $R$ is the set of the other random variables in the ID. This probability can be computed using standard inference techniques for BNs; however, it will require an exponential number of inference steps, one for each permutation of $D$.

Given that in a *regular* ID the decision nodes are ordered, a more efficient evaluation can be done by evaluating the decisions in (inverse) order [12]. That is, instead of maximizing $P(u = 1 \mid D, R)$, we maximize $P(D_j \mid u = 1, R)$. We can recursively optimize each decision node, $D_j$, starting from the last decision, continuing with the previous decision, and so on, until we reach the first decision. This gives a much more efficient evaluation procedure. Additional improvements have been proposed to the previous algorithms based on *decomposable* IDs (see the additional readings section).

Traditional techniques for solving IDs make two important assumptions:

Total ordering: all the decisions follow a total ordering according to a directed path in the graph.
Non forgetting: all previous observations are remembered for future decisions.

These assumptions limit the applicability of IDs to some domains, in particular temporal problems that involve several decisions at different times. For instance, in medical decision making, a total ordering of the decisions is an unrealistic assumption since there are situations in which the decision maker does not know in advance what decision should be made first to maximize the expected utility. For a system that evolves over a large period of time, the number of observations grows linearly with the passing of time, so the non-forgetting requirement implies that the size of policies grows exponentially.

## 10.4.3 Extensions

### 10.4.3.1 Limited Memory Influence Diagrams

In order to avoid the previous limitations of IDs, Lauritzen and Nilsson [7] proposed limited-memory influence diagrams (LIMIDs) as an extension of influence diagrams. The term limited-memory reflects the property that a variable known when making a decision is not necessarily remembered when making a posterior decision. Eliminating some variables reduces the complexity of the model so it is solvable with a computer, although at the price of obtaining a sub-optimal policy.

### 10.4.3.2 Dynamic Decision Networks

Another extension is applied for sequential decision problems, that involve several decisions over time. As with BNs, we can consider decision problems in which a series of decisions have to be made at different time intervals; this type of problem is known as a *sequential decision problem*. A sequential decision problem can be modeled as a *dynamic decision network* (DDN)—also known as a *dynamic influence diagram*—which can be seen as an extension of a DBN, with additional decision and utility nodes for each time step, see Fig. 10.9.

In principle, we can evaluate a DDN in the same way as an ID, considering that the decisions have to be ordered in time. That is, each decision node $D_t$, has incoming informational arcs from all previous decision nodes, $D_{t-1}$, $D_{t-2}$, etc. However, as the number of time epochs increases, the complexity increases and can become computationally intractable. Additionally, in some applications we do not know in advance the number of decision epochs, so in principle there might be an *infinite* number of decisions.

DDNs are closely related to *Markov decision processes* which are the topic of the next chapter.

## 10.5 Applications

We present two applications: (i) an influence diagram for the treatment of non-small cell lung cancer, and (ii) a dynamic decision network for assisting elderly or handicapped persons in washing their hands.

### 10.5.1 Decision Support System for Lung Cancer

M. Luque and J. Dıez [8] developed a decision support system to determine the most efficient selection of tests and therapy for the treatment of non-small cell lung cancer.

Lung cancer is a very frequent tumor in the world and the leading cause of cancer death. It is classified into two major types: small-cell lung cancer (SCLC) and non-small cell lung cancer (NSCLC). A correct assessment in an early stage of NSCLC and an appropriate selection of patients for surgery is very important to avoid dangerous and unnecessary surgery in bad prognosis patients. Determining whether malignant mediastinal lymph nodes are present or absent is the most important prognostic factor in patients with NSCLC to determine the therapeutic strategy. There are different techniques to study the mediastinum and potentially malignant lymph

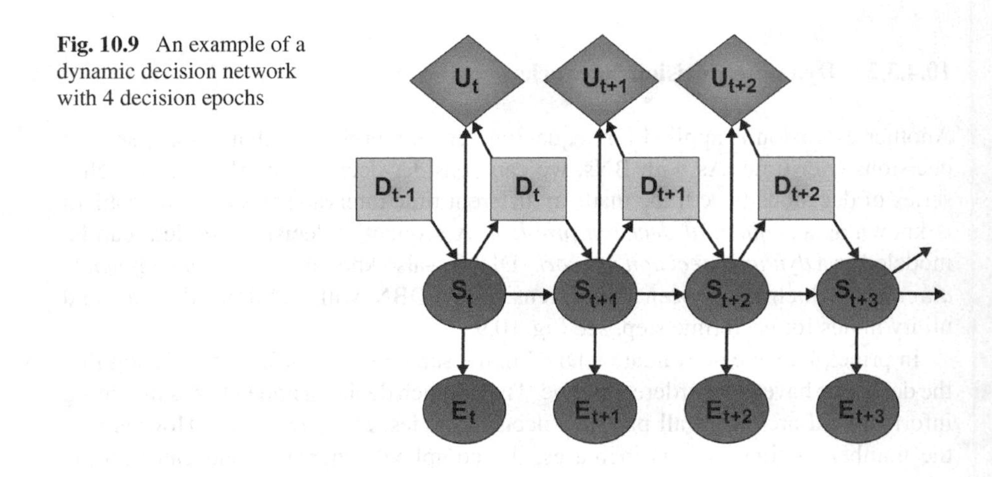


**Fig. 10.9** An example of a dynamic decision network with 4 decision epochs

nodes. Because of this variety of available tests, each one having pros and cons, it is difficult to decide which technologies should be used and in what order.

As an aid to select the *best* strategy for the treatment of NSCLC an influence diagram was constructed named *Mediastined*, which is shown in Fig. 10.10. The model includes three types of variables: random, decision and utility.

The random variables correspond to possible causes and risk factors of the disease, including also the results of laboratory tests:

N factor: indicates whether or not the cancer has reached lymph nodes. N0 indicates that there is no cancer in any lymph nodes; while N1, N2 and N3 indicate there is cancer in different sites. In the model these have been grouped into two states, N0-N1which means the cancer is operable, and N2-N3 which means it is inoperable (this is the variable N2-N3 in the ID, with values present/absent).
CT-scan: the result of computer tomography (positive or negative).
TBNA: the result of transbronchial needle aspiration (positive, negative, or no-result).
PET: the result of positron emission tomography (positive, negative, or no-result).
EBUS: the result of endobronchial ultrasound (positive, negative, or no-result).
EUS: the result of endoscopic ultrasound (positive, negative, or no-result).
MED: the result of mediastinoscopy—a process in which a small incision is made in the neck—(positive, negative, or no-result).
MED-Sv: indicates if the patient has survived mediastinsocopy (yes or no).

Note the variables that represent different tests have three posible values: positive, if cancer is detected; negative, if cancer is not detected; and no-result, if the test was not performed; except for CT-scan which is always performed. An important aspect of this model are the decisions to perform or not a test, as these tests have risks.

The decision variables are whether to perform or not certain test and what possible treatments to apply to the patient:

Treatment: indicates the set of possible treatments: thoracotomy, chemotherapy, and palliative.
Decision-TBNA: decision to perform the TBNA test (yes or no).
Decision-PET: decision to perform the PET test (yes or no).
Decision-MED: decision to perform the MED test (yes or no).
Decision-EBUS-EUS: decision to perform EBUS or EUS or both or none (four possible values).

In this model there are several utility nodes, divided into *ordinary utility nodes* and *super value utility nodes*. The ordinary utility nodes represent the decision maker's preferences (the medical specialists in this case). Several of these utilities are measured in "QALEs", which measure the quality-adjusted life expectancy. The ordinary utility nodes are:

Survivors-QALE: QALE of the survivors to medical tests and treatment.
TBNA-Morbidity: morbidity represented in QALEs due to TBNA.
EBUS-Morbidity: morbidity represented in QALEs due to EBUS.

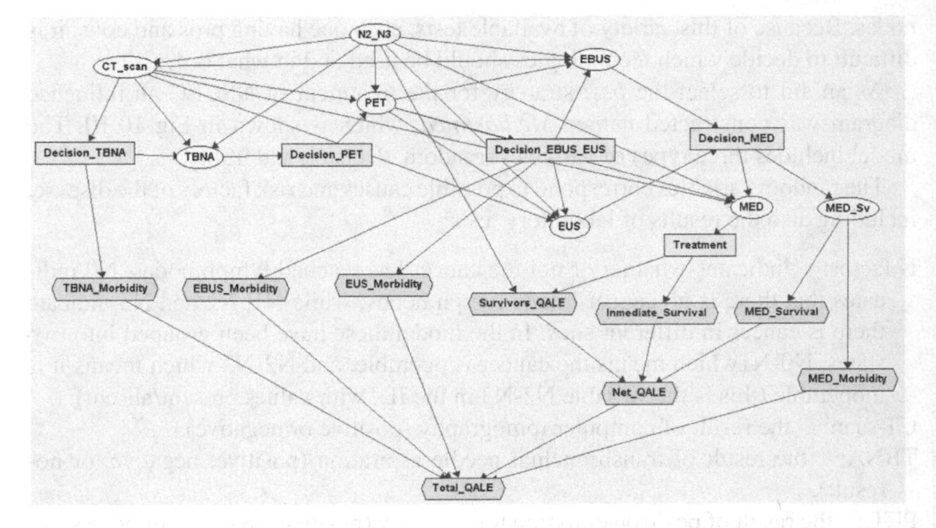


**Fig. 10.10** Influence diagram for the treatment of non-small cell lung cancer. See text for a description of all the variables in the model

EUS-Morbidity: morbidity represented in QALEs due to EUS.
Med-Morbidity: morbidity represented in QALEs due to mediastenoscopy.
Med-Survival: indicates if the patient survived the mediastenoscopy.
Immediate-Survival: probability of survival to the treatment.

The super value utility nodes combine the ordinary utility nodes, so at the end there is a single utility node which is maximized (Total-QALE):

Net-QALE: combines via a product the Survivors-QALE, Med-Survival and Immediate-Survival; it is a product as it multiplies the QALE of survival to the tests (except mediastenoscopy) by the probability of survival to mediastenoscopy and the probability of survival to the treatment.
Total-QALE: combines the nodes Net-QALE, TBNA-Morbidity, EBUS-Morbidity, EUS-Morbidity and Med-Morbidity via addition. A sum is used as the morbidities (that will take negative values of QALE) tend to reduce the Net-QALE.

The parameters of the model, CPTs for the random variables and values for the utility nodes, were elicited from domain experts. There are 61 independent parameters: one parameter is the prior probability of metastasis (N2-N3), 46 parameters are the conditional probabilities of the tests, one is the probability of surviving MED, 10 are utilities measured in QALEs, and 3 parameters are utilities that represent the probability of surviving the treatment.

A strategy (policy) that maximized the effectiveness (Total-QALE) was obtained by solving the ID. The optimal policy according to the model is the following:

- Decision-TBNA: perform TBNA only when the CT-scan is positive.
- Decision-PET: perform always PET.

- Decision-EBUS-EUS: perform EBUS if there is a negative CT-scan and a positive PET; or if there is a positive CT-scan and contrary results in the PET and TBNA. It recommends never to perform the EUS.
- Decision-MED: never perform mediastenoscopy.
- Treatment: apply chemotherapy if the last performed test is positive; otherwise, it is better to apply thoracotomy.

The previous strategy was presented to an expert. The expert agreed with most parts of the policy, except on the part if the tests EBUS, EUS and MED should be performed. He agreed in the case the CT-scan is positive; however, if the CT-scan is negative, he would vary the decisions after performing the TBNA: (i) if the result is positive, do not perform additional tests; (ii) otherwise, perform EBUS instead of PET, and if EBUS is negative perform mediastenoscopy. The expert commented that the system does not take into account that performing the PET causes a delay and forces the patient to go again to the hospital a different day (something not represented in the model).

The previous model does not take into account the economic costs of the diagnostic tests and the treatments, which in medical decision making cannot be ignored. One way to take costs into account is to transform the problem into a multi-objective optimization problem. Another is to combine benefits (QALEs) and costs into a single variable to optimize, known as the net benefit (NE). NE can be estimated as effectiveness (Total-QALE) minus the costs (C) multiplied by the *net effectiveness* (medical benefit minus economic costs). A second ID was developed that takes into account the economic costs, see [8] for details.

## 10.5.2 Decision-Theoretic Caregiver

The objective of the *caregiver* is to guide a person in completing a task using an adequate selection of prompts. We consider the particular task of *cleaning one's hands*. The system acts as a caregiver that guides an elderly or handicapped person in performing this task correctly [9].

The relevant objects in the wash stand environment are: soap, faucet, and towel. The system detects the behavior of the user when interacting with these objects. Then it chooses an action (we use audible prompts) to guide the user to complete a task, or it may simply say nothing (null action) if the user is performing the sequence of steps required correctly.

### 10.5.2.1 Model

A dynamic decision network (DDN) is used to model the user behavior and make the optimal decisions at each time step based on the user's behavior (observations) and the objectives of the system (utilities). As optimal actions might involve evaluating

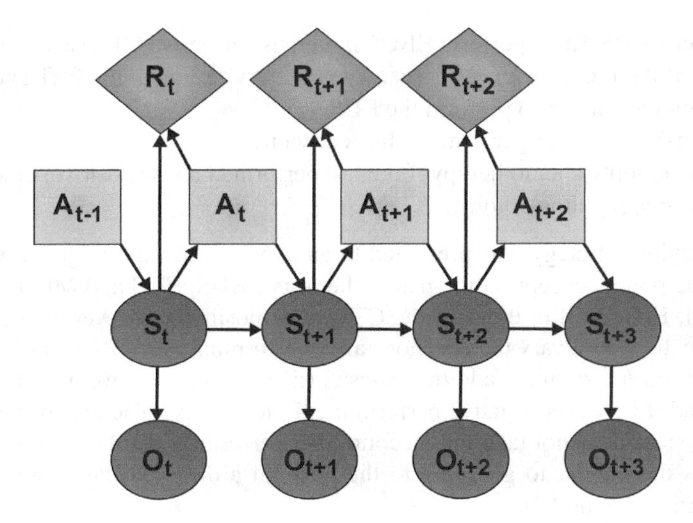


**Fig. 10.11** A 4 stage dynamic decision network that models the caregiver scenario. $S_t$ represents the activity of the user, $O_t$ corresponds to the observed activity, $A_t$ is the action selected, and $R_t$ the immediate reward

the model *many* steps in advance until the task of washing the hands is completed, a compromise between optimality and efficiency is obtained by analyzing $k$ steps in advance (lookahead). For this, the optimal decisions for $k$ decision nodes are obtained by solving the DDN with one of the techniques presented previously in this chapter. In this case the DDN was transformed to a DBN and solved using BN inference.

The DDN for this scenario is illustrated in Fig. 10.11. In this model, the state variables ($S$) represent the activity of the user at each time step, which are not directly observable. The observation nodes ($O$) represent the information obtained from a vision system used to recognize the activity being performed by the user. The action nodes ($A$) correspond to the different actions (prompts) that can be selected by the controller at each time. Finally, the rewards ($R$) represent the immediate reward that depends on the current state and preferred action. These elements of the model are described in detail below.

**States**. The state space is characterized by the activities (hand gestures) carried out by a person. In this case, the state variable has 6 possible values: $s_1$ = *opening the faucet*, $s_2$ = *closing the faucet*, $s_3$ = *using the soap*, $s_4$ = *drying the hands*, $s_5$ = *taking the towel* and $s_6$ = *washing the hands*.

**Observations**. These correspond to the information obtained by a visual gesture recognition system [9] that tries to recognize the activity performed by the person while washing their hands. The observation values are the same as the state values.

**Actions**. Actions are audible prompts to help the person complete the task. There are 8 actions that correspond to the possible prompts considered by the system: $a_1$ = *open the faucet*, $a_2$ = *close the faucet*, $a_3$ = *put soap on the hands*, $a_4$ = *wash the hands*, $a_5$ = *take the towel*, $a_6$ = *dry the hands*, $a_7$ = *Null* and $a_8$ = *call for help*.

**Rewards**. Rewards are associated with the preferences of the different actions selected by the system. In the caregiver setting, we must consider the prompts, clarity of the prompts and user's response. Three different reward values were used: $+3$ indicates a preference, $-3$ indicates a penalty, and $-6$ is used for selecting the action *call for help*. The idea is that asking for help should be the last option.

Additionally, the model requires two conditional probability tables: the transition function, $P(S_{t+1} \mid S_t)$, and the observation function, $P(O_t \mid S_t)$:

**Transition function**. The transition function defines the probability of the next state (next gesture) given the current state and action. In this setting this is predictable, with some degree of uncertainty. The transition functions were defined subjectively.

**Observation function**. In this case, the observation function consists of the probability of the observed gesture given the state, that is the *actual* gesture. It can be easily obtained from the confidence (confusion matrix) of the gesture recognition system.

It is assumed that the model is time invariant, that is that the previous CPTs do not change over time.

#### 10.5.2.2 Evaluation

The model was evaluated in terms of its: (i) sensitivity relative to the number of stages or lookahead, (ii) efficiency in terms of the time required to solve the model for selecting the next action, and (iii) performance, comparing the actions selected by the system with a human caregiver.

The DDN was solved with 2 to 11 time stages, and the actions selected under different scenarios were compared for the different model sizes. The expected utility increases as the lookahead is increased, tending to stabilize after 6 or 7 stages. However, the selected actions do not vary after a lookahead of 4, so this value was selected.

The model with 4 time stages (4 decision nodes) was evaluated in terms of its response time, and it could be solved in about 3 seconds with a standard personal computer. Thus, this number of stages provides a good compromise between performance and efficiency.

To evaluate the optimality (or near optimality) of the actions selected by the system, its decisions were compared to those of a human performing the same task. A preliminary evaluation was done with normal persons simulating that they had problems washing their hands. Ten adults participated in the experiment, divided in two groups, 5 each (test and control). The first group was guided to complete the task of washing their hands by verbal prompts given by the system, the control group was guided by verbal instructions given by a human assistant. The aspects evaluated in a questionnaire given to each participant were: (i) clarity of the prompt, (ii) detail of the prompt, and (iii) effectiveness of the system. Detail of the prompt refers to the level of specificity. Clarity considers the user's ease of understanding the message. Effectiveness evaluates the system's guidance for the successful completion of the task.

The results obtained are summarized in Table 10.4. The results indicate a small advantage when the best prompt is selected by a human, however the difference between the system and human controller is not considerable. Two aspects show a small difference (0.6 or less), and one shows a more significant difference (detail of the prompt). This last aspect has to do with the verbal phrase recorded to be associated with each prompt, which could be easily improved; and not so much with the decisions of the system.

## 10.6 Additional Reading

An Introduction to Decision Theory by Martin Peterson is a good first introduction to decision theory [10]. The original reference on influence diagrams is the book by Howard and Matheson [5]. The book by Jensen includes decision networks [6]. Decision trees are described in [2]. The elimination algorithm for evaluating influence diagrams is presented in [11]; the alternative evaluation technique based on a transformation to a BN is described in [3]. Limited memory influence diagrams are introduced in [7] and extended to dynamic models in [13].

## 10.7 Exercises

1. Define a function for utility in terms of monetary value considering risk aversion, such that the optimal decision for the example of Sect. 10.2.1.1 is to keep the money (D1) when calculating the expected utility for both possible decisions.
2. Consider that you have $100,000 dollars in savings, and have the opportunity to invest half of this in stocks, with the possibility of a return of 50% (you gain $25,000 ) with probability 0.5, and the possibility of a lose of 50% (you loose $25,000 ) with probability 0.5. What will be the optimal decision between invest or no invest if you are (a) risk averse, (b) risk seeking, or (c) neutral?
3. Consider the decision tree in Fig. 10.2. The futures on the stock market have changed, and now the potential gain has increased to $3000. Also, the insurance

**Table 10.4** Results obtained in the caregiver setting, where a person simulating memory problems is guided to complete the activity of cleaning their hands. It compares the decisions of a human assistant and the decision-theoretic system based on a DDN. The evaluation scale is from 1 (worst) to 5 (best)

| | Human | System |
|---|---|---|
| Clarity | 4.4 | 3.9 |
| Detail | 4.6 | 3.6 |
| Effectiveness | 4.2 | 3.6 |

price for gold has increased to \$600. After applying these changes to the decision tree, reevaluate it. Have the decisions changed with respect to the original example?
4. Repeat the previous problem based on the ID representation and applying variable elimination.
5. Define the required CPTs for the influence diagram of Fig. 10.3. Consider two possible airport locations, $Loc_A$ and $Loc_B$, and that all the random variables are binary: $traffic = \{low, high\}$, $demand = \{no, yes\}$, $construction = \{simple, complex\}$, $accidents = \{very-low, low\}$, $noise=\{low, medium\}$, $costs = \{medium, high\}$. Define the parameters according to your intuition and following the axioms of probability.
6. For the ID of Fig. 10.3 define a utility function in terms of the accidents, noise and costs, using the same values as in the previous exercise: (a) Define it as a mathematical function, $U = f(accidents, noise, costs)$. (b) Define it as a table.
7. Based on the parameters and utilities of the previous two exercises, evaluate the ID of Fig. 10.3 by calculating the utility function for each possible location, considering there is no evidence. Which is the best location according to the model?
8. Repeat the previous exercise for different scenarios, for instance: $traffic = low, demand = no, construction = simple$; and $traffic = high, demand= yes, construction = complex$. Does the optimal location change under the different scenarios?
9. Transform the the influence diagram of Fig. 10.3 to a decision tree.
10. Obtain the optimal decision for the DT of the previous problem, using the same parameters of Problems 5 and 6. (a) Without evidence. (b) With evidence.
11. Consider the problem of deciding whether to take an umbrella according to the weather forecast. There are two decisions for this problem: watch the weather forecast (no,yes) and take the umbrella (no, yes); and one random variable: $weather = \{sunny, light-rain, heavy-rain\}$. Model this problem as a decision tree, establishing costs / utilities for the different scenarios.
12. Define an influence diagram for the umbrella decision problem of the previous exercise.
13. *** Develop a program that transforms an ID to a BN. Then use probabilistic inference to evaluate the BN. Apply it to solve the airport location model using the parameters and utilities defined in the previous exercises.
14. *** Develop a program that transforms an ID to a DT. Apply it to solve the airport location model.
15. *** Investigate how to implement LIMIDs, and apply them to a complex decision problem.

## References


1. Borrás, R.: Análisis de Incertidumbre y Riesgo para la Toma de Decisiones. Comunidad Morelos, Orizaba, Mexico (2001) (In Spanish)
2. Cole, S., Rowley, J.: Revisiting decision trees. Manag. Decis. **33**(8), 46–50 (1995)
3. Cooper, G.: A method for using belief networks as influence diagrams. In: Proceedings of the Twelfth Conference on Uncertainty in Artificial Intelligence (UAI), pp. 55–63 (1988)
4. González-Ortega, J., Radovic, V., Ríos-Insua, D.: Utility Elicitation. In: Dias, L., Morton, A., Quigley, J. (eds.) Elicitation. International Series in Operations Research & Management Science, vol. 261. Springer, Cham (2018)
5. Howard, R., Matheson, J.: Influence Diagrams. In: Howard, R., Matheson, J. (eds.) Readings on the Principles and Applications of Decision Analysis. Strategic Decisions Group, Menlo Park (1984)
6. Jensen, F.V.: Bayesian Networks and Decision Graphs. Springer, New York (2001)
7. Lauritzen, S., Nilsson, D.: Representing and solving decision problems with limited information. Manage. Sci. **47**, 1235–1251 (2001)
8. Luque-Gallego, M.: Probabilistic graphical models for decision making in medicine. Ph.D. Thesis, Universidad Nacional de Educación a Distancia, Spain (2009)
9. Montero, A., Sucar, L.E., Martínez, M.: Decision-theoretic assistants based on contextual gesture recognition. In: Gupta, A. et al. (eds.) Advances in Healthcare Informatics and Analytics, Annals of Information Systems, vol. 19, pp. 157–185 (2016)
10. Peterson, M.: An Introduction to Decision Theory. Cambridge University Press, Cambridge (2009)
11. Shachter, R.D.: Evaluating influence diagrams. Oper. Res. **34**(6), 871–882 (1986)
12. Shachter, R.D., Peot, M.: Decision making using probabilistic inference methods. In: Proceedings of the Eight Conference on Uncertainty in Artificial Intelligence (UAI), pp. 276–283 (1992)
13. Van Gerven, M.A.J., Díez, F.J., Taal, B.G., Lucas, P.J.F.: Selecting treatment strategies with dynamic limited-memory influence diagrams. Artif. Intell. Med. **40**(3), 171–186 (2007)
14. Von Neumann, J., Morgenstern, O.: Theory of Games and Economic Behavior. Princeton University Press, Princeton (1944)

# Chapter 11
# Markov Decision Processes

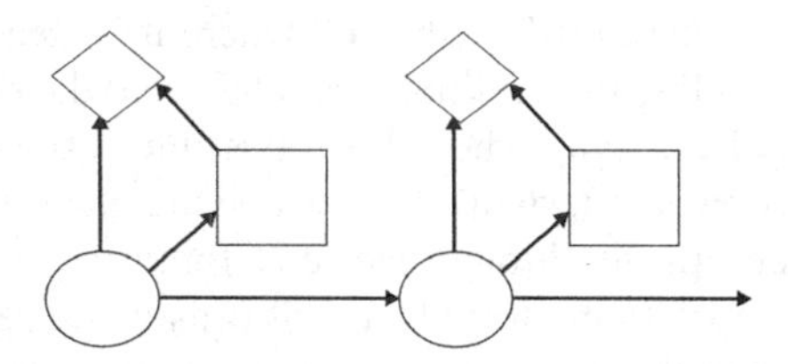

## 11.1 Introduction

In this chapter we will see how to solve sequential decision problems; that is, those that involve a series of decisions over time. It is assumed that the decision agent is rational, so the objective is to maximize the expected utility for the long term. Considering that there is uncertainty in the results of the agent's decisions, these type of problems can be modeled as Markov decision processes (MDPs). By solving the MDP model we obtain what is known as a *policy*, which indicates to the agent which action to select at each time step based on its current state; the optimal policy is the one that selects the actions so that the expected value is maximized.

We will first formalize the MDP model, and then present two standard ways to solve it: value iteration and policy iteration. Although the complexity of solving an MDP is quadratic in terms of the number of state–actions, it could still become impractical in terms of memory and time when the number of state–actions is too large. Factored MDPs provide a representation based on graphical models to solve very large MDPs. Other alternatives to solve complex problems are: *abstraction*, in which the number of states is reduced by grouping similar states; and *decomposition*, in which a large problem is decomposed in several smaller problems.

Partially observable MDPs (POMDPs), represent not only uncertainty in the results of the actions but also in the state; these are the topic of the next chapter.



L. E. Sucar, *Probabilistic Graphical Models*, Advances in Computer Vision and Pattern Recognition, https://doi.org/10.1007/978-3-030-61943-5_11

## 11.2 Modeling

A Markov decision process (MDP) [16] models a sequential decision problem, in which a system evolves over time and is controlled by an agent. The system dynamics are governed by a probabilistic transition function $\Phi$ that maps states **S** and actions **A** to new states **S'**. At each time, an agent receives a reward $R$ that depends on the current state $s$ and the applied action $a$. By solving an MDP representation of the problem, we obtain a recommendation strategy or *policy* that maximizes the expected reward over time and that also deals with the uncertainty of the effects of an action.

For example, consider an agent (a simulated robot) that *lives* in a grid world, so the state of the robot is determined by the cell where it is; see Fig. 11.1. The robot wants to go to the goal (cell with a smiling face) and avoid the obstacles and dangers (filled cells and a forbidden sign). The robot's possible actions are to move to the neighboring cells (up, down, left, right). We assume that the robot receives a certain immediate reward when it passes through each cell, for instance $+100$ when it arrives to the goal, $-100$ if it goes to the forbidden cell (this could represent a dangerous place), and $-1$ for all the other cells (this will motivate the robot to find the shortest route to the goal).

Consider that there is uncertainty in the result of each action taken by the robot. For example, if the selected action is *up* the robot goes to the upper cell with a probability of 0.8 and with probability of 0.2 to other cells (in the example in Fig. 11.1, it will have a probability of 0.1 off staying in the same cell and of 0.1 off going the right). This is illustrated with the width of the arrows in Fig. 11.1. Having defined the states, actions, rewards and transition function, we can model this problem as an MDP.

The objective of the robot is to go to the goal cell as fast as possible and avoid the dangers. This will be achieved by solving the MDP that represents this problem, and maximizing the expected reward.[1] The solution will provide the agent with a policy,

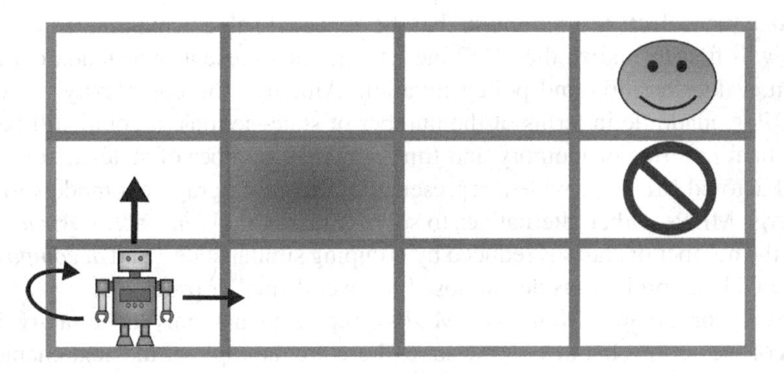

**Fig. 11.1** A robot in the grid world. Each cell represents the possible states of the robot, with a smiling face for the goal and a forbidden sign for danger. The robot is shown in a cell with the width of the arrows illustrating the probability for the next state given the action *up*

[1] This assumes that the defined reward function correctly models the desired objective.

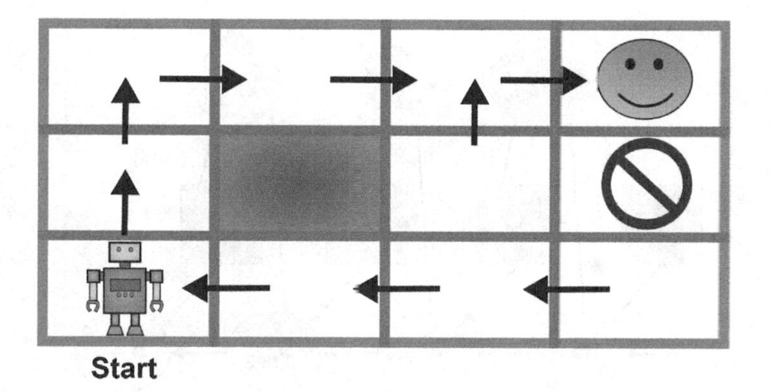


**Fig. 11.2** The policy resulting from solving the MDP model is illustrated, with an arrow for each cell indicating the optimal action

that is, what is the best action to perform in each state, as illustrated graphically (by the arrows) for this small example in Fig. 11.2.

Formally, an MDP is a tuple $M =< S, A, \Phi, R >$, where $S$ is a finite set of states $\{s_1, \ldots, s_n\}$. $A$ is a finite set of actions $\{a_1, \ldots, a_m\}$. $\Phi : A \times S \times S \rightarrow [0, 1]$ is the state transition function specified as a probability distribution. The probability of reaching state $s'$ by performing action $a$ in state $s$ is written as $\Phi(a, s, s')$. $R : S \times A \rightarrow \Re$ is the reward function. $R(s, a)$ is the reward that the agent receives if it takes action $a$ in state $s$.

A graphical model of an MDP is shown in Fig. 11.3. As its represented in the figure, it is in general assumed that the process is Markovian (the future state only depends on the present state) and stationary (the parameters remain the same overtime).

Markov property:

$$P(S_{t+1} \mid S_t, A_t, S_{t-1}, A_{t-1}, ...) = P(S_{t+1} \mid S_t, A_t), \forall S, A \tag{11.1}$$

$$P(R_t \mid S_t, A_t, S_{t-1}, A_{t-1}, ...) = P(R_t \mid S_t, A_t), \forall R, S, A \tag{11.2}$$

Stationary property:

$$P(S_{t+1} \mid S_t, A_t) = P(S_{t+2} \mid S_{t+1}, A_{t+1}), \forall t \tag{11.3}$$

$$P(R_t \mid S_t, A_t) = P(R_{t+1} \mid S_{t+1}, A_{t+1}), \forall t \tag{11.4}$$

Depending on how much into the future (horizon) we consider there are two main types of MDPs: (i) finite horizon and (ii) infinite horizon. Finite horizon problems consider that there exists a fixed, predetermined number of time steps for which we want to maximize the expected reward (or minimize the cost). For example, consider an investor that buys or sells actions each day (time step) and wants to maximize his profit for a year (horizon). Infinite horizon problems do not have a

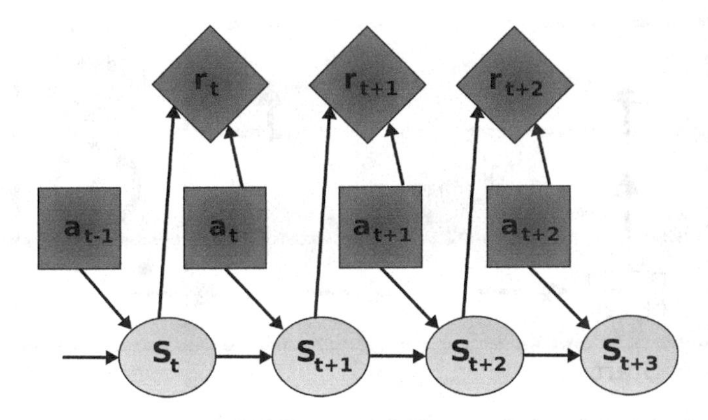


**Fig. 11.3** A graphical model representation of an MDP with four time steps. $s_t$ represents the state at time $t$, $a_t$ the action, and $r_t$ the reward

fixed, predetermined number of time steps, these could vary and in principle could be infinite. This is the case for the robot planning problem in general, as initially the number of movements (actions) that the robot will require to reach its goal or goals is unknown.

In the rest of this chapter we will focus on infinite horizon problems, as these are more common in practice. They also have the advantage that under certain conditions the optimal policy is stationary; that is, it only depends on the state and not on the time step.

A *policy*, $\pi$, for an MDP is a function $\pi : S \rightarrow A$ that specifies for each state, $s_i$, the action to be executed, $a_i$. Given a certain policy, the expected accumulated reward for a certain state, $s$, is known as the *value* for that state according to the policy, $V^\pi(s)$; it can be calculated using the following recursive equation:

$$V^\pi(s) = R(s,a) + \sum_{s' \in \mathbf{S}} \Phi(a,s,s')V^\pi(s') \tag{11.5}$$

where $R(s,a)$ represents the immediate reward given action $a$, and $\sum_{s' \in \mathbf{S}} \Phi(a,s,s')$ $V^\pi(s')$ is the expected value of the next states according to the chosen policy.

For the infinite horizon case, a parameter known as the *discount factor*, $0 \leq \gamma < 1$, is included so that the sum converges. This parameter can be interpreted as giving more value to the rewards obtained at the present time than those obtained in the future.[2]

Including the discount factor, the value function is written as:

$$V^\pi(s) = R(s,a) + \gamma \sum_{s' \in \mathbf{S}} \Phi(a,s,s')V^\pi(s') \tag{11.6}$$

[2]This has a clear interpretation in the case of financial investments, related to the inflation or interest rates. For other applications there usually isn't an easy way to determine the discount factor, and in general, a value close to one, such as 0.9, is used.

What is desired is to find the policy that maximizes the expected reward; that is, the policy that gives the highest value for all states. For the discounted infinite-horizon case with any given discount factor $\gamma$, there is a policy $\pi^*$ that is optimal regardless of the starting state and that satisfies what is known as the *Bellman* equation [2]:

$$V^{\pi}(s) = max_a\{R(s,a) + \gamma \sum_{s' \in \mathbf{S}} \Phi(a,s,s')V^{\pi}(s')\} \quad (11.7)$$

The policy that maximizes the previous equation is then the optimal policy, $\pi^*$:

$$\pi^*(s) = argmax_a\{R(s,a) + \gamma \sum_{s' \in \mathbf{S}} \Phi(a,s,s')V^{\pi}(s')\} \quad (11.8)$$

The *Bellman* equation is a recursive equation that can not be solved directly. However, there are several methods to solve it efficiently; these will be covered in the next section.

## 11.3 Evaluation

There are three basic methods for solving an MDP and finding an optimal policy: (a) value iteration, (b) policy iteration, and (c) linear programming [16]. The first two techniques solve the problem iteratively, improving an initial value function or policy, respectively. The third one transforms the problem to a linear program which can then be solved using standard optimization techniques such as the simplex method. We will cover the first two approaches for more on the third one see the additional reading section.

### *11.3.1 Value Iteration*

Value iteration consists in iteratively estimating the value for each state, $s$, based on Bellman's equation. Note that this is actually a set of $N$ equations, one for each state, $s_1, s_2, \ldots, s_N$. It starts by assigning an initial value to each state; usually this value is the immediate reward for that state. That is, at iteration 0, the $\mathbf{V_0}(s) = R(a, s)$. Then these estimates of the values are improved in each iteration by maximizing using the *Bellman* equation. The process is terminated when the value for all states *converges*, this is when the difference between the values in the previous and current iterations is less than a predefined threshold. The actions selected in the last iteration correspond to the optimal policy. The method is shown in Algorithm 11.1.

The time complexity of the algorithm is quadratic in terms of the number of state–actions, per iteration.

**Algorithm 11.1** The Value Iteration Algorithm.

1: $\forall_s V_0(s) = R(s, a)$ {Initialization}
2: $t = 1$
3: **repeat**
4: $\forall_s V_t(s) = max_a\{R(s, a) + \gamma \sum_{s' \in S} \Phi(a, s, s')V_{t-1}(s')\}$ {Iterative improvement}
5: **until** $\forall_s \mid V_t(s) - V_{t-1}(s) \mid < \epsilon$
6: $\pi^*(s) = argmax_a\{R(s, a) + \gamma \sum_{s' \in S} \Phi(a, s, s')V_t(s')\}$ {Obtain optimal policy}

Usually the policy converges before the values converge; this means that there is no change in the policy even if the value has not converged yet. This gives rise to the second approach, policy iteration.

### *11.3.2 Policy Iteration*

Policy iteration starts by selecting a random, initial policy (if we have certain domain knowledge, this can be used to seed the initial policy). Then the policy is iteratively improved by selecting the action for each state that increases the most the expected value. The algorithm terminates when the policy converges, that is, the policy does not change from the previous iteration. The method is shown is Algorithm 11.2.

**Algorithm 11.2** The Policy Iteration Algorithm.

1: $\pi_0 : \forall_s a_0(s) = a_k$ {Initialize the policy}
2: $\forall_s V_0(s) = R(s, a)$ {Initialize the values}
3: $t = 1$
4: **repeat**
5: {Iterative improvement}
6: $\forall_s V_t^{\pi_{t-1}}(s) = \{R(s, a) + \gamma \sum_{s' \in S} \Phi(a, s, s')V_{t-1}(s')\}$ {Calculate values for the current policy}
7: $\forall_s \pi_t(s) = argmax_a\{R(s, a) + \gamma \sum_{s' \in S} \Phi(a, s, s')V_t(s')\}$ {Iterative improvement}
8: **until** $\pi_t = \pi_{t-1}$

Policy iteration tends to converge in fewer iterations than value iteration, however the computational cost of each iteration is higher, as the values have to be updated.

### *11.3.3 Complexity Analysis*

Value iteration and policy iteration have been shown to perform in polynomial time for fixed $\gamma$, but value iteration can take a number of iterations proportional to $1/(1 - \gamma)log(1/(1 - \gamma))$ in the worst case [10].

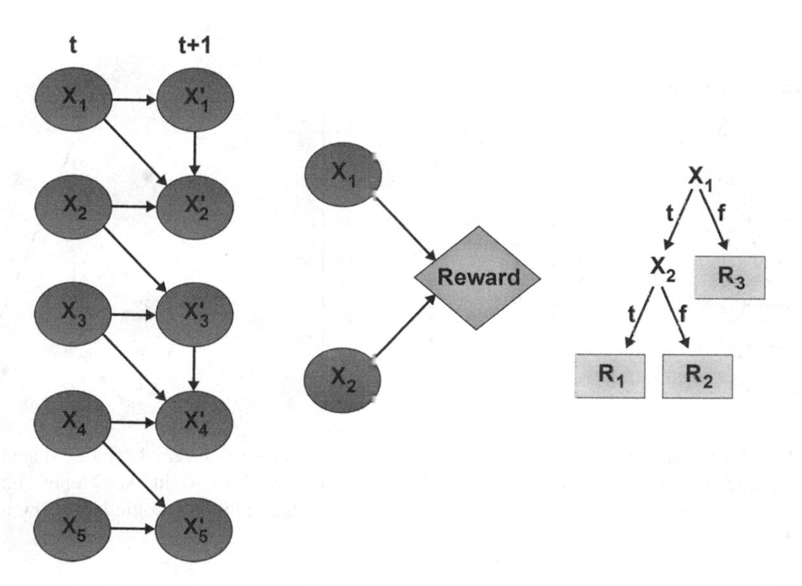


**Fig. 11.4** Left: A DBN with 5 state variables that represents the transition function for one action. Center: Influence diagram denoting a reward function. Right: Structured conditional reward (CR) represented as a binary decision tree

Solving *small* MDPs with the previous algorithms is very efficient; however it becomes difficult when the state–actions space is very large. Consider, for instance, a problem with 10, 000 states and 10 actions, which is common in applications such as robot navigation. In this case the space required to store the transition table will be $10,000 \times 10,000 \times 10 = 10^9$; and updating the value function will require in the order of $10^8$ operations per iteration. So solving very large MDPs could become problematic even with current computer technology. An alternative is to decompose the state space and take advantage of the independence relations to reduce the memory and computation requirements, using a graphical model-based representation of MDPs known as *Factored* MDPs.

## 11.4 Factored MDPs

In a factored MDP, the set of states is described via a set of random variables $\mathbf{X} = \{X_1, \ldots, X_n\}$, where each $X_i$ takes on values in some finite domain $Dom(X_i)$. A state $s$ defines a value $x_i \in Dom(X_i)$ for each variable $X_i$. The transition model and reward function can become exponentially large if they are explicitly represented as matrices, however, the frameworks of dynamic Bayesian networks (see Chap. 7) and decision trees [13] give us the tools to describe the transition model and the reward function concisely.

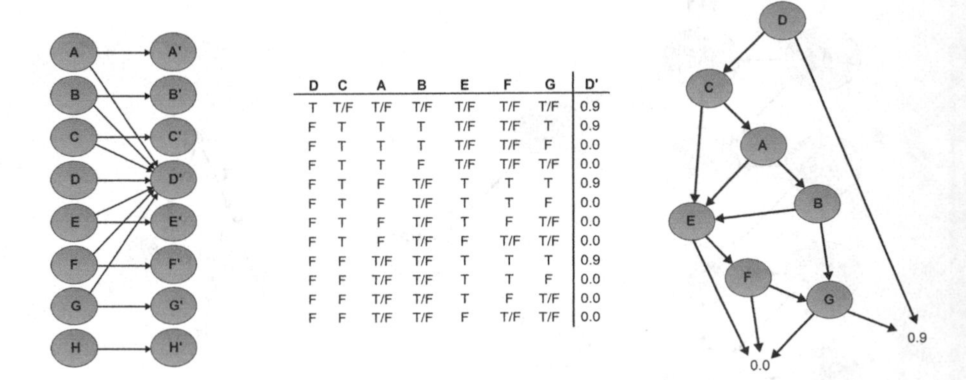


| D | C | A | B | E | F | G | D' |
|---|---|---|---|---|---|---|---|
| T | T/F | T/F | T/F | T/F | T/F | T/F | 0.9 |
| F | T | T | T | T/F | T/F | T | 0.9 |
| F | T | T | T | T/F | T/F | F | 0.0 |
| F | T | T | F | T/F | T/F | T/F | 0.0 |
| F | T | F | T/F | T | T | T | 0.9 |
| F | T | F | T/F | T | T | F | 0.0 |
| F | T | F | T/F | T | F | T/F | 0.0 |
| F | T | F | T/F | F | T/F | T/F | 0.0 |
| F | F | T/F | T/F | T | T | T | 0.9 |
| F | F | T/F | T/F | T | T | F | 0.0 |
| F | F | T/F | T/F | T | F | T/F | 0.0 |
| F | F | T/F | T/F | F | T/F | T/F | 0.0 |

**Fig. 11.5** An example of a CPT represented as an algebraic decision diagram. Left: two stage DBN representing the transition function. Center: CPT for one of the variables. Right: ADD representation for the CPT—for each node, the right arrow corresponds to the value $T$ and the left arrow to the value $F$

Let $X_i$ denote a variable at the current time and $X_i'$ the variable at the next step. The transition function for each action, $a$, is represented as a two–stage dynamic Bayesian network, that is a two–layer directed acyclic graph $G_T$ whose nodes are $\{X_1, \ldots, X_n, X_1', \ldots, X_n'\}$; see Fig. 11.4 (left). Each node $X_i'$ is associated with a *conditional probability distribution* $P_\Phi(X_i' \mid Parents(X_i'))$, which is usually represented by a matrix (*conditional probability table*) or more compactly by a decision tree. The transition probability $\Phi(a, s_i, s_i')$ is then defined to be $\Pi_i P_\Phi(X_i' \mid \mathbf{u_i})$ where $\mathbf{u_i}$ represents the values of the variables in $Parents(X_i')$.[3]

The reward associated with a state often depends only on the values of certain features of the state. The relationship between rewards and state variables can be represented with value nodes in an influence diagrams, as shown in Fig. 11.4 (center). The conditional reward table (CRT) for such a node is a table that associates a reward with every combination of values for its parents in the graph. This table is exponential in the number of relevant variables. Although in the worst case the CRT will take exponential space to store the reward function, in many cases the reward function exhibits structure allowing it to be represented compactly using decision trees or graphs, as shown in Fig. 11.4 (right).

In many cases, the conditional probability tables (CPTs) in the DBN exhibit a certain structure; in particular some probability values tend to be repeated many times (such as a zero probability). Taking advantage of these properties, the representation can be compacted even more by representing a CPT as a tree or a graph, such that repeated probability values appear only once in the leaves of these graphs. A particular presentation that is very efficient is an *algebraic decision diagram* or ADD. An example of a CPT represented as an ADD is depicted in Fig. 11.5.

[3]In this case, *synchronic* arcs (that connect variables in the same time step) are not considered for time $t$, and could be included or not at time $t + 1$.

The representation of the transition functions in MDPs as two-stage DBNs, and the reward function as a DT with the further reduction of these based on trees or ADDs, implies, in many cases, huge savings in memory for storing very large MDPs. An example of this will be shown in the applications section of this chapter.

Additionally, based on this compact representation, very efficient versions of the value and policy iteration algorithms have been developed that also reduce the computational time required to solve complex MDP models. An example of this is the SPUDD algorithm [8].

Further reduction in computational complexity can be achieved using other techniques, such as *abstraction* and *decomposition*, which are summarized below.

### 11.4.1 Abstraction

The idea of abstraction is to reduce the state space by creating an abstract model where states with similar features are grouped together [9].

*Equivalent states* are those that have the same transition and reward functions; these can be grouped together without altering the original model, so that the optimal policy will be the same for the reduced model. However, the state space reduction achieved by only joining equivalent states is in general not significant; further reductions can be achieved by grouping *similar* states; this results in approximate models and creates a trade-off between the precision of the model (the resulting policy) and its complexity.

Different strategies have been used to create reduced, approximate models. One is to partition the state space into a set of blocks such that each block is stable; that is, it preserves the same transition probabilities as the original model [5]. Another alternative is to partition the state space into *qualitative* states that have similar reward functions [17]; which is described below.

The basic idea of *qualitative MDPs* [18] is to partition the state space into several qualitative states that share similar rewards. This scheme applies to discrete MDPs with a large state space and also to continuous state spaces. For example, in robot navigation, the state represents the position $(X, Y)$ of a mobil robot, and it could be codified as a large number of cells that provide a partition of the environment (discrete) or as the actual position in $X$, $Y$ coordinates (continuous).

Based on expert knowledge or data, the reward for different parts of the state space can be estimated, and clustered in regions of similar reward. This can be represented by a *reward decision tree* (RDT), which codifies the reward function based on partitions of the different state variables. The leafs of this tree correspond to state regions with the same reward, so this are the initial abstract states (q-states), obtained by transforming the RDT to a *Q-tree*, see Fig. 11.6.

This strategy provides a coarse, initial partition, as it is based on the immediate reward; ideally it should be according to the value function. So in a second phase, the initial partition is refined according to the following procedure:

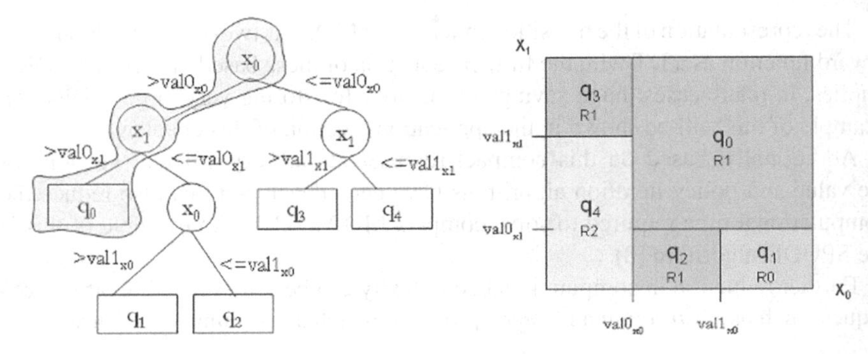


**Fig. 11.6** Left: a *Q-tree* that corresponds to a reward decision tree according to two state variables, $X_0$ and $X_1$; the leafs of this tree correspond to the q-states for the MDP. Right: partition of the state space in 5 qualitative states according to the Q-tree. The state $q_0$ si highlighted in the tree

1. Solve the abstract MDP and obtain the value for each q-state.
2. Estimate the variance of the value of each state with respect to its neighboring states.
3. Partition the state with highest variance (not marked) into two new q-states, in the dimension with highest value-difference with its neighbors.
4. Solve the new MDP that includes the new partition.
5. If the policy for the new MDP is the same as before, return to the previous MDP and mark the q-state so it is not partitioned again; otherwise keep the new partitions.
6. If the dimension of all states is equal or below a predefined hyper-volume terminate; otherwise return to step 2.

Although this procedure implies solving an MDP several times, it could still obtain important savings in storage and computation time if the original problem has a very large or continuous state space. It has been applied to the operation of power plants with good results [18].

### 11.4.2 Decomposition

Decomposition consists in dividing the global problem into smaller subproblems that are solved independently and their solutions combined [4, 11]. There are two main types of decomposition: (i) serial or hierarchical, and (ii) parallel or concurrent.

Hierarchical MDPs provide a sequential decomposition, in which different subgoals are solved in sequence to reach the final goal. That is, at the execution phase, only one task is active at a given time. Hierarchical MDPs accelerate the solution of complex problems by defining different subtasks that correspond to intermediate goals, solving for each subgoal, and then combining these subprocesses to solve the overall problem; examples of hierarchical approaches are HAM [14] and MAXQ [6].

In concurrent or parallel MDPs, the subtasks are executed in parallel to solve the global task. In general these approaches consider that the task can be divided in several relatively independent subtasks that can be solved independently and then the solutions are combined to solve the global problem. When the subtasks are not completely independent some additional considerations are required. For instance, Loosely Coupled MDPs [11] consider several independent subprocesses which are coupled due to common resource constraints. To solve them they use an iterative procedure based on a heuristic allocation of resources to each task. An alternative approach is taken by [4], which initially solves each subtask independently, and when the solutions are combined, they take into account potential conflicts between the partial policies, and solve these conflicts to obtain a global, approximately optimal policy. An example of the application of this last technique in robotics is given below in Sect. 11.5.

## 11.5 Applications

The application of MDPs is illustrated in two different domains. One is for assisting power plant operators in the operation of a power plant under difficult situations. The other is for coordinating a set of modules to solve a complex task for service robots.

### 11.5.1 Power Plant Operation

MDPs have been applied to several problems in power plants and power systems [19], including dam management in hydroelectric power plants, inspection of electric substations, and optimization of steam generation in a combined-cycle power plant. This last application is presented.

The steam generation system of a combined-cycle power plant provides superheated steam to a steam turbine. It is composed by a recovery steam generator, a recirculation pump, control valves and interconnection pipes. A *heat recovery steam generator* (*HRSG*) is a process machinery capable of recovering residual energy from a gas turbine's exhaust gases to generate high pressure (*Pd*) steam in a special tank (*steam drum*). The *recirculation pump* is a device that extracts residual water from the steam drum to keep a water supply in the HRSG (*Ffw*). The result of this process is a high-pressure steam flow (*Fms*) that keeps running a *steam turbine* to produce electric energy (*g*) in a *power generator*. The main control elements associated are the feedwater valve (*fwv*) and the main steam valve (*msv*). A simplified diagram of the steam generation system is shown in Fig. 11.7.

During normal operation, a three-element feedwater control system commands the feed-water control valve (*fwv*) to regulate the level (*Ld*) and pressure (*Pd*) in the drum. However, this traditional controller does not consider the possibility of failures in the control loop (valves, instrumentation, or any other process devices).

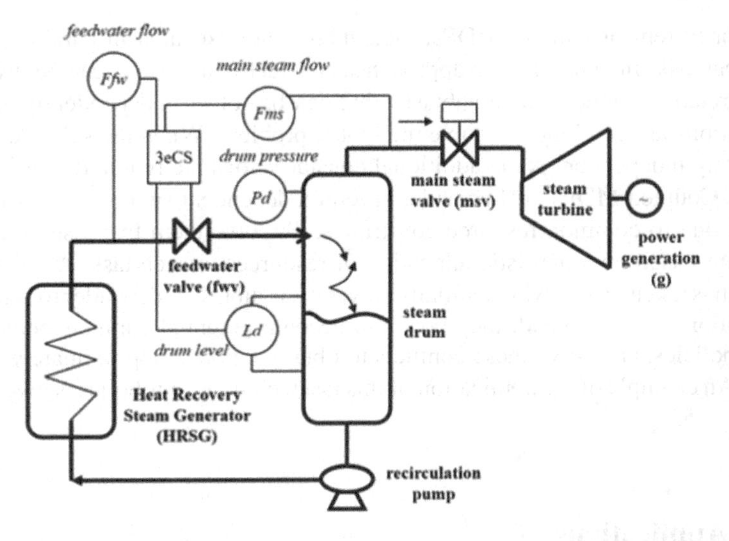


**Fig. 11.7** Main elements of the steam generation system of a combined-cycle power plant [19]

Furthermore, it ignores whether the outcomes of executing a decision will help, in the future, to increase the steam drum lifetime, security, and productivity. The problem is to obtain a function that maps plant states to recommendations for the power plant operator; this function should consider all these aspects.

This problem was modeled as an MDP, which served as the basis for developing a tool for training and assistance for power plant operators.

#### 11.5.1.1 Power Plant Operator Assistant

*AsistO* [18] is an intelligent assistant that provides recommendations for training and on-line assistance in the power plant domain. The assistant is coupled to a power plant simulator capable of partially reproducing the operation of a combined cycle power plant.

AsistO is based on a decision-theoretic model that represents the main elements of the steam generation system of a combined-cycle power plant. The main variables in the steam generation system represent the state in a factored form. Some of these variables are continuous, so they are discretized into a finite number of intervals. The actions correspond to the control of the main valves in this subsystem of the power plant, in particular those that have to do with controlling the level of the drum (a critical element of the plant): feed-water valve (*fwv*) and main steam valve (*msv*). The reward function is defined in terms of a recommended operation curve for the relation between the drum pressure and steam flow (see Fig. 11.8). The idea is to maintain a balance between the efficiency and safety of the plant. As such, the control actions should try to maintain the plant within this recommended operation

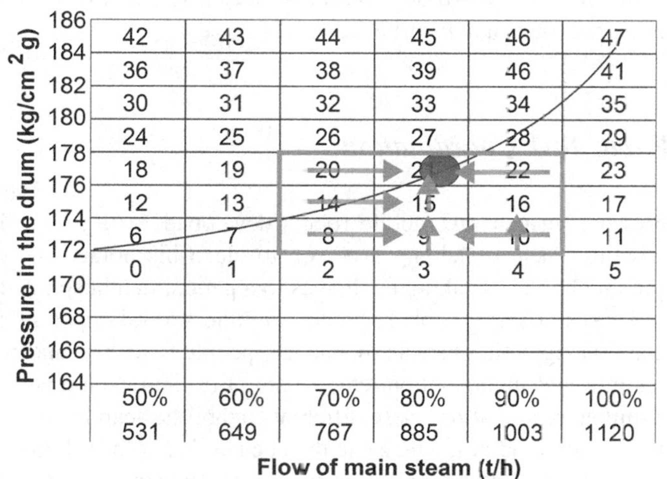


**Fig. 11.8** Recommended operation curve for the drum of a combined-cycle power plant, defining the desirable relation between drum pressure and steam flow. The arrows illustrate schematically the idea of the recommender system—returning the operation to a safe relation (circle) when the plant deviates from the recommended operation curve

curve; if it deviates they should return it to a *safe* point; this is shown schematically with arrows in Fig. 11.8.

The last element to be defined to complete the MDP model is the transition function. In this application it can be learned by using the power plant simulator and sampling the state and action spaces. Once the MDP is built, it can be solved to obtain the optimal policy, and from this, the system can give recommendations to the operator under different plant conditions.

#### 11.5.1.2 Experimental Results

A relatively simple example was considered with five state variables: $Fms$, $Ffw$, $Pd$, $g$, $d$; and four actions: open/close the feed-water (*fwv*) and main steam (*msv*) valves a certain amount. The reward function was defined based on the recommended operation curve. To learn the transition function, samples of the system dynamics were gathered using simulation.

The memory requirements for a *flat* MDP representation and a factored representation were compared. The flat MDP required $589,824$ parameters (probability values) while the factored MDP only 758. The optimal solution for the factored MDP was obtained in less than two minutes on a standard personal computer.

The recommended actions of the MDP controller and the traditional automatic control were compared using the power plant simulator. The actions taken by both are similar, however the MDP-based controller takes less time to return the plant to a safe operating point when a disturbance occurs.

### 11.5.2 Robot Task Coordination

Service robots are, in general, mobile robots developed for assisting humans in different activities, such as helping a person to clean his home, assisting senior citizens in nursing homes or taking medicines to a patient in a hospital. To perform these types of tasks, a service robot should combine several capabilities, such as localization and navigation, obstacle avoidance, people detection and recognition, object recognition and manipulation, etc. To simplify the development of service robots and promote reutilization, these different capabilities can be implemented as independent software modules, which can then be *combined* for solving a particular task, such as delivering messages or objects between people in an office environment. In this case, it is necessary to coordinate the different modules to perform a task, ideally in an optimal way.

Markov decision processes provide an appropriate framework for task coordination for service robots. The state space can be defined in terms of a number of variables that define the high–level situation of the tasks. The actions correspond to commands (calls) to the different software modules, for instance indicating to the navigator to move the robot to a certain position in the environment. The reward function can be defined based on the objectives of the task, for example for the message delivery robot, a certain reward for when it receives a message from the sender, and another, higher reward when it delivers it to the recipient. Once a task is modeled as an MDP, the MDP can be solved to obtain a policy to perform the task. This is in general better than a traditional plan, as it provides a *plan* for any initial state (a kind of general plan) and it is robust with respect to the uncertainty in the results of the different actions.

Under this framework, based on general software modules and an MDP–based coordinator, it is in principle relatively easy for a service robot to solve different tasks. We just need to modify the MDP reward function according to the new task objectives, and solve the modified MDP to obtain a policy for the other task.

Additionally, it is desirable for the robot to perform several actions *simultaneously*, such as navigation to a certain location, avoiding obstacles and looking for people; all at the same time. However, if we represent the robot task coordination problem as a single MDP, we have to consider all possible combinations of all the possible simultaneous actions. This implies an explosion in the action—state space and thus an important increase in the complexity for solving the MDP. It also becomes much more difficult to specify or learn the model.

An alternative is to model each subtask as an independent MDP, then solve each MDP to obtain its optimal policy, and then execute these policies *concurrently*. This approach is known as *concurrent MDPs* [4].

#### 11.5.2.1 Concurrent Markov Decision Processes

Based on functional decomposition, a complex task is partitioned into several subtasks. Each subtask is represented as an MDP and solved independently, and the policies are executed in parallel assuming no conflicts. All the subtasks have a common goal and can share part of the state space, that is represented in a factored form. However, conflicts may arise between the subtasks.

There are two main types of conflicts: (i) *resource conflicts*, and (ii) *behavior conflicts*. Resource conflicts occur when two actions require the same physical resource (e.g., to control the wheels of a robot) and cannot be executed concurrently. This type of conflict is solved off–line by a two-phase process [3]. In the first phase we obtained an optimal policy for each subtask (MDP). An initial global policy is obtained by combining the local policies, such that if there is a conflict between the actions selected by each MDP for a certain state, the one with maximum value is considered, and the state is marked as a *conflict* state. This initial solution is improved in a second phase using policy iteration. By taking the previous policy as its initial policy and considering only the states marked as conflicts, the time complexity is drastically reduced and a near-optimal global policy is obtained.

Behavior conflicts arise in situations in which it is possible to execute two (or more) actions at the same time but it is not desirable given the application. For example, it is not desirable for a mobile robot to be navigating and handling an object to a person at the same time (this situation is also difficult for a person). Behavior conflicts are solved on–line based on a set of restrictions specified by the user. If there are no restrictions, all the actions are executed concurrently; otherwise, a constraint satisfaction module selects the set of actions with the highest expected utility.

#### 11.5.2.2 Experiments

An experiment was done with *Markovito*, a service robot, which performed a delivery task; behavior conflicts were considered. Markovito is a service robot based on an ActivMedia PeopleBot robot platform, which has laser, sonar and infrared sensors; a camera, a gripper and two computers (see Fig. 11.9) [1].

In the task considered for Markovito the goal for the robot is to receive and deliver a message, an object or both, under a user's request. The user gives an order to send a message/object and the robot asks for the name of the sender and the receiver. The robot either records a message or uses its gripper to hold an object, and navigates to the receiver's position for delivery. The task is decomposed into five subtasks, each represented as an MDP:

1. *navigation*, the robot navigates safely in different scenarios;
2. *vision*, for looking and recognizing people and objects;
3. *interaction*, for listening and talking with a user;
4. *manipulation*, to receive and deliver an object safely; and

**Fig. 11.9** Markovito, a service robot

5. *expression*, to show *emotions* using an animated face.

Each subtask is represented as a factored MDP:

Navigation: States: 256 decomposed into 6 state variables (*located, has-place, has-path, direction, see-user, listen-user*). Actions: 4 (*go-to, stop, turn, move*).
Vision: States: 24 decomposed into 3 state variables (*user-position, user-known, user-in-db*). Actions: 3 (*look-for people, recognize-user, register-user*).
Interaction: States: 9216 decomposed into 10 state variables (*user-position, user-known, listen, offer-service, has-petition, has-message, has-object, has-user-id, has-receiver-id, greeting*). Actions: 8 (*hear, offer-service, get-message, deliver-message, ask-receiver-name, ask-user-name, hello, bye*).
Manipulation: States: 32 decomposed into 3 state variables (*has-petition, see-user, has-object*). Actions: 3 (*get-object, deliver-object, move-gripper*).
Expression: States: 192 decomposed into 5 state variables (*located, has-petition, see-user, see-receiver, greeting*). Actions: 4 (*normal, happy, sad, angry*).

Note that several state variables are common to two or more MDPs. If we represent this task as a single, flat MDP, there are 1,179,648 states (considering all the non-duplicated state variables) and 1,536 action combinations, giving a total of nearly two billion state-actions. Thus, to solve this task as a single MDP will be difficult even for state of the art MDP solvers.

The MDP model for each subtask was defined using a structured representation [8]. The transition and reward functions were specified by the user based on task knowledge and intuition. Given that the MDP for each subtask is relatively simple, its manual specification is not too difficult. In this task, conflicts might arise between the different subtasks, so we need to include conflict resolution strategies. The conflicts considered are behavior conflicts, so these are solved based on a set of restrictions, which are summarized in Table 11.1.

**Table 11.1** Restriction set for the messenger robot

| Action(s) | Restriction | Action(s) |
|---|---|---|
| Get message | **Not_during** | Turn OR advance |
| Ask_user_name | **Not_before** | Recognize_user |
| Recognize_user | **Not_start** | Avoid_obstacle |
| Get_object OR deliver_object | **Not_during** | Directed towards OR turn OR moving |

For comparison, the delivery task was solved under two conditions: (i) without restrictions and (ii) with restrictions.

In the case without restrictions, all actions can be executed concurrently, but the robot performs some undesirable behaviors. For example, in a typical experiment, the robot is not able to identify the person who wants to send a message for a long time. This is because the *vision MDP* cannot get a good image to analyze and recognize the user because the *navigation MDP* is moving trying to avoid the user. Also, because of this, the user has to follow the robot to provide an adequate input to the microphone.

In the case where restrictions were used, these allowed a more fluid and efficient solution. For example, in a similar experiment as without restrictions, the *vision MDP* with the restriction set is now able to detect and recognize the user much earlier. When the *interaction MDP* is activated, the *navigation MDP* actions are not executed, allowing an effective interaction and recognition.

On average, the version with restrictions takes about 50% of the time steps required by the version without restrictions to complete the task. In summary, by considering conflicts via restrictions, not only does the robot perform the expected behavior, but also it has a more robust performance, avoids conflicts, and displays a significant reduction in the time required to complete the task.

## 11.6 Additional Reading

Puterman [16] is an excellent reference for MDPs, including their formal definition and the different solution techniques. A recent overview of decision theoretic models and their applications is given in [20]. Reference [15] reviews MDPs and POMDPs. The representation of MDPs using ADDs and SPUDD are described in [8].

## 11.7 Exercises

1. For the grid world example of Fig. 11.1, consider that each cell is a state and that there are four possible actions: *up, down, left, right*. Complete the specification of the MDP, including the transition and reward functions.
2. Solve the MDP for the previous exercise by value iteration. Initialize the values for each state to the immediate reward, and show how the values change with each iteration.
3. Solve the grid world MDP using policy iteration. Initialize the policy to *up* for all states. Show how the policy changes with each iteration.
4. Define a factored representation for the grid world example, considering that a state is represented in terms of two variables, *row* and *column*. Specify the transition function as a two-stage DBN and the reward function as a decision tree.
5. Consider a grid world scenario in which the grid is divided in several rooms and a hallway that connects the rooms. Develop a hierarchical solution to the robot navigation for this scenario, considering one MDP to go from any cell in a room to the door that connects to the hallway, other MDP that navigates through the hallway from one door to another, and a third MDP to navigate from the door entering a room to the goal in the room. Specify each MDP and then a high-level controller that coordinates the low-level policies.
6. Develop a solution to the grid navigation problem based on two concurrent MDPs, one that navigates towards the goal (without considering obstacles) and another that avoids obstacles. Specify the model for each of these two subtasks and how to combine their resulting policies to navigate to the goal and avoid obstacles. Does the solution always reach the goal or can it get stuck in local maxima?
7. Given a large 10 x 10 meters flat scenario divided in 10 cm x 10 cm cells, there are two goals of one square meter each, one has a reward of $+100$ and its at coordinates are $X = 1, Y = 1$, and the other a reward of $+50$ at coordinates $X = 5, Y = 5$; all others cells have a reward of $-1$. A robot navigates in this scenario, the state is the cell location, and it has four possible actions: *up, down, left, right*; with 0.7 probability reaches the next cell according to the action, and with 0.1 probability it reaches another neighboring cell in the grid. (a) Define a flat MDP model for this scenario. (b) What is the number of state-actions?
8. For the previous problem consider two state variables, $X$ and $Y$, define the reward function as a reward decision tree.
9. Solve problem 7 based on the RDT of the previous problem, using the abstract MDP approach considering a minimum abstract state size of $1\ x\ 1$ meters. (a) What are the initial and the final partitions of the state space? (b) What is the final policy?
10. Consider the power plant operation example. Given the recommended operation curve in Fig. 11.8, define a reward function (as a table) such that it has a positive reward for the states (cells) over the recommended curve and a negative reward

for the other cells. Transform the reward function to a RDT based on the two variables, drum pressure and steam flow. Compare the complexity and storage requirement of both representations.

11. Prove that the solution to the Bellman equation using value iteration always converges.
12. *** Develop a general program that implements the value iteration algorithm.
13. *** Develop a program to solve the grid world robot navigation problem, including a graphical interface to define the size of the grid, the obstacles and the position of the goal. Considering a high positive reward for the goal cell, high negative rewards for the obstacles, and small negative rewards for the other cells, represent the problem as an MDP and obtain the optimal policy using value iteration.
14. *** Investigate how to transform an MDP to a linear optimization problem such that it can be solved via the simplex method.
15. *** When the model for an MDP is not known, an alternative is to learn the optimal policy by trial and error with *reinforcement learning*. Investigate the basic algorithms for reinforcement learning, such as *Q-learning*, and implement a program to learn a policy to solve the robot navigation problem for the grid world example of Fig. 11.1 (use the same rewards as in the example). Is the policy obtained the same as the solution to the MDP model?

## References


1. Avilés-Arriaga, H.H., Sucar, L.E., Morales, E.F., Vargas, B.A., Corona, E.: Markovito: a flexible and general service robot. In: Liu, D., Wang, L., Tan, K.C. (eds.) Computational Intelligence in Autonomous Robotic Systems, pp. 401–423. Springer, Berlin (2009)
2. Bellman, R.: Dynamic Programming. Princeton University Press, Princeton (1957)
3. Corona, E., Morales, E.F., Sucar, L.E.: Solving policy conflicts in concurrent markov decision processes. In: ICAPS Workshop on Planning and Scheduling Under Uncertainty. Association for the Advancement of Artificial Intelligence (2010)
4. Corona, E., Sucar, L.E.: Task coordination for service robots based on multiple Markov decision processes. In: Sucar, L.E., Hoey, J., Morales, E. (eds.) Decision Theory Models for Applications in Artificial Intelligence: Concepts and Solutions. IGI Global, Hershey (2011)
5. Dean, T., Givan, R.: Model minimization in Markov decision processes. In: Proceedings of the 14th National Conference on Artificial Intelligence (AAAI), pp. 106–111 (1997)
6. Dietterich, T.: Hierarchical reinforcement learning with the MAXQ value function decomposition. J. Artif. Intell. Res. **13**, 227–303 (2000)
7. Elinas, P., Sucar, L., Reyes, A., Hoey, J.: A decision theoretic approach for task coordination in social robots. In: Proceedings of the IEEE International Workshop on Robot and Human Interactive Communication (RO-MAN), pp. 679–684 (2004)
8. Hoey, J., St-Aubin, R., Hu, A., Boutilier, C.: SPUDD: stochastic planning using decision diagrams. In: Proceedings of the International Conference on Uncertainty in Artificial Intelligence (UAI), pp. 279–288 (1999)
9. Li, L., Walsh, T.J., Littman, M.L.: Towards a unified theory of state abstraction for MDPs. In: Proceedings of the Ninth International Symposium on Artificial Intelligence and Mathematics, pp. 21–30 (2006)

10. Littman, M., Dean, T., Kaelbling, L.: On the complexity of solving Markov decision problems. In: Proceedings of the Eleventh Conference on Uncertainty in Artificial Intelligence, pp. 394–402 (1995)
11. Meuleau, N., Hauskrecht, M., Kim, K.E., Peshkin, L., Kaelbling, L.P., Dean, T., Boutilier, C.: Solving very large weakly coupled Markov decision processes. In: Proceedings of the Association for the Advancement of Artificial Intelligence (AAAI), pp. 165–172 (1998)
12. Montero, A., Sucar, L.E.: Decision-theoretic assistants based on contextual gesture recognition. Annals of Information Systems (to be published)
13. Quinlan, J.R.: Induction of decision trees. Mach. Learn. **1**(1), 81–106 (1986)
14. Parr, R., Russell, S.J.: Reinforcement learning with hierarchies of machines. In: Proceeding of the Advances in Neural Information Processing Systems (NIPS) (1997)
15. Poupart, P.: An introduction to fully and partially observable Markov decision processes. In: Sucar, L.E., Hoey, J., Morales, E. (eds.) Decision Theory Models for Applications in Artificial Intelligence: Concepts and Solutions. IGI Global, Hershey (2011)
16. Puterman, M.L.: Markov Decision Processes: Discrete Stochastic Dynamic Programming. Wiley, New York (1994)
17. Reyes, A., Sucar, L.E., Morales, E.F., Ibargüngoytia, P.: Hybrid Markov decision processes. Lecture Notes in Computer Science, vol. 4293, pp. 227–236. Springer, Berlin (2006)
18. Reyes, A., Sucar, L.E., Morales, E.F.: AsistO: a qualitative MDP-based recommender system for power plant operation. Computacion y Sistemas **13**(1), 5–20 (2009)
19. Reyes, A., Sucar, L.E., Ibargüngoytia, P., Morales, E.: Planning under uncertainty applications in power plants using factored Markov decision processes. Energies **13**, (2020)
20. Sucar, L.E., Hoey, J., Morales, E.: Decision Theory Models for Applications in Artificial Intelligence: Concepts and Solutions. IGI Global, Hershey (2011)

# Chapter 12
# Partially Observable Markov Decision Processes

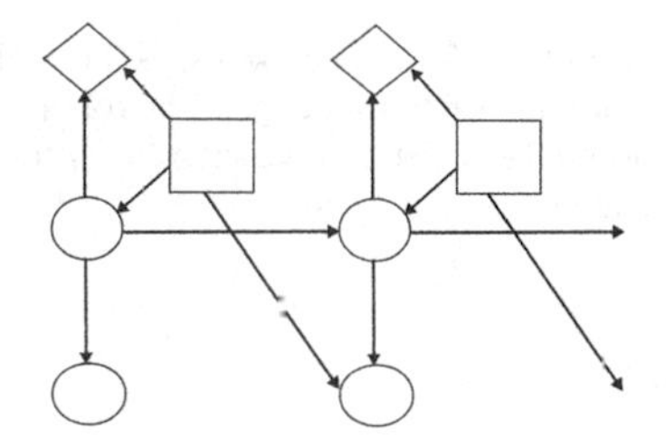

## 12.1 Introduction

The previous chapter introduced sequential decision problems, and a technique to solve this type of problems when the state is completely observable: Markov decision processes. However, in many practical applications such as robotics, finance, health, among others; the state can not be observed completely, there is only partial information about the state of the system. This type of problems are known as *partially observable Markov decision processes* (POMDPs). In this case, there are certain observations from which the state can be estimated probabilistically. For instance, consider the example of the robot in the grid world. It could be that the robot can not determine precisely the cell where it is (its state), but can estimate the probability of being in each cell by observing the surrounding environment. Such is the case in real mobile robot environments, where the robot can not know with precision its localization in the environment, only probabilistically by using its sensors and certain *landmarks*. Figure 12.1 illustrates a POMDP in the grid world; the robot is not certain of its current state, it could be in different cells with different probabilities (depicted according to the size of the robot).

In a POMDP the current state is not known with certainty, only the probability distribution of the state, which is known as the *belief state*. Thus, since the states are not directly observable, the action selection has to be based on the past observations. This make solving a POMDP, that is obtaining an optimal policy, much more difficult than solving an MDP.



L. E. Sucar, *Probabilistic Graphical Models*, Advances in Computer Vision and Pattern Recognition, https://doi.org/10.1007/978-3-030-61943-5_12

**Fig. 12.1** A robot in the grid world. Each cell represents the possible states of the robot, with a smiling face for the goal and a forbidden sign for danger. The robot is shown in different cells, the probability of each one is proportional to the size of the robot. This corresponds to a POMDP where there is uncertainty about the state

## 12.2 Representation

Formally, a POMDP is a tuple $M =< S, A, \Phi, R, O, \Omega, \Pi >$. The first four elements are the same as in an MDP: $S$ is a finite set of states $\{s_1, \ldots, s_n\}$. $A$ is a finite set of actions $\{a_1, \ldots, a_m\}$. $\Phi : A \times S \times S \rightarrow [0, 1]$ is the state transition function specified as a probability distribution. $R : S \times A \rightarrow \Re$ is the reward function. $R(s, a)$ is the reward that the agent receives if it takes action $a$ in state $s$.

The three new elements are the Observations, the Observation Function and the Initial State Distribution. $O$ is a finite set of observations $\{o_1, \ldots, o_l\}$. $\Omega : S \times O \rightarrow [0, 1]$ is the observation function specified as a probability distribution, which gives the probability of an observation $o$ given that the process is in state $s$ and it executed action $a$, $P(o \mid s, a)$. $\Pi$ is the initial state distribution that specifies the probability of being in state $s$ at $t = 0$.

Note that the observation values are not necessarily the same as the state values. In practice observations are the information that the agent can obtain from its environment to estimate its state. For example, in the case of the robot in the grid world, it could be a range sensor that estimates the distances to the walls, and from these measurements, the robot can estimate in which cell (state) it is currently located.

A graphical model of a POMDP is shown in Fig. 12.2. It is similar to an MDP, with the addition of the observation variables, which depend on the current state and previous actions.

As in the case of MDPs, the process is assumed to be Markovian and stationary. This applies to the transition, reward and observation functions.

Markov property:

$$P(S_{t+1} \mid S_t, A_t, S_{t-1}, A_{t-1}, \ldots) = P(S_{t+1} \mid S_t, A_t), \forall S, A \qquad (12.1)$$

$$P(O_t \mid S_t, A_{t-1}, S_{t-1}, A_{t-2}, \ldots) = P(O_t \mid S_t, A_{t-1}), \forall O, S, A \qquad (12.2)$$

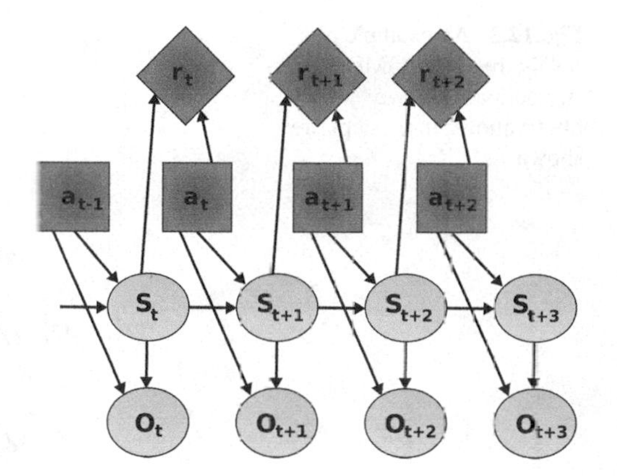


**Fig. 12.2** A graphical model representation of a POMDP with four time steps. $s_t$ represents the state at time $t$, $a_t$ the action, $o_t$ the observation, and $r_t$ the reward

Stationary property:

$$P(S_{t+1} \mid S_t, A_t) = P(S_{t+2} \mid S_{t+1}, A_{t+1}), \forall t \tag{12.3}$$

$$P(O_t \mid S_t, A_{t-1}) = P(O_{t+1} \mid S_{t+1} A_t), \forall t \tag{12.4}$$

Once a sequential decision problem is modeled as a POMDP, we usually want to *solve* the POMDP to obtain the optimal policy; this is the topic of the next section.

## 12.3 Solution Techniques

In a POMDP the current state is not known with certainty, so we cannot specify a policy as a mapping from states to actions. Given that we do know the past actions and observations, a policy can be specified based on the *history* of past observations and actions. That is, an action can be selected based on this history $H$:

$$H : A_0, O_1, A_1, O_2, \ldots A_t, O_{t+1} \rightarrow A_{t+1}$$

This can be represented as what is known as a *policy tree* that depicts the alternative actions and observations starting from the initial action as the root of the tree, see Fig. 12.3. The policy tree in the figure shows a simple example with two observations and two actions; after each action the possible observations are shown in the arrows, and then the actions selected based on the previous actions–observations. That is the actions in the second level need to consider the previous action and observation, the actions in the third level need to consider the two previous actions and observations, and so on.

Although the process is Markovian in terms of the states, it is not so based only on actions and observations; so an optimal policy needs to consider all the actions

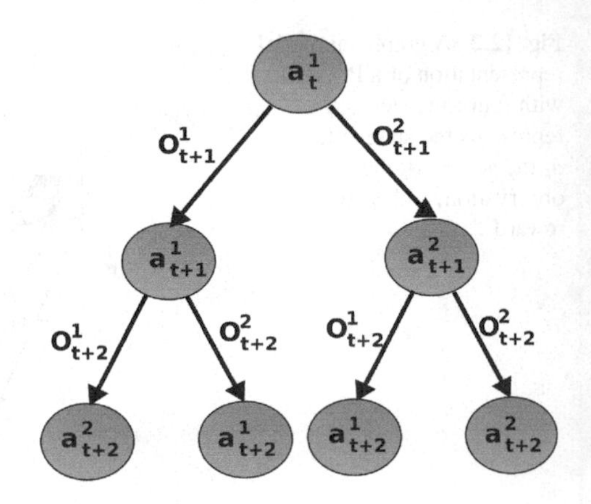


**Fig. 12.3** An example of a policy tree of a POMDP with two actions and two observations; three steps are shown

and observations from the past. As the policy grows exponentially with the length of the history, this strategy is computationally and memory wise very expensive, in particular for long histories.

An alternative to consider the whole history of actions and observations, is to use the probability distribution of the state, that is known as the *belief state* ($B$). A belief state is the probability distribution over all possible states given the previous history of actions–observations (or the initial distribution). For example, in the case of the robot in the grid world, Fig. 12.1, there are 12 states (cells). If we number the states from the top-left by rows, then the belief illustrated in the figure could be:

$$0, 0, 0, 0, 0.7, 0, 0, 0, 0.2, 0.1, 0, 0$$

So solving a POMDP requires finding a mapping from the belief space to the action space, such that the optimal action is selected:

$$\pi : B \rightarrow A$$

Given the belief state the process is Markovian, that is, it gives as much information as the entire history of actions and observations. Although the state can not be observed directly, the belief state can be estimated based on the previous actions and observations. In particular, knowing the belief state at time $t$, $b_t$ (which could be the initial belief), and after executing action $a_t$ and observing $o_{t+1}$, the belief state at time $t + 1$ can be estimated by applying Bayes rule ($s$ is the state at time $t$ and $s'$ at time $t + 1$):

$$b_{t+1}(s') = P(s' \mid b_t, a_t, o_{t+1}) = \frac{P(s', b_t, a_t, o_{t+1})}{P(b_t, a_t, o_{t+1})}$$

$$b_{t+1}(s') = \frac{P(o_{t+1} \mid b_t, a_t, s')P(s' \mid b_t, a_t)P(b_t, a_t)}{P(o_{t+1} \mid b_t, a_t, )P(b_t, a_t)}$$

$$b_{t+1}(s') = \frac{P(o_{t+1} \mid s', a_t)\sum_s P(s' \mid b_t, a_t, s)P(s \mid b_t, a_t)}{P(o_{t+1} \mid b_t, a_t, )}$$

$$b_{t+1}(s') = \frac{P(o_{t+1} \mid s', a_t)\sum_s P(s' \mid a_t, s)b_t(s)}{P(o_{t+1} \mid b_t, a_t, )}$$

The denominator can be treated as a normalizing factor, thus:

$$b_{t+1}(s') \propto P(o_{t+1} \mid s', a_t)\sum_s P(s' \mid a_t, s)b_t(s) \tag{12.5}$$

So the next belief state can be calculated in terms of the current belief, the transition probabilities and the observation probabilities.

Given that beliefs states can be considered as states, the original POMDP is equivalent to a continuous state space MDP or belief space MDP. Thus, solving a POMDP is more difficult than solving an MDP, as the belief space is in principle infinite. However, as the belief MDP is a fully observable MDP and has an optimal policy that is Markovian and stationary, it is easier to solve the belief MDP than the POMDP. Several techniques for solving POMDPs are based on the equivalent belief MDP.

### 12.3.1 Value Functions

Recall that the value function specifies the expected sum of expected rewards according to certain policy, $\pi$. Based on the belief states, the value function of a POMDP (or belief MDP) can be written in a similar way as for an MDP:

$$V^{\pi}(b) = R(b, a) + \gamma \sum_o P(o \mid b, a)V^{\pi}(b_o^a), \forall b \tag{12.6}$$

where $a$ is the action selected for the current belief state according to the policy, $\pi(b)$; and $b_o^a$ is the next belief state according to Eq. 12.5 after executing action $a$ and observing $o$.

The optimal policy, as for MDPs, is the one that maximizes the value function for all beliefs and therefore satisfies Bellman's equation:

$$V^{\pi *}(b) = max_a \left[ R(b, a) + \gamma \sum_o P(o \mid b, a)V^{\pi *}(b_o^a) \right], \forall b \tag{12.7}$$

Although in theory this is fine, in practice it is not clear how to solve Bellman's equation as the belief space is continuous, so in principle it will represent an infinite number of equations! Fortunately it can be shown that the finite-horizon value function is piece-wise linear and convex [9], which is key to several solution techniques.

#### 12.3.1.1 Representation of the Value Function

A finite policy tree induces a value function that is linear in the belief state, $b$; so several alternative policy trees can be represented as a collection of linear functions. The upper surface of this collection of linear functions will represent the optimal value function, $V^*$, which is piece-wise linear and convex. To illustrate this we will use a simple example.

Consider a POMPD with two states, $s_1, s_2$, two actions, $a_1, a_2$, and three observations, $o_1, o_2, o_3$. Since that there are only two states, the probability distribution (belief state) can be represented by a single value, for example for $s_1$, given that they must add one (i.e., $P(s_1) = 0.4$, then $P(s_2) = 0.6$). Therefore the belief space can be represented by a line segment [0, 1], see Fig. 12.4. Assuming a certain initial belief, $b_1$, there are a finite number of possible next belief states, $b_2$, depending on the actions and observations, that can be estimated as was shown in the previous section. In this example, there will be three possible values for $b_2$ if action $a_1$ is selected, one for each observation (these must add to one); and other three if action $a_2$ is chosen. This is illustrated in Fig. 12.4.

Let see now how we can represent the value function for different planning horizons. For horizon 1, the value will be the same as the reward; since there are two states and two actions in the example, we have four values:

$$V_1(a_1, s_1) = 2,\ V_1(a_1, s_2) = 1,\ V_1(a_2, s_1) = 0,\ V_1(a_1, s_2) = 3$$

For instance, for action $a_1$, the value for some belief states:

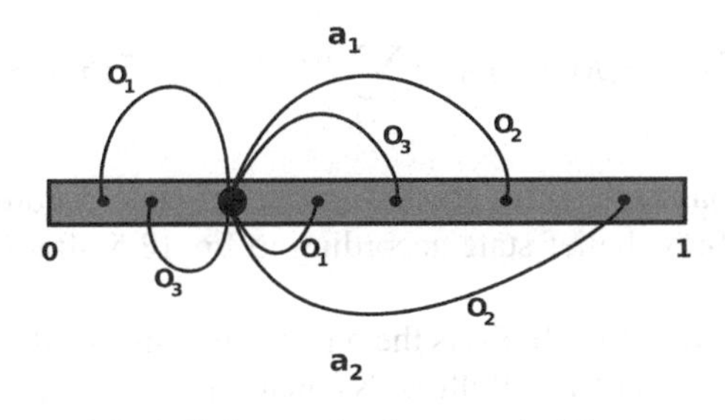


**Fig. 12.4** An illustration of the belief space for the example. The bar represents the probability of $s_1$ ($P(s_2) = 1 - P(s_1)$). Graphically we show the possible transitions from $b_1$ to $b_2$ for each possible action–observation: the large black dot represents the initial belief state and the smaller black dots the resulting belief states

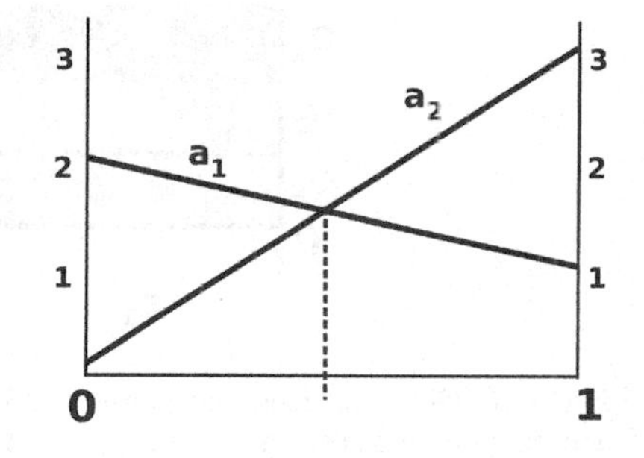


**Fig. 12.5** The value function for horizon 1 of the two states example. Given that we want to maximize the value, we will choose $a_1$ for a belief state to the left of the dashed vertical line, and $a_2$ for a belief state to the right

$$b = [1, 0],\ V_1(a_1) = 1 \times 2 + 0 \times 1 = 2$$

$$b = [0.2, 0.8],\ V_1(a_1) = 0.2 \times 2 + 0.8 \times 1 = 1.2$$

$$b = [0, 1],\ V_1(a_1) = 0 \times 2 + 1 \times 1 = 1$$

The value is a linear combination of the reward for the different states, weighted by their belief, so it can be represented by a line in this example with two states (or an hyperplane in general). Similarly we can obtain the values for $a_2$ and we obtain another line. The resulting value function for horizon 1 is depicted in Fig. 12.5. We can see that the value function is piece-wise linear and convex, and should choose the action that maximizes the value according to the current belief state: $b(s_1) < 0.5 \Rightarrow a_1, b(s_1) > 0.5 \Rightarrow a_2$ (if $b(s_1) = 0.5$ any action can be selected).

The next step is to obtain the value function for horizon 2; that is, given the initial belief and after performing an action $a$ and observing $o$. The value for horizon 2 will be the value of the immediate action plus the value of the next action. Given the current belief state, $b$, according to the action and observations, we can estimate the next belief state, $b'$. As the observation is not known a priori, we need to estimate the expected value of each action, considering the probabilities of each possible observation given $b$. For instance, if we consider one of the actions, $a_1$, we will obtain three value functions, one for each potential observation. Each one is also piece-wise linear (similar to the one for horizon 1, Fig. 12.5), and partitions the belief state in two parts, so that in each part the preferred next action will be $a_1$ or $a_2$. By combining the three functions, the next best action can be selected according to the observation, see Fig. 12.6.

In a similar way we can obtain the value functions for the other action, $a_2$, which will give the best next action according to the observation. Finally we combine the value functions for both actions, each one a set of linear functions, as depicted in Fig. 12.7. To select the optimal action, we maximize the value function for each belief state, giving us the upper surface, as shown in Fig. 12.7. This process can be repeated to obtain the value functions and optimal actions for horizon 3, 4, and so on, to horizon $H$.

The representation of the value as a combination of linear functions applies to higher dimensions (more states), and in general they can be represented as vectors,

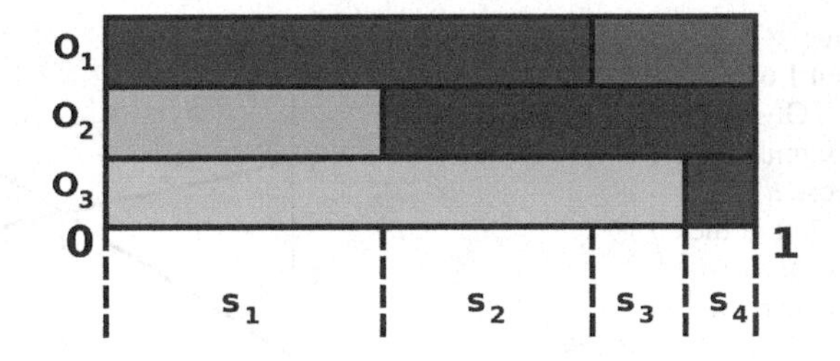


**Fig. 12.6** Graphical representation of the best next action $a(t = 2)$ based on the belief state, depending on the observation, $o_1, o_2, o_3$. The lighter regions represent when $a_1$ is preferred, and the darker ones when $a_2$ is preferred. Although in principles there could be 8 different strategies combining actions and observations, in this example we can partition the belief space in four strategies, shown in the bottom of the figure, $s_1, \ldots s_4$. For example, the strategy for $s_1$ is: $o_1 \Rightarrow a_2, o_2 \Rightarrow a_1, o_3 \Rightarrow a_1$

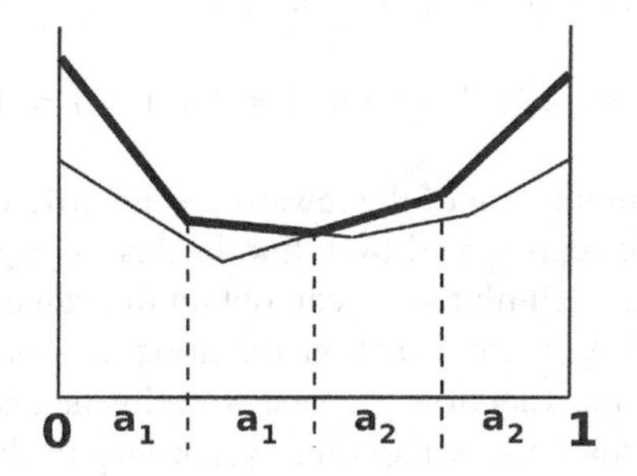


**Fig. 12.7** The value function for horizon 2 of the two states example. The figure shows the value functions combined for both initial actions; maximizing them we obtain the upper surface drawn with a bold line. Based on the upper value, the regions for selecting the next action are shown in the bottom

known as $\alpha$-*vectors*. As we saw in the example, the optimal value as a function of the belief state corresponds to the upper surface of a set of linear functions, so it is piecewise linear and convex. The convexity of the optimal value makes intuitive sense, as it has higher values in the extremes of the belief space (i.e, in the two states example, when $P(s_1) = 1$ or $P(s_2) = 1$) where the uncertainty is lower (low-entropy) and the agent takes actions more likely to be appropriate. In the "middle" of the belief state the entropy is higher and it is more difficult for the agent to select the best actions.

### *12.3.2 Solution Algorithms*

Solving a POMDP is a computationally hard problem, finding the optimal policy for finite horizon problems is PSPACE-Complete [4]. As a result, several approximate solution techniques have been developed. The main types of approaches for finding approximate solutions to a POMDP are the following: (i) Value iteration, (ii) Policy search, and (iii) Forward search. Next we describe the value iteration algorithm; see the additional reading section for references of the alternative approaches.

#### 12.3.2.1 Value Iteration

The value iteration algorithm for POMDPs computes the value function for increasing horizons, considering a finite horizon of length $H$. The algorithm constructs the value function as this combination of linear functions or $\alpha$-vectors in a recursive way. The $\alpha$-vectors at step $n$ are built by adding the reward vector for the action at the root (of the policy tree) and the $\alpha$-vectors of the step $n-1$ policy trees for each observation. The algorithm returns a set of $\alpha$-vectors whose upper surface represents the maximum value for each belief state. The optimum policy that gives a mapping from belief states to actions is derived from these $\alpha$-vectors by computing a one step lookahead according to Bellman's equation:

$$\pi^*(b) = argmax_a \left[ R(b,a) + \gamma \sum_o P(o \mid b,a) max_\alpha \alpha(b^a_o) \right], \forall b \qquad (12.8)$$

The execution of the policy is done by alternating between updating the belief state (Eq. 12.5) and selecting the best action according to Bellman's equation. Assuming an initial belief state the following steps are repeated:

1. Select the best action for the current belief state.
2. Execute the action.
3. Receive an observation.
4. Update the belief state.

The computational complexity of POMDP value iteration is $O(|S|^2|A|^{|O^H|})$; where$|S|$ is the dimension of the state space, $|A|$ the dimension of the action space, $|O|$ the number of observations and $H$ the horizon [6]. This is intractable except for very small problems, as it is exponential in $|O|$ and double exponential in $H$. In practice, for many problems the initial belief $b_0$ is known, so the algorithm can restrict the attention to only those belief points that are reachable from $b_0$. This results in a practical variant of value iteration, known as *point-based value iteration* [5]; which computes in each step a set $\alpha$-vectors bounded by a fixed number of belief points, $B$ (one optimal $\alpha$-vector per belief). The set of $\alpha$-vectors may not contain all those necessary to represent the optimal value function, but in practice a reasonable number of $\alpha$-vectors is sufficient for a good policy.

**Point-Based Value Iteration**

The point-based value iteration (PBVI) algorithm solves a POMDP for a finite set of belief points $B = \{b_0, b_1, \dots b_q\}$. It initializes a separate $\alpha$-vector for each selected point $\{\alpha_0, \alpha_1, \dots, \alpha_q\}$. Given a solution set $\mathcal{P}^n$ only one $\alpha$-vector per belief point is maintained. By maintaining an $\alpha$-vector for each belief point, PBVI preserves the piece-wise linearity and convexity of the value function, and defines a value function over the entire belief simplex. The complete PBVI algorithm is designed as an anytime algorithm, interleaving steps of value iteration and steps of belief set expansion.

To plan for a finite set of belief points, PBVI modifies the backup operator such that only one $\alpha$-vector per belief point is maintained. When performing point-based updates, the backup creates projections as in exact value iteration. However the final solution $V^*$ is limited to containing only $|B|$ components (in time $|S||A||V||O||B|$). Thus a full point value update takes only polynomial time, and even more crucial, the size of the solution set $V$ remains constant. As a result, the pruning of vectors (and solving of linear programs), so crucial in exact POMDP algorithms, is now unnecessary. The only pruning step is to refrain from adding any vector already included, which arises when two nearby belief points support the same vector.

For the expansion phase it initializes the set $B$ to contain the initial belief $b_0$ and expand $B$ by greedily choosing new reachable beliefs that improve the worst-case density as rapidly as possible. PBVI stochastically simulates a single- step forward trajectory using each action to produce new beliefs $B = \{b_{a0}, b_{a1}, \ldots\}$. Given a solution set $\mathscr{P}^n$, the value function is modified so only one $\alpha$-vector per belief point is maintained. The modified algorithm now gives an $\alpha$-vector which is valid over a region around $b$. It assumes that the other belief points in that region have the same action choice and lead to the same facets of $V_{t-1}$ as the point $b$.

Finally it keeps only the new belief, $b_{ai}$, which is farthest away from any point already in $B$. PBVI tries to generate one new belief from each previous belief; so $B$ at most doubles in size on each expansion. The complete algorithm of PBVI terminates once a fixed number of expansions have been completed. The basic algorithm selects belief points at random. More sophisticated approaches for selecting beliefs are: stochastic simulation with random action, stochastic simulation with greedy action, stochastic simulation with exploratory action.

PBVI is summarized in Algorithm 12.1 [6].

---

**Algorithm 12.1** The Point-Based Value Iteration Algorithm.

---

**Require:** POMPD, $\epsilon$
1: $n \leftarrow 0$, $\mathscr{P}^0 \leftarrow \{0\}$
2: **repeat**
3: $\quad n \leftarrow n+1$
4: $\quad \mathscr{P}^n \leftarrow \{\}$
5: $\quad$ **for** $b \in B$ **do**
6: $\quad\quad \alpha_{ao} \leftarrow argmax_{\alpha \in \mathscr{P}^{n-1}} \alpha(b), \forall a, o$
7: $\quad\quad a^* = argmax_a \{R(b, a) + \gamma \sum_o P(o \mid b, a) \alpha_{ao}(b_o^a)\}$
8: $\quad\quad \alpha(b) = R(b, a^*) + \gamma \sum_o P(o \mid b, a^*) \alpha_{a^*o}(b_o^a)$
9: $\quad\quad \mathscr{P}^n \leftarrow \mathscr{P}^n \cup \{\alpha\}$
10: $\quad$ **end for**
11: **until** $\max_{b \in B} \mid \max_{\alpha \in \mathscr{P}^n} \alpha(b) - \max_{\alpha' \in \mathscr{P}^{n-1}} \alpha'(b) \mid \leq \epsilon$
12: **return** $\mathscr{P}^n$

---

## 12.4 Applications

We describe two applications of POMDPs. The first one is for automatic adaptation of the game difficulty in virtual rehabilitation, which starts from a general policy obtained from a POMDP, and then is refined for each user based on reinforcement learning. The second one addresses the problem of task planing in robotics, and in particular the challenge of large state spaces via hierarchical POMDPs.

### *12.4.1 Automatic Adaptation in Virtual Rehabilitation*

Motor impairment subsequent to a stroke leaves survivors functionally disabled, in many cases affecting the movement of one of the upper limbs (arm and hand). Rehabilitation therapies aim at alleviating motor sequels and returning the patient some of its former independence. Virtual rehabilitation is among the available therapies for the physically impaired; it consists on the use of virtual reality scenarios to afford training environments with a great and versatile capacity for feedback and customization. A virtual environment is a simulation of the real world that is generated through computer software and it is experienced by the user through a human-machine interface.

Gesture Therapy [11] is a virtual platform for the rehabilitation of the upper limb. It consists of several serious games specially designed for rehabilitation, and a gripper that the patient holds with the affected arm, see Fig. 12.8. The gripper includes a colored ball that facilitates the tracking of the hand of the patient using a monocular camera, and a pressure sensor used for detecting the closing and opening of the hand. A patient interacts with the games by moving her arm and closing/opening her hand, and in this way she performs the rehabilitation exercises that will help her recover the movements of the affected limb. The games have different difficulty levels by

**Fig. 12.8** A user interacting with Gesture Therapy; the interface of one of the games is shown in the computer screen. The hand gripper facilitates tracking of arm movements as well as incorporates a sensor for measuring gripping force

varying, for instance, the distance or velocity at which the targets appear in the screen, or the number of targets. According to the patient's initial capacities and progress in the therapy, the difficulty level should be adjusted to provide the optimal level of challenge. In a hospital setting this adjustment is performed by an experienced therapist; however, if the system is used by the patient at home without a therapist present, the system should adjust automatically the level of difficulty of the games.

An intelligent adaptation system [1] for a virtual rehabilitation platform, Gesture Therapy, was developed to control the games difficulty during a rehabilitation session. This adaptation model is based on POMDPs. The model deals with within-game adaptation; after each game is played for a certain amount of time, the speed and control exhibited by the patient during the game are evaluated. Depending on the evaluation the decision-theoretical adaptation system proposes a new level of challenge for the next game to match the patient's progress through the therapy. The objective is to favor improvement in both, speed and control, so the reward function was defined accordingly.

The model has two major elements. First, a POMDP that was specified based on expert knowledge, is solved to obtain an initial, general policy. Second, a reinforcement learning algorithm progressively tailors this initial model to a particular patient. The improvement of the initial policy comes from sensing the performance of the user and also from educative rewards from the therapist. Thus, the new dynamic policy built by means of reinforcement learning may vary from patient to patient, as well as for the same patient as the therapy progresses. In other words, an initial, general model to the patient is iteratively refined ensuring the policy remains optimal to fit the actual circumstances, and in turn the actions chosen by the dynamic policy permit adaptation of the virtual environment.

#### 12.4.1.1 Model

The initial model is a POMDP in which the system's states, $S$, are described by a discretized bivariate performance space relying on the subject *speed* and *control*. Speed is measured based on the on-screen avatar trace path length over the time needed to fulfill the task, and expressed relatively against empirically found values for a healthy subject, with three possible intervals: *low*, *medium* and *high*. Control is calculated as the deviation in pixels from a straight path in the screen going from the user avatar location at task onset to the location of the task target. Control is also expressed relatively against ranges for healthy subjects, with three possible intervals: *poor*, *fair* and *good*.

The actions considered by the POMDP are either to *increase*, *maintain* or *decrease* the challenge level of the game, the increments or decrements of the difficulty levels are limited by each game capacity. The reward function was defined to favor a balance between speed and control, and the transition function was defined subjectively based on experts' experience. The full details can be found in [1]. The optimal policy is obtained with the value iteration algorithm.

#### 12.4.1.2 Policy Improvement

Reinforcement learning is a machine learning paradigm inspired by how a human learns. An agent evolves to maximize expected rewards as it receives feedback from the environment [13]. A common reinforcement learning algorithm is Q-*Learning*. The quality function $Q(s, a)$ specifies the value (expected accumulated reward) of executing action $a$ whilst in state $s$. $Q(s_t, a_t)$ evaluates the rewards obtained through the course of actions. In Q-*Learning*, the value of $Q(s_t, a_t)$ is updated considering the action that maximizes the expected utility. The goal of the learning agent is consequently to maximize its accumulated reward and it does so by learning which action is optimal for each state.

The original version of the Q-*Learning* considers only a single source of rewards, i.e. those given by the environment. In order to accelerate the learning process, a second reward based on the therapist's feedback is incorporated, using an strategy known as *reward shaping*. The new learning algorithm, $Q+$, is essentially the same as Q-*Learning*, except for the reward considered when the $Q$ values are updated is the sum of the environment reward, $r$, and the shaping reward, $f$. In the case of virtual rehabilitation, the shaping reward is provided by the therapist, by indicating if the action suggested by the current policy is correct (positive reward) or incorrect (negative reward). In this way the initial general policy is improved and adapted for each particular patient.

#### 12.4.1.3 Preliminary Evaluation

If the virtual therapy is to be deployed to the patient's home without continuous expert supervision, it is critical that the adaptation model makes the system to behave intelligently to replicate what the expert would recommend at any given time. For this, the initial policy should be optimized for a particular patient in a few sessions at the clinic under the therapist's supervision. Ideally, the policy updating period should not last more than 2–3 therapeutics sessions. Considering a common rehabilitation session of about 45 min, and assuming that during each session the patient performs in average 10 to 15 tasks (games) of about 3 min per game; the policy shall be adjusted within 30 to 45 feedback iterations.

An initial experimental evaluation of the adaptation system was conducted through a small feasibility study in laboratory controlled conditions. A few subjects played the role of patients and two experts, representing therapists, assessed the model decisions. The objective of this experiment was to analyze how rapidly the policy is improved to match the therapists decisions. Four subjects participated in the study. A physician and an experienced researcher played the roles of the therapists. The experiment was carried out in a single session, in which each participant played 25 blocks of 2 tasks (50 learning policy updates episodes) selected among a set of the five rehabilitation games available in Gesture Therapy.

Table 12.1 summarizes the congruence between the model and the expert decisions expressed as a percentage of agreements over the total decisions. Congruence levels

**Table 12.1** Agreement of the expert with the decisions taken by the adaptation model per participant

| Subject | 1 | 2 | 3 | 4 |
|---|---|---|---|---|
| Congruence | 56% | 92% | 96% | 100% |

surpassed 90% for 3 of the subjects. Congruence in the remaining subject (1) was lower following some initial disagreements at the beginning of the session, strongly influencing subsequent decisions. For more details on the adaptation system and experimental results see [1, 12].

### *12.4.2 Hierarchical POMDPs for Task Planning in Robotics*

Two important challenges when applying POMDPs are: (i) designing a representation that models as best as possible a particular problem, and (ii) the complexity of finding a good policy with large state spaces. To address both challenges, an architecture for task planning oriented towards service robotics is proposed, that combines a knowledge representation scheme and hierarchical POMDPs to automatically build a hierarchy of actions that enables the decomposition of the problem [8]. The knowledge representation defines a list of parameters, so that domain specific information can be encoded by a designer, and used by the architecture to automatically generate and execute plans to solve tasks.

In order to solve task planning problems, the proposed architecture follows three steps: (a) knowledge base construction (KBC), (b) architecture initialization (AI), and (c) architecture operation (AO). In the KBC step, a human designer is required to encode, into the architecture's knowledge base (KB), domain specific information that describes the robot's skill set and the particular scenario in which it will operate. Next, in the AI step, the architecture builds a POMDP from the information in the KB and uses a hierarchical description of the environment to automatically build a hierarchy of POMDPs. Finally, in the AO step, the architecture is ready to receive task requests, for which it builds and executes *Hierarchical Policies* based on the hierarchy of POMDPs.

#### 12.4.2.1 Knowledge Base

The encoding of the KB consists in providing general and specific knowledge. The general knowledge is specified by three components: basic modules, domain dynamics, and a hierarchical function. Whilst the specific knowledge is defined by four lists of elements: concrete values, abstract values, neighborhood and hierarchical function pairs. With regards to the general knowledge, for each skill set the robot has, e.g. navigation, object manipulation, etc., a basic module must be specified. The domain dynamics must contain a description of how the robot's skills can interact with the environment, while the hierarchical function is employed to abstract the state space. The specific knowledge describes the particular environment in which the robot will operate. This description consists of a list of objects that characterize the whole scenario (concrete values), the objects in which the domain can be abstracted (abstract values), the pairs of objects that are neighbors by some action (neighborhood pairs) and pairs of objects that exist in the hierarchical function (hierarchical function pairs).

For each basic module, for example, navigation, the following elements are specified: (i) actions, (ii) state variables, (iii) observations, (iv) state transitions, (v) observation transitions. Additionally, a set of constraints, relations and rules are defined for the dynamics of the environment, that include: (a) neighborhood relations, (b) causal laws, (c) state constraints, (d) executability conditions. To build de hierarchy of POMDPs, the system requires a function that describes a hierarchy, whose leaf nodes are the values for one of the state variables, while internal nodes are abstract values the designer must provide in the specific knowledge. Such function should map a value to its parent, within the hierarchy.

#### 12.4.2.2 Architecture Initialization

After the general and specific knowledge has been specified, the architecture uses the domain dynamics and the set of concrete values to build a stochastic transition diagram. The sets of state and observation transitions are used to assign probabilities in the transition diagram, leading to the definition of the bottom POMDP. Next, a recursive definition for abstract actions is introduced to build a hierarchy of POMDPs from the bottom POMDP and hierarchical function.

The construction of the bottom POMDP from the KB proceeds as follows. Let a POMDP be defined by a tuple $M = \langle S, A, \Phi, R, O, \Omega, B_0 \rangle$ where $S$, $A$, and $O$ are the sets of states, actions and observations, respectively, $\Phi$, $\Omega$ and $R$ are the transition, observation and reward function, respectively, and $B_0$ the initial belief distribution. The purpose of the bottom POMDP is to describe the dynamics of the environment. Thus, the basic modules, domain dynamics and specific knowledge (specified in the knowledge base) are employed to define all the parameters of $M$, with the exception of $R$ and $B_0$.

The hierarchical function, defined in the knowledge base, is used to build a hierarchical representation of $S$. This representation is employed along with the bottom POMDP to build a hierarchy of POMDPs that will represent abstract actions, that the architecture can later use to generate plans. See [8] for details.

#### 12.4.2.3 Architecture Operation

During operation phase, the agent is ready to receive task requests, which must be passed as an $n$-tuple that specifies a value for each one of the $n$ state variables (defined in the knowledge base). The architecture assumes that the agent's state is

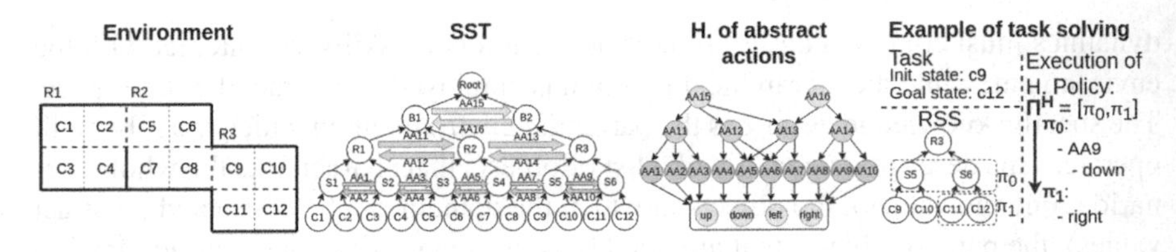


**Fig. 12.9** Navigation example. The **Environment** is composed by 12 cells, which are abstracted into 6 sections, 3 rooms and 2 buildings to define the environment's state-space tree (**SST**). Abstract actions (thick arrows) are computed to transit between specific pairs of neighbor abstract states and build the **Hierarchy of abstract actions**. An **Example of task solving** is shown, defined as $c9 \rightarrow c12$, its RSS is specified and a hierarchical policy is computed to reach $c12$. The execution of $\Pi^H$ consists in first transiting to $s6$ and then to $c12$

known with certainty at the moment a task request is received. Thus, for a pair of initial and goal states, a *hierarchical policy* is computed, and executed in a top-down way to gradually bring the agent to the goal state.

A hierarchical policy is built to operate over a sub-region of the state space, called *Relevant Sub-Space* (RSS). For a pair $\langle s_0, s_{goal} \rangle$, an RSS is a sub-tree from the state-space tree (SST), whose root node is the deepest common ancestor of $s_0$ and $s_{goal}$ in the SST. Also, in order to represent the goal state at several levels within the RSS, a *hierarchical state* is defined, which is a path from the RSS root to the goal state.

The execution of a hierarchical policy $\Pi^H$ is performed in a top-down manner, i.e. when $\Pi^H[i]$ reaches $G^H[i+1]$ (the state it was designed to reach), $\Pi^H[i]$ terminates and $\Pi^H[i+1]$ starts its execution. In the example from Fig. 12.9, the hierarchical policy $\Pi^H$ is employed to reach cell $c12$ from $c9$. First, $\pi_0$ is executed, which invokes $AA9$ that in turn is a policy and executes $down$. By reaching $c11$, both $AA9$ and $\pi_0$ have reached their goal states simultaneously, a children of $S6$ and $S6$, respectively. Finally, $\pi_1$ starts and invokes $right$ to reach the task's goal state.

#### 12.4.2.4 Example: Robot Navigation

To illustrate the proposed architecture, a mobile robot navigation domain is employed, see Fig. 12.9. In this domain, an environment is modeled as a sequence of interconnected buildings, each one discretized as a grid of square uniform cells. The bottom POMDP, obtained from the description of the domain encoded in the knowledge base, has a state and an observation for each cell in the environment, while the set of actions is constituted by four actions: *up,down,left,right*. The transition distribution of each action assigns probabilities of 0.1 and 0.9 to staying in the current cell and transiting to the target cell, respectively. The observation distribution of each action is modeled as a $3 \times 3$ Gaussian kernel centered in the reached cell, whose standard deviation is specified differently for several experimental configurations. For the hierarchical function, four levels are provided (from bottom to the top of the hierarchy): cells, subsections, rooms, buildings.

In simulation experiments it is shown that the hierarchical POMDP has a higher success ratio of reaching the goal against a flat POMDP for different environment dimensions, and it is one order of magnitude faster [8].

## 12.5 Additional Reading

A review of different approaches for solving POMDPs is presented in [6, 7]. The point-based value iteration algorithm is described in detail in [5]. Different applications of POMDPs are included in [10]. Several resources for POMDPs, including a tutorial, code, etc., can be found at: https://www.pomdp.org.

## 12.6 Exercises

1. For the example of the robot in the grid world, Fig. 12.1, specify the POMDP, including the states, actions, rewards, transition probabilities, observation probabilities and initial state probabilities. Assume the observation gives the cell in which the robot is with high probability, and one of the neighboring cells with low probabilities. Define the parameters according to your intuition.
2. Based on the POMDP of the previous problem, assume the that the observation variable can take only two values, $o_1$ and $o_2$. Draw three levels of the policy tree, assuming the initial state is known.
3. According to the parameters of the POMDP in Problem 1, if the initial belief state ($t = 0$) is $(1, 0, 0, 0, 0, 0, 0, 0, 0, 0, 0, 0)$ (the robot is in the top-left cell), estimate the belief state for $t = 1$ for the action $right$.
4. Continue the previous problem, estimating the belief state for $t = 2$, for the action $right$. Define the required parameters.
5. Repeat the previous two problems, if now the actions are $down, down$.
6. Figure 12.5 shows the lines ($\alpha$-$vectors$) for the two states example for horizon 1, obtain the $\alpha$-$vectors$ for horizon 2 and the resulting value function.
7. According to the solution of the previous problem, specify the optimal action for each region of the belief space.
8. Specify a possible POMDP model for the automatic adaptation system described in Sect. 12.4.1.
9. *** Implement the point-based value iteration algorithm.
10. *** Investigate alternative algorithms to solve POMDPs.

## References

1. Ávila-Sansores, S., Orihuela-Espina, F., Sucar, L.E., Álvarez-Cárdenas, P.: Adaptive virtual rehabilitation environments. In: ICML Workshop: Role of Machine Learning in Transforming Health (2013)
2. Bellman, R.: Dynamic Programming. Princeton University Press, Princeton (1957)
3. Meuleau, N., Hauskrecht, M., Kim, K.E., Peshkin, L., Kaelbling, L.P., Dean, T., Boutilier, C.: Solving very large weakly coupled Markov decision processes. In: Proceedings of the Association for the Advancement of Artificial Intelligence (AAAI), pp. 165–172 (1998)
4. Papadimitriou, C.H., Tsitsilis, J.N.: The complexity of Markov decision processes. Math. Oper. Res. **12**(3), 441–450 (1987)
5. Pineau, J., Gordon, G., Thrun, S.: Anytime point-based approximations for large POMDPs. J. Artif. Intell. Res. **27**, 335–380 (2006)
6. Poupart, P.: An introduction to fully and partially observable Markov decision processes. In: Sucar, L.E., Hoey, J., Morales, E. (eds.) Decision Theory Models for Applications in Artificial Intelligence: Concepts and Solutions. IGI Global, Hershey (2011)
7. Ross, S., Pineau, J., Paquet, S., Chaib-draa, B.: Online planning algorithms for POMDPs. J. Artif. Intell. Res. **32**, 663–704 (2008)
8. Santiago, E., Serrano, S., Sucar, L.E.: A knowledge and probabilistic based task planning architecture for service robotics. In: 18th Mexican International Conference on Artificial Intelligence, pp. 646–657. Springer (2019)
9. Smallwood, R.D., Sondik, E.J.: The optimal control of partially observable Markov decision processes over a finite horizon. Oper. Res. **21**, 1071–1088 (1973)
10. Sucar, L.E., Hoey, J., Morales, E.: Decision Theory Models for Applications in Artificial Intelligence: Concepts and Solutions. IGI Global, Hershey (2011)
11. Sucar, L.E., Orihuela-Espina, F., Luis-Velazquez, R., Reinkensmeyer, D.J., Leder, R., Hernández-Franco, J.: Gesture therapy: an upper limb virtual reality-based motor rehabilitation platform. IEEE Trans. Neural Syst. Rehabil. Eng. **22**(13), 634–643 (2014)
12. Sucar, L.E., Ávila-Sansores, S., Orihuela-Espina, F.: User modelling for patient tailored virtual rehabilitation. In: Lucas, P., Hommerson, A. (eds.) Foundations of Biomedical Knowledge Representation, pp. 281–302. Springer, Berlin (2016)
13. Sutton, R.S, Barto, A.G.: Reinforcement Learning: An Introduction. The MIT Press, Cambridge (1998)

# Part IV
# Relational, Causal and Deep Models

The fourth and last part of the book describes two interesting extensions to probabilistic graphical models: relational probabilistic models and causal graphical models. Relational probabilistic models increase the representational power of *standard* PGMs, by combining the expressive power of first-order logic with the uncertain reasoning capabilities of probabilistic models. Causal graphical models go beyond representing probabilistic dependencies, to express cause and effect relations, based on the same framework of graphical models. Additionally, we include a chapter on deep neural network models, their bases and interactions with graphical models.

# Chapter 13
# Relational Probabilistic Graphical Models

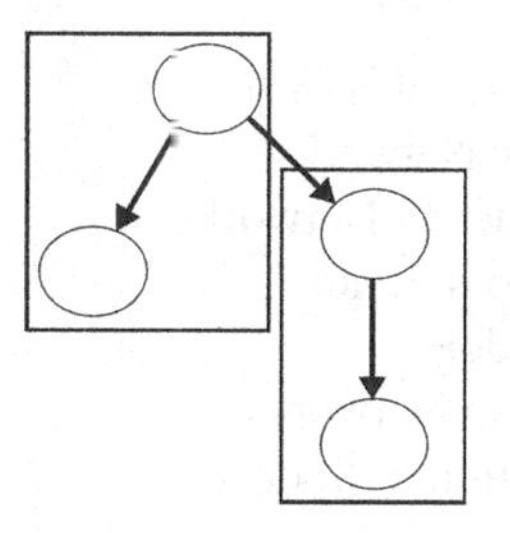

## 13.1 Introduction

The *standard* probabilistic graphical models that have been covered until now, have to represent explicitly each object in the domain, so they are equivalent in terms of their logical expressive power to propositional logic. However, there are problems in which the number of objects (variables) could increase significantly, so a more expressive (compact) representation is desirable. Consider, for instance, that we want to model a student's knowledge of a certain topic (this is known as student modeling or, in general, user modeling), and that we want to include in the model all of the students in a college, where each student is enrolled in several topics. If we model this explicitly with a PGM such as a Bayesian network, it could become a huge model, difficult to acquire and store. Instead, it would be more efficient if in some way we could have a general model that represents the dependency relations for any student, $S$ and any course, $C$, which could then be parameterized for particular cases. This can be done using predicate logic; however standard logical representations do not consider uncertainty. Thus, a formalism is required that combines predicate logic and probabilistic models.

Relational probabilistic graphical models (RPGMs) combine the expressive power of predicate logic with the uncertain reasoning capabilities of probabilistic graphical models. Some of these models extend PGMs such as Bayesian networks or



L. E. Sucar, *Probabilistic Graphical Models*, Advances in Computer Vision and Pattern Recognition, https://doi.org/10.1007/978-3-030-61943-5_13

Markov networks by representing objects, their attributes, and their relations with other objects. Other approaches extend the logic–based representations, in order to incorporate uncertainty, by describing a probability distribution over logic formulas.

Different formalisms have been proposed to combine logic and PGMs. We propose a taxonomy for the classification of RPGMs (inspired and partially based on [5]):

1. Extensions of logic models
    a. Undirected graphical models
        i. Markov Logic Networks
    b. Directed graphical models
        i. Bayesian Logic Programs
        ii. Bayesian Logic Networks
2. Extensions of probabilistic models
    a. Undirected graphical models
        i. Relational Markov Networks
        ii. Relational Dependency Networks
        iii. Conditional Random Fields
    b. Directed graphical models
        i. Relational Bayesian Networks
        ii. Probabilistic Relational Models
3. Extensions of programming languages
    a. Stochastic Logic Programs
    b. Probabilistic Inductive Logic programming
    c. Bayesian Logic (BLOG)
    d. Probabilistic Modeling Language (IBAL)

In this chapter we will review two of them. One, called *probabilistic relational models* [3], extends Bayesian networks to incorporate objects and relations, as in a relational data base. The other are *Markov logic networks* [12], which add weights to logical formulas, and can be considered as an extension of Markov networks. First a brief review of first-order logic is presented, then we describe the two relational-probabilistic approaches, and finally we illustrate their application in two domains: student modeling and visual object recognition.

## 13.2 Logic

Logic is a very well studied representation language with well defined syntax and semantic components. Here we only include a concise introduction, for more extensive references see the additional reading section. We will start by defining propositional logic, and then go into first-order logic.

### 13.2.1 Propositional Logic

Propositional logic allows us to reason about expressions or *propositions* that are *True* (T) or *False* (F). For instance, *Joe is an engineering student*. Propositions are denoted with capital letters, such as $P$, $Q$, ..., known as atomic propositions or *atoms*. Propositions are combined using logic connectives or operators, obtaining what are known as *compound propositions*. The logic operators are:

- Negation: $\neg$
- Conjunction: $\wedge$
- Disjunction: $\vee$
- Implication: $\rightarrow$
- Double implication $\leftrightarrow$

For example, if $P$ ="Joe is an engineering student" and $Q$ ="Joe is young", then $P \wedge Q$ means "Joe is an engineering student AND Joe is young".

Atoms and connectives can be combined according to a set of syntactic rules; valid combinations are known as *well formed formulas* (WFF). A well formed formula in propositional logic is an expression obtained according to the following rules:

1. An atom is a WFF.
2. If $P$ is a WFF, then $\neg P$ is a WFF.
3. If $P$ and $Q$ are WFFs, then $P \wedge Q$, $P \vee Q$, $P \rightarrow Q$, and $P \leftrightarrow Q$ are WFFs.
4. No other formula is a WFF.

For instance, $P \rightarrow (Q \wedge R)$ is a WFF; $\rightarrow P$ and $\vee Q$ are not WFFs.

The meaning (semantics) of a logical formula can be expressed by a function which gives a *True* or *False* value to the formula for each possible *interpretation* (truth values of the atoms in the formula). In the case of propositional logic, the interpretation can be represented as a *truth table*. Table 13.1 depicts the truth tables for the basic logic operators. An interpretation that assigns a truth value to a formula $F$ is a *model* of $F$.

A formula $F$ is a *logical consequence* of a set of formulas $\mathbf{G} = \{G_1, G_2, \ldots, G_n\}$ if for every interpretation for which $\mathbf{G}$ is True, $F$ is True. It is denoted as $G \models F$.

Propositional logic can not express general properties, such as *all engineering students are young*. For this we need first-order logic.

**Table 13.1** Truth tables for the logical connectives

| $P$ | $Q$ | $\neg P$ | $P \wedge Q$ | $P \vee Q$ | $P \rightarrow Q$ | $P \leftrightarrow Q$ |
|---|---|---|---|---|---|---|
| T | T | F | T | T | T | T |
| T | F | F | F | T | F | F |
| F | T | T | F | T | T | F |
| F | F | T | F | F | T | T |

### 13.2.2 First-Order Predicate Logic

Consider the two statements: "All students in a technical university are engineering majors" and "A particular person $t$ of a particular university $\mathscr{I}$ is a computer science major". Whereas the first statement declares a property that applies to all persons in a technical university, the second applies only to a specific person in a particular university. First order logic lets us deal with these differences.

Expressions or formulas in first-order predicate logic are constructed using four types of symbols: *constants, variables, functions*, and *predicates*. Constant symbols represent objects in the domain of interest (e.g., persons, courses, universities, etc.). Variable symbols range over the objects in the domain. Variables, which could be arguments for a predicate or function, are represented with lower case letters, i.e., *x*, *y*, *z*.

Predicates are expressions that can be True or False; they contain a number of arguments. Predicates can represent relations among objects in the domain (e.g., *Above*) or attributes of objects (e.g., *IsRed*). If the number of arguments is zero then it is an atom as in propositional logic. Predicates are represented with capital letters, for instance $P(x)$.

Function symbols (e.g., $neighborOf$) represent mappings from tuples of objects to objects. They are represented with lower case letters and can also have a number of arguments. For example, $f(x)$ will represent a function with one argument, $x$. A function with no arguments is a constant.

Predicate logic includes the logical connectives of propositional logic, plus other operators known as *quantifiers*:

Universal quantifier: $\forall x$ (for all $x$).
Existential quantifier: $\exists x$ (exists an $x$).

A *term* is a constant, variable or a function of terms. An *atomic formula* or *atom* is a predicate that has as arguments *N* terms.

Now we can rewrite the example at the beginning of the section in first-order logic. If the formula $M(t_x, \mathscr{I})$ represents the major of a person $t_x$ in a technical university $\mathscr{I}$, the statement "All students in a technical university are engineering majors" can be rewritten as:

$$\forall t_x \forall \mathscr{I} : M(t_x, \mathscr{I}) = engineering. \tag{13.1}$$

Similarly to predicate logic, in first-order predicate logic there are a set of syntactic rules that define which expressions are well formed formulas:

1. An atom is a WFF.
2. If $P$ is a WFF, then $\neg P$ is a WFF.
3. If $P$ and $Q$ are WFFs, then $P \wedge Q$, $P \vee Q$, $P \rightarrow Q$, and $P \leftrightarrow Q$ are WFFs.
4. If $P$ is a WFF and $x$ is a free variable in $P$, then $\forall x P$ and $\exists x P$ are WFFs.
5. No other formula is a WFF.

Roughly speaking, a first-order knowledge base (*KB*) is a set of sentences or formulas in first-order logic [4].

An $\mathscr{L}$-interpretation specifies which objects, functions, and relations in the domain are represented by which symbols. Variables and constants may be typed, in which case variables range only over objects of the corresponding type, and constants can only represent objects of the corresponding type. For example, the variable $x$ might range over universities (e.g., public universities, private universities, etc.), and the constant $C$ might represent a university or a specific set of universities (e.g., $u_2$, $(u_6 \cup u_2)$, etc.). An *atomic formula* (or simply *atom*) is a predicate symbol applied to a tuple of terms: $Near(u_4, u_1)$.

In standard predicate logic all predicates are $True$ or $False$, so it can not deal with probabilistic uncertainty directly. For example, if we do not know if a person is in a certain university, we can only say that $Univ(p) = u_1$ OR $Univ(p) = u_2$; but we can not specify if it is more probable that she is in $u_1$ than in $u_2$.

To have the expressive power of predicate logic and at the same time be able to represent and reason under uncertainty (in probabilistic terms) we need to combine predicate logic with probabilistic models under a single framework. Next we will describe two of these frameworks.

## 13.3 Probabilistic Relational Models

Probabilistic relational models (PRMs) [6] are an extension of Bayesian networks that provide a more expressive, object-oriented representation facilitating knowledge acquisition. They also makes it easier to extend a model to other domains. For the case of a very large model, only part of it is considered at any time, so the inference complexity is reduced.

The basic entities in a PRM are objects or domain entities. Objects in the domain are partitioned into a set of disjoint classes $X_1, \ldots, X_n$. Each class is associated with a set of attributes $A(X_i)$. Each attribute $A_{ij} \in A(X_i)$ (that is, attribute $j$ of class $i$) takes on values in some fixed domain of values $V(A_{ij})$. $X.A$ denotes the attribute $A$ of an object in class $X$ [6]. A set of relations, $R_j$, are defined between the classes. A binary relation $R(X_i, X_j)$ between classes $X_i$ and $X_j$ can be viewed as a *slot* of $X_i$. The classes and relations define the *schema* of the model. Then, a PRM defines a probability distribution over a set of instances of a schema; in particular, a distribution over the attributes of the objects in the model.

The dependency model is defined at the class level, allowing it to be used for any object in the class. PRMs explicitly use the relational structure of the model, so an attribute of an object will depend on some attributes of related objects. A PRM specifies the probability distribution using the same underlying principles used in Bayesian networks. Each of the random variables in a PRM, the attributes $x.a$ of the individual objects $x$, is directly influenced by other attributes, which are its parents. A PRM, therefore, defines for each attribute, a set of parents, which are the directed influences on it, and a local probabilistic model that specifies probabilistic parameters.

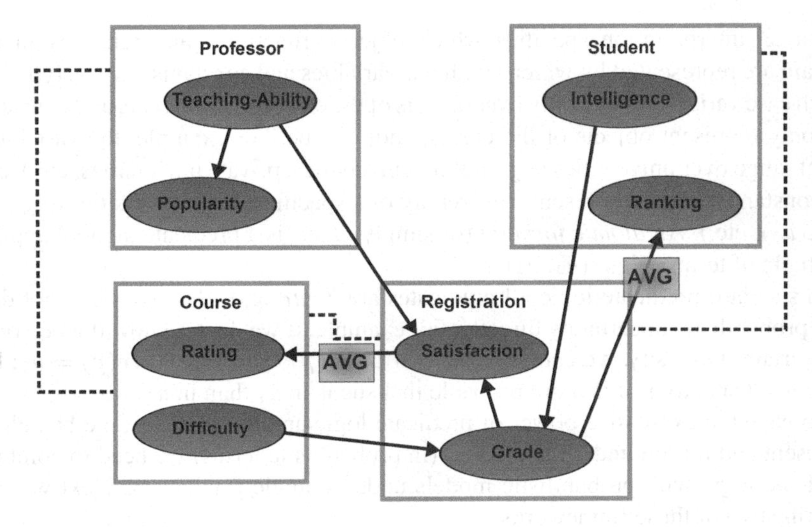


**Fig. 13.1** An example of a PRM structure for the school domain. Dashed edges represent relations between classes, and arrows correspond to a probabilistic dependency. The $AVG$ in a link indicates that the conditional probabilities depend on this variable

There are two basic differences between PRMs and Bayesian networks [6]: (i) In a PRM the dependency model is specified at the class level, allowing it to be used for any object in the class. (ii) A PRM explicitly uses the relational structure of the model, allowing an attribute of an object to depend on attributes of related objects.

An example of a PRM for the school domain, based on [6], is shown in Fig. 13.1. There are 4 classes, with 2 attributes each, in this example:

Professor: teaching-ability, popularity
Student: intelligence, ranking
Course: rating, difficulty
Registration: satisfaction, grade

This representation allows for two types of attributes in each class: (i) information variables, (ii) random variables. The random variables are the ones that are linked in a kind of Bayesian network that is called a *skeleton*. From this skeleton, different Bayesian networks can be generated, according to other variables in the model. For example, in the student model described in Sect. 13.5, we define a general skeleton for an experiment, from which particular instances for each experiment are generated. This gives the model a greater flexibility and generality, facilitating knowledge acquisition. It also makes inference more efficient, because only part of the model is used in each particular case.

The probability distribution for the skeletons are specified as in Bayesian networks. A PRM, therefore, defines for each attribute $x.a$, a set of parents, which are the directed influences on it, and a local probabilistic model that specifies the conditional probability of the attribute given its parents. To guarantee that the local models define a coherent global probability distribution, the underlying dependency structure should be acyclic, as in a BN. Then, the joint distribution is equal to the product of the local distributions, similar to Bayesian networks.

### *13.3.1 Inference*

The inference mechanism for PRMs is the same as for Bayesian networks, once the model is instantiated to particular objects in the domain. However, PRMs can take advantage of two properties to make inference more efficient. These properties are made explicit by the PRM representation, in contrast to Bayesian networks where they are only implicit.

One property is the locality of influence, most attributes will tend to depend mostly on attributes of the same class, and there are few interclass dependencies. Probabilistic inference techniques can take advantage of this locality property by using a divide and conquer approach.

The other aspect which can make inference more efficient is reuse. In a PRM there are usually several objects of the same class, with similar structure and parameters. Then, once inference is performed for one object, this can be reused for the other similar objects.

### *13.3.2 Learning*

Given that PRMs share the same underlying principles of BNs, the learning techniques developed for BNs can be extended for PRMs. The expectation maximization algorithm has been extended to learn the parameters of a PRM, and structure learning techniques have been developed to learn the dependency structure from a relational database [3].

## 13.4 Markov Logic Networks

In contrast to PRMs, Markov logic networks (MLN) start from a logic representation, adding *weights* to formulas to incorporate uncertainty.

In logic, a $\mathscr{L}$-interpretation which violates a formula given in a knowledge base (KB) has zero probability. It means that its occurrence is impossible, because all the *possible worlds* must be consistent with the KB. In Markov Logic Networks, this assumption is relaxed. If the interpretation violates the KB, it has *less* probability than others with no violations. Less probability means that it has a non zero probability. In a MLN, a weight to each formula is added in order to reflect how strong the constraint is: higher weights entail higher changes in probability whether that interpretation satisfies that formula or not.

Before we formally define a MLN, we need to review Markov networks, also known as Markov random fields (see Chap. 6). A Markov Network is a model for the joint distribution of a set of variables $\mathbf{X} = (X_1, X_2, \ldots, X_n) \in \mathcal{X}$. It is composed of an undirected graph $G$ and a set of potential functions $\phi_k$. The associated graph has a node for each variable, and the model has a potential function for each clique in the graph. The joint distribution represented by a Markov network is given by

$$P(\mathbf{X} = \mathbf{x}) = \frac{1}{z} \prod_k \phi_k(x_{\{k\}}) \tag{13.2}$$

where $x_{\{k\}}$ is the state of the $k$th clique, and $z$ is the partition function.

Normally, Markov Networks can also be represented using *log-linear* models, where each clique potential function is replaced by an exponential weighted sum:

$$P(\mathbf{X} = \mathbf{x}) = \frac{1}{z} \exp \sum_j w_j f_j(x) \tag{13.3}$$

where $w_j$ is a weight (real value) and $f_j$ is, for our purposes, a binary formula $f_j(x) \in \{0, 1\}$.

We now provide a formal definition for a MLN [1].

An MLN $L$ is a set of pairs $(F_i, w_i)$, where $F_i$ is a formula in first-order logic and $w_i$ is a real number. Together with a finite set of constants $C = \{c_1, c_2, \ldots, c_{|C|}\}$, it defines a Markov network $M_{L,C}$:

1. $M_{L,C}$ contains one binary node for each possible grounding of each formula appearing in the MLN $L$. The value of the node is 1 if the ground atom is true, and 0 otherwise.
2. $M_{L,C}$ contains one feature for each possible grounding of each formula $F_i$ in $L$. The value of this feature is 1 if the ground formula is true, and 0 otherwise. The weight of the feature is the $w_i$ associated with $F_i$ in $L$.

MLNs are a generalization of Markov networks, so they can be seen as templates for constructing Markov networks. Given a MLN and a set of different constants, different Markov networks can be produced; these are known as ground Markov networks. The joint probability distribution of a ground Markov network is defined in a similar way as a Markov network, using Eq. 13.3.

The graphical structure of a MLN is based on its definition; there is an edge between two nodes of the MLN if the corresponding ground atoms appear together in at least one grounding of one formula in the knowledge base, $L$. For example [1], consider the following MLN consisting of two logical formulas:

$$\forall x Smoking(x) \rightarrow Cancer(x)$$

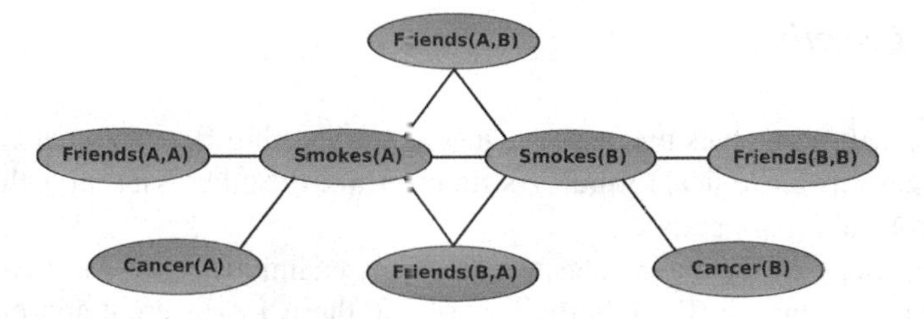


**Fig. 13.2** Structure of the ground MLN obtained from the two logical formulas of the example. (Figure based on [1].)

$$\forall x \forall y Friends(x, y) \rightarrow (Smoking(x) \leftrightarrow Smoking(y))$$

If the variables $x$ and $y$ are instantiated to the constants $A$ and $B$, we obtain the structure depicted in Fig. 13.2.

## 13.4.1 Inference

Inference in MLNs consists in estimating the probability of a logical formula, $F_1$, given that another formula (or formulas), $F_2$, are true. That is, calculating $P(F_1 \mid F_2, L, C)$, where $L$ is a MLN consisting of a set of weighted logical formulas, and $C$ is a set of constants. To compute this probability, we can estimate the proportion of possible worlds in which $F_1$ and $F_2$ are true, over the possible worlds in which $F_2$ holds. For this calculation, the probability of each possible world is considered according to the weights of the formulas and the structure of the corresponding grounded Markov network.

Performing the previous calculations directly is, computationally, very costly; thus, it becomes prohibitive but for very small models. Several alternative probabilistic inference techniques can be used to make this computation more efficient. One alternative is using stochastic simulation; by sampling the possible worlds we can obtain an estimate of the desired probability; for instance, using Markov chain Montecarlo techniques.

Another alternative is to make certain reasonable assumptions about the structure of the logical formulas that simplify the inference process. In [1], they develop an efficient inference algorithm for the case that $F_1$ and $F_2$ are conjunctions of ground literals.

### 13.4.2 Learning

Learning a MLN involves two aspects, just as for learning Bayesian networks. One aspect is learning the logical formulas (structure), and the other is learning the weights for each formula (parameters).

For learning the logical formulas, we can apply techniques from the area of inductive logic programming (ILP). In the area of ILP there are different approaches that can induce logical relations from data, usually considering some background knowledge. For more information, see the additional reading section.

The weights of the logical formulas can be learned from a relational database. Basically, the weight of a formula is proportional to its number of true groundings in the data with respect to its expectation according to the model. Counting the number of true groundings of a formula could be computationally expensive for large domains. An alternative is make an estimate based on sampling the groundings of a formula and verifying if they are true in the data.

An example of a MLN in the context of an application for visual object recognition is presented in Sect. 13.5.

## 13.5 Applications

We will illustrate the application of the two classes of RPGMs presented in the previous sections in two different domains. First we will see how we can build a kind of *general* student model for virtual laboratories based on PRMs. Then we will use MLNs for representing visual grammars for object recognition.

### 13.5.1 Student Modeling

A particularly challenging area for student modeling are virtual laboratories. A virtual lab provides a simulated model of some equipment, so that students can interact with it and learn by doing. A tutor serves as a virtual assistant in this lab, providing help and advice to the user, and setting the difficulty of the experiments, according to the student's level. In general, it is not desirable to trouble the student with questions and tests to update the student model. So the cognitive state should be obtained based solely on the student's interactions with the virtual lab and the results of the experiments. For this a student model is required. The model infers, from the student's interactions with the laboratory, the cognitive state; and based on this model, an intelligent tutor can give personalized advice to the student [14].

#### 13.5.1.1 Probabilistic Relational Student Model

PRMs provide a compact and natural representation for student modeling. Probabilistic relational models allow each attending student to be represented in the same model. Each class represents the set of parameters of several students, like in databases, but the model also includes the probabilistic dependencies between classes for each student.

In order to apply PRMs to student modeling we have to define the main objects involved in the domain. A general student model oriented to virtual laboratories was designed, starting from a high level structure at the class level, and ending with specific Bayesian networks for different experiments at the lower level. As shown in Fig. 13.3, the main classes, related with students and experiments, were defined. In this case there are 8 classes, with several attributes for each class, as listed below:

Student: student-id, student-name, major, quarter, category.
Knowledge Theme: student-id, knowledge-theme-id, knowledge-theme-known.
Knowledge Sub-theme: student-id, knowledge-sub-theme-id, knowledge-sub-theme-known.
Knowledge Items: student-id, knowledge-item-id, knowledge-item-known.
Academic background: previous-course, grade.
Student behavior: student-id, experiment-id, behavior-var1, behavior-var2, …
Experiment results: student-id, experiment-id, experiment-repetition, result-var1, result-var2, …
Experiments: experiment-id, experiment-description, exp-var1, exp-var2, …

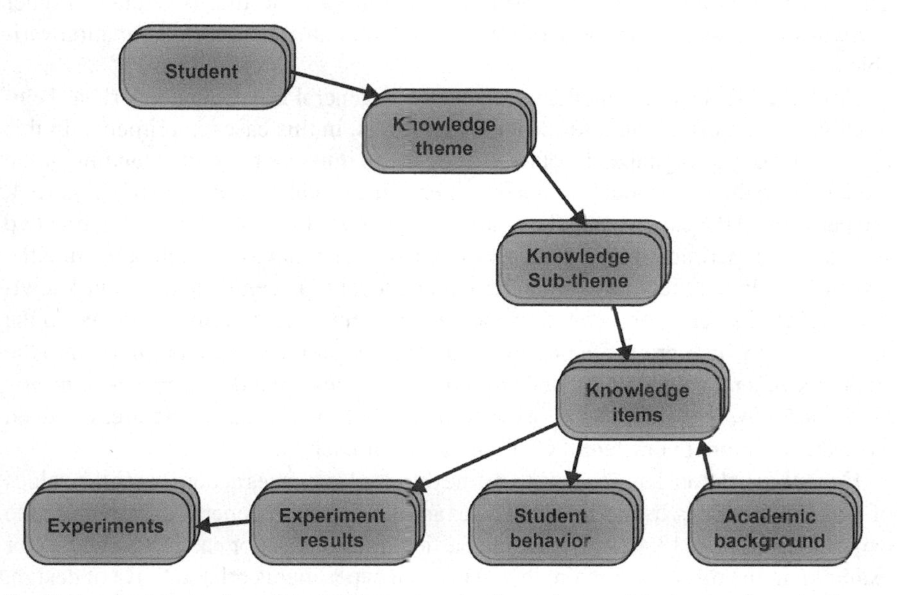


**Fig. 13.3** A high level view of the PRM structure for the student model, showing the main classes and their relations

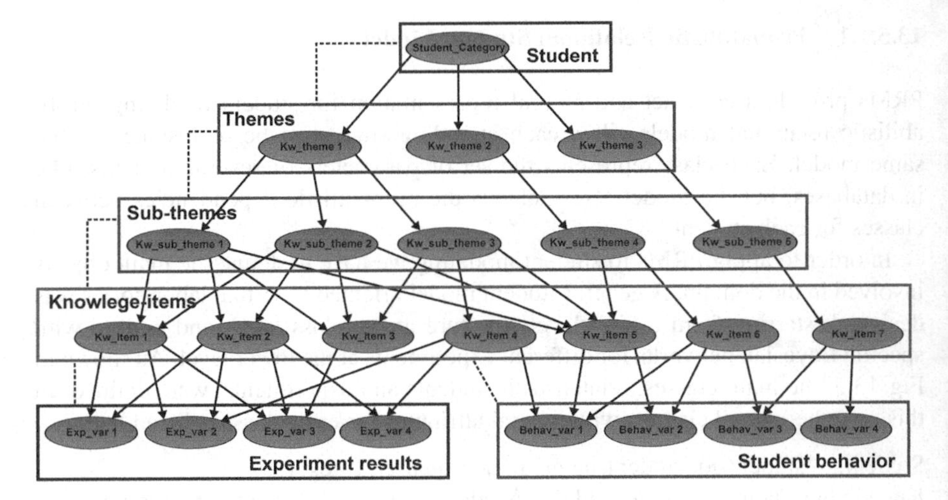


**Fig. 13.4** A general skeleton for an experiment derived from the PRM student model for virtual laboratories. Basically the network has a hierarchical structure, starting form a node that represents the student category at the top, and then with three layers of variables that represent the knowledge the student has of the domain at different levels of abstraction: themes, sub-themes and items. At the bottom level there are two sets of variables that correspond to the results of the student's interaction with an experiment in the virtual lab, divided in experimental results and student behavior [14]

The dependency model is defined at the class level, allowing it to be used for any object in the class. Some attributes in this model represent probabilistic variables. This means that an attribute represents a random variable that is related to other attributes in the same class or in other classes. Other attributes are information variables.

From the PRM student model we can define a general Bayesian network, a skeleton, that can be instantiated for different scenarios, in this case experiments. In this model it is easy to organize the classes by levels to improve the understanding of the model. From the class model we obtain a hierarchical skeleton, as shown in Fig. 13.4. We partitioned the experiment class, according to our object of interest, creating two subclasses: experiment performance and experiment behavior, which constitute the lowest level in the hierarchy. The intermediate level represents the different knowledge items (concepts) associated to each experiment. These items are linked to the highest level which groups the items in sub-themes and themes, and finally into the student's general category. We defined three categories of students: *novice, intermediate and advanced*. Each category has the same Bayesian net structure, obtained from the skeleton, but different CPTs are used for each one.

From the skeleton, it is possible to define different instances according to the values of specific variables in the model. For example, from the general skeleton for the experiments of Fig. 13.4, we can define particular instances for each experiment (for example, in the robotics domain, there could be experiments related to robot design, control, motion planning, etc.) and student level (novice, intermediate, advanced).

Once a specific Bayesian network is generated, it can be used to update the student model via standard probability propagation techniques. In this case, it is used to propagate evidence from the experiment evaluation to the knowledge items, and then to the knowledge sub-themes and to the knowledge themes. After each experiment, the knowledge level at each different level of granularity—items, sub-themes and themes, is used by the tutor to decide if it should provide help to the student, and at what level of detail. For example, if in general the experiment was successful, but some aspect was not very good, a lesson on a specific concept (item) is given to the student. While if the experiment was unsuccessful, a lesson on a complete theme or sub-theme is recommended. Based on the student category, the tutor decides the difficulty of the next experiments to be presented to the student.

### *13.5.2 Visual Grammars*

A visual grammar describes objects hierarchically. It can represent a diagram, a geometric drawing or an image. For example, the description of a flowchart is made by decomposition: complex elements are decomposed into simple elements (from the complete image to arrows or simple boxes).

For visual object recognition, we need a grammar that allows us to model the decomposition of a visual object into its parts and how they relate with each other [13]. One interesting kind of relational grammar are Symbol-Relation Grammars (*SR-grammars*) [2], because they can provide this type of description and also incorporate the possibility of adding rewriting rules to specify relationships between terminal and non-terminal symbols once a decomposition for all the non-terminal symbols has taken place.

#### **13.5.2.1 Representing Objects with SR Grammars**

Classes of objects are represented based on symbol-relational grammars. This includes three basic parts: (i) the basic elements of the grammar or lexicon, (ii) the spatial relations, (iii) the transformation rules. Next we briefly describe them.

The idea is to use simple and general features as basic elements so they can be applied to different classes of objects. These regions define a visual dictionary. The visual features considered are: *uniform color regions* (color is quantized in 32 levels) and edges at different orientations (obtained with Gabbor filters). These features form a set of training examples that are clustered and the centroids of these clusters constitute the *Visual Lexicon*.

The spatial relations include topological and order relationships. The relationships used are: $Inside\_of(A, B)$ (A region is *within* B region), $Contains(A, B)$ (A region *covers* completely B region), $Left(A, B)$ (A is *touched* by B and A is located *left* from B), $Above(A, B)$ (A is *touched* by B and A is located *above* from B),

$Invading(A, B)$ (A is *partially covering* B more than $Above$ and $Left$ but less than $Contains$).

The next step is to generate the rules that make up the grammar. Using training images for the class of the object of interest, the most common relationships between clusters are obtained. Such relationships become candidate rules to build the grammar. This is an iterative process where the rules are subsumed and converted to new non-terminal elements of the grammar. This process is repeated until a threshold (in terms of the minimum number of elements) is reached; the starting symbol of the grammar represents the class of objects to be recognized.

Visual object recognition involves uncertainty: noise in the image, occlusions, imperfect low-level processing, etc. SR grammars do not consider uncertainty, so to incorporate it, they can be extended using RPGMs, in particular MLNs.

#### 13.5.2.2 Transforming a SR Grammar into a Markov Logic Network

The SR grammar for a class of objects is transformed *directly* to formulas in the MLN language. In this way the structural aspect of the MLN is obtained. The parameters—weights associated to each formula—, are obtained from the training image set.

Consider a simple example, a SR grammar to recognize faces based on high-level features: eyes, mouth, nose, head. The productions of this simple SR-grammar for faces are:

$$1 : FACE^0 \rightarrow < \{eyes^2, mouth^2\}, \{above(eyes^2, mouth^2)\} >$$
$$2 : FACE^0 \rightarrow < \{nose^2, mouth^2\}, \{above(nose^2, mouth^2)\} >$$
$$3 : FACE^0 \rightarrow < \{eyes^2, head^2\}, \{inside\_of(eyes^2, head^2)\} >$$
$$4 : FACE^0 \rightarrow < \{nose^2, head^2\}, \{inside\_of(nose^2, head^2)\} >$$
$$5 : FACE^0 \rightarrow < \{mouth^2, head^2\}, \{inside\_of(mouth^2, head^2)\} >$$

The transformation into a MLN is nearly straightforward. First, we need to declare the formulas:

```
aboveEM(eyes,mouth)
aboveNM(nose,mouth)
insideOfEH(eyes,head)
insideOfNH(nose,head)
insideOfMH(mouth,head)
isFaceENMH(eyes,nose,mouth,head)
```

Subsequently, we need to declare the domain:

```
eyes={E1,E2,E3,E4}
nose={N1,N2,N3,N4}
mouth={M1,M2,M3,M4}
head={H1,H2,H3,H4}
```

Finally we need to write the weighted first-order formulas. We used a validation image dataset and translated the probabilities into weights:

**Fig. 13.5** Several examples of the detection of the terminal elements in images. From top-left, clockwise: mouth, face, eyes, nose. Correct detections are shown as green rectangles and incorrect ones as red rectangles [9]

```
1.58 isFaceENMH(e,n,m,h) => aboveEM(e,m)
1.67 isFaceENMH(e,n,m,h) => aboveNM(n,m)
1.16 isFaceENMH(e,n,m,h) => insideOfEH(e,h)
1.25 isFaceENMH(e,n,m,h) => insideOfNH(n,h)
1.34 isFaceENMH(e,n,m,h) => insideOfMH(m,h)
```

To recognize a *face* in an image, the relevant aspects of the image are transformed into a first-order $KB$. For this, the terminal elements are detected (in the example the eyes, mouth, nose and head) in the image, as well as the spatial relations between these elements. Then, the particular image $KB$ is combined with the *general* model represented as a MLN; and from this combination a *grounded Markov network* is generated. Object (face) recognition is performed using standard probabilistic inference (see Chap. 6) over the Markov network.

This method was implemented to recognize faces in images [9]. Although the detectors of the terminal elements are not very good, generating many false positives (see Fig. 13.5), the restrictions provided by the grammar eliminate most false positives and can recognize faces with high accuracy. In particular, this approach is more robust than traditional face detectors for cases with large occlusions as those depicted in Fig. 13.5.

## 13.6 Additional Reading

There are several introductory books to logic, such as [8, 10]. From an artificial intelligence perspective, [4] provides a good introduction to predicate logic and logic-based representations. Most of the current relational probabilistic models are described in [5], which includes a chapter on each approach. Probabilistic relational models are introduced in [6]; and how to learn PRMs from data is presented in [3]. A review of Markov logic networks is included in [1]. References [7] and [11] are introductory books to inductive logic programming.

## 13.7 Exercises

1. If $p$ and $r$ are false, and $q$ and $s$ are true, determine the truth values of the following expressions:
   - $p \vee q$
   - $\neg p \wedge \neg(q \wedge r)$
   - $p \rightarrow q$
   - $(p \rightarrow q) \wedge (q \rightarrow r)$
   - $(s \rightarrow (p \wedge \neg r)) \wedge ((p \rightarrow (r \vee q)) \wedge s)$
2. Which of the following expressions are true?
   - $\forall x((x^2 - 1) > 0)$
   - $\forall x(x^2 > 0)$
   - $\exists x(1/(x^2 + 1) > 1)$
   - $\neg \exists x((x^2 - 1) \leq 0)$
3. Determine if the next expressions are WFFs:
   - $\forall x \neg(p(x) \rightarrow q(x))$
   - $\exists x \forall x(p(x) \leftrightarrow q(x))$
   - $\exists x \vee q(x) \wedge q(y)$
   - $\forall x \exists y p(x) \vee (p(x) \leftrightarrow q(y))$
4. Write the following sentences in first-order predicate logic: (i) all persons have a father and a mother, (ii) some persons have brothers or sisters or brothers and sisters, (iii) if a person has a brother, then his father is also the father of his brother, (iv) no person has two natural fathers or two natural mothers.
5. Given the PRM of Fig. 13.1, assume that all variables are binary (i.e., Teaching-Ability={Good, Average}, Popularity={High, Low}, etc.). According to the structure of the model, specify the required conditional probability tables.
6. Based on the PRM of the previous exercise, assume there are two professors, three students, three courses and five registrations (one student registers for two courses and the other to three courses). Expand the PRM according to the previous objects and generate the corresponding Bayesian network.
7. According to the BN and parameters of the two previous exercises, we want to estimate the probability of satisfaction of one student given the teaching ability of one professor and the rating of one course. (a) Define the minimum subset of the BN required for this query. (b) Calculate the posterior probabilities of satisfaction based on the reduced model.

8. Consider the MLN example of Sect. 13.4 with two logical formulas, and an additional third formula: $\forall x Unhealthy Diet(x) \rightarrow Cancer(x)$. Given that the variables $x$ and $y$ are instantiated to $Tim$, $Sue$, and $Maria$, obtain the graphical dependency structure of the MLN.
9. Determine the weights for the two logical formulas of the MLN example of Sect. 13.4, assuming that you have extracted the following statistics from a database: (i) 19 persons smoked and have cancer, (ii) 11 persons smoked and do not have cancer, (iii) 10 persons did not smoke and do not have cancer, (iv) 5 persons did not smoke and have cancer, (v) 15 pairs of persons were friends and both smoked, (vi) 10 pairs of persons were friends and both did not smoked, (vii) 5 pairs of persons were friends, one smoked and the other did not.
10. Based on the model of the previous problem, and given that the variables are instantiated to $John$ and $Linda$: (a) depict the resulting graph of the grounded MLN; (b) given that $John$ and $Linda$ are friends and $John$ smokes, estimate the probability that $Linda$ has cancer.
11. Estimate the weights for the formulas of the MLN for face detection, Sect. 13.5.2.2, considering the following data. In 1000 samples, 920 eyes above mouth, 860 nose above mouth, 950 eyes inside head, 850 nose inside head, 800 mouth inside head.
12. According to the formulas of the MLN for face detection, Sect. 13.5.2.2, depict the grounded network considering the following elements detected in an image: 2 eyes, 3 mouths, 2 noses, 1 head.
13. *** Investigate alternative formalisms that combine logic and probability. Analyze their advantages and disadvantages in terms of expressive power and computational efficiency.
14. *** Develop a program for doing inference with PRMs. Given the PRM described at the class level (you can use an object-oriented database), and a set of objects, transform it to a BN. Then perform inference over the BN (use the algorithms developed previously).
15. *** Implement the system to detect faces in images based on MLNs using state of the art detectors for eyes, mouth, nose and face. Extend this system to detect persons based on different body parts (face, arms, torso, legs) by defining a grammar for "persons" and the corresponding MLN. Test both systems in images with occlusions.

## References


1. Domingos, P., Richardson, M.: Markov logic: a unifying framework for statistical relational learning. In: Getoor, L., Taskar, B. (eds.) Introduction to Statistical Relational Learning, pp. 339–371. MIT Press, Cambridge (2007)
2. Ferrucci, F., Pacini, G., Satta, G., Sessa M.I., Tortora, G., Tucci, M., Vitiello, G.: Symbol-relation grammars: a formalism for graphical languages. Inf. Comput. **131**(1), 1–46 (1996)

3. Friedman, N., Getoor, L., Koller, D., Pfeffe, A.: Learning probabilistic relational models. In: Proceeding of the International Joint Conference on Artificial Intelligence (IJCAI), pp. 1300–1309 (1999)
4. Genesereth, M.R., Nilsson, N.J.: Logical Foundations of Artificial Intelligence. Morgan Kaufmann, Burlington (1988)
5. Getoor, L., Taskar, B.: Introduction to Statistical Relational Learning. MIT Press, Cambridge (2007)
6. Koller, D.: Probabilistic Relational Models. In: Proceedings of the 9th International Workshop on Inductive Logic Programming. Lecture Notes in Artificial Intelligence, vol. 1634, pp. 3–13. Springer (1999)
7. Lavrac, N., Dzeroski, S.: Inductive Logic Programming: Techniques and Applications. Ellis Horwood, New York (1994)
8. Lemmon, E.J.: Beginning Logic. Hackett Publishing Company, Indianapolis (1978)
9. Meléndez, A., Sucar, L.E., Morales, E.: A visual grammar for face detection. In: Proceedings of IBERAMIA. Lecture Notes in Artificial Intelligence, vol. 6433, pp. 493–502. Springer (2010)
10. Newton-Smith, W.H.: Logic: An Introductory Course. Routledge, Milton Park (1985)
11. Nienhuys-Cheng, S., de Wolf, R.: Foundations of Inductive Logic Programming. Springer, Berlin (1991)
12. Richardson, M., Domingos, P.: Markov Logic Networks. Mach. Learn. **62**(1–2), 107–136 (2006)
13. Ruiz, E., Sucar, L.E.: An object recognition model based on visual grammars and Bayesian networks. In: Proceedings of the Pacific Rim Symposium on Image and Video Technology. LNCS, vol. 8333, pp. 349–359. Springer (2014)
14. Sucar, L.E., Noguez, J.: Student Modeling. In: Pourret, O., Naim, P., Marcot, B. (eds.) Bayesian Belief Networks: A Practical Guide to Applications, pp. 173–186. Wiley, Hoboken (2008)

# Chapter 14
# Graphical Causal Models

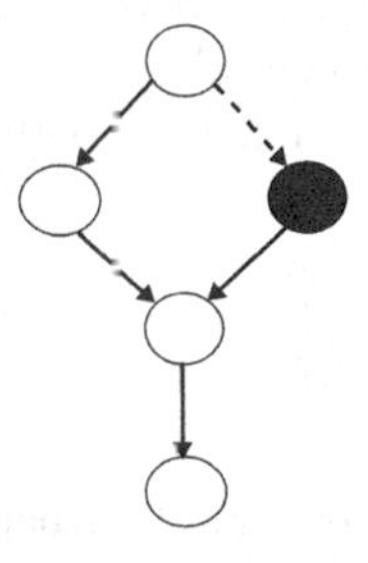

## 14.1 Introduction

Causal reasoning is a constant element in our lives as it is in human nature to constantly ask why [7, 9]. Acting in the world is conceived by human beings as intervening the world, and humans are able to learn and use causal relations while making choices. Causality has to do with cause–effect relations; that is, identifying when there are two (or more) related phenomena, which is the *cause* and which is the *effect*. However, there could be a third explanation, namely, there is another phenomena which is the *common cause* of the original two phenomena of interest.

Probabilistic models do not necessarily represent causal relations. For instance, consider the following two Bayesian networks: BN1: $A \rightarrow B \rightarrow C$ and BN2: $A \leftarrow B \leftarrow C$. From a probabilistic perspective, both represent the same set of dependency and independency relations: direct dependencies between $A$ and $B$, $B$ and $C$; and $I(A, B, C)$, that is, $A$ and $C$ are independent given $B$. However, if we define that a directed link $A \rightarrow B$ means $A$ causes $B$, both models represent very different causal relations.

Given that we can model and solve complex phenomena with probabilistic graphical models, and also that causality is for many a complex and controversial concept, we may ask ourselves, Why do we need causal models? There are several advantages to causal models, in particular, graphical causal models.



L. E. Sucar, *Probabilistic Graphical Models*, Advances in Computer Vision and Pattern Recognition, https://doi.org/10.1007/978-3-030-61943-5_14

First, a causal model provides a deeper understanding of a domain, by identifying the direct and indirect causes of certain events, which is not necessarily true for associative models such as Bayesian networks. Secondly, with causal models we can perform other types of reasoning that are not possible, at least in a direct way, with PGMs such as Bayesian networks. These other inference situations are: (i) *interventions*, in which we want to find the effects of setting a certain variable to a specific value by an external agent (note that this is different from observing a variable); and (ii) *counterfactuals*, where we want to reason about what would have happened if certain information had been different from what actually happened. We will discuss these two situations in more detail in the rest of the chapter.

Several causal modeling techniques have been developed; for example, functional equations, path diagramas and structural equation models, among others. In this chapter we will focus on graphical models, in particular causal Bayesian networks.

### 14.1.1 Definition of Causality

There are several interpretations of causality, here we adopt the *manipulationist* interpretation [6, 9]: "manipulation of a cause will result in a manipulation of the effect". For example: *if we force a rooster to crow it will not cause the sun to rise, instead if the sun rises it causes the rooster to crow.*

We adopt the formal definition of causality provided by Spirtes et al. [9] which states that causation is a stochastic relation between events in a probability space; this is, some event (or events) causes another event to ocurr. Let $\Omega, F, P$ be a finite probability space, and consider a binary relation $\rightarrow \subseteq F \times F$ which is:

- Transitive: If $A \rightarrow B$ and $B \rightarrow C$, $\forall A, B, C \in F$, then $A \rightarrow C$.
- Irreflexive: For all $A \in F$ it doesn't hold that $A \rightarrow A$.
- Antisymmetric: For $A, B \in F$ such that $A \neq B$ if $A \rightarrow B$ then it does not hold that $B \rightarrow A$.

$A$ is a cause of $B$ if $A \rightarrow B$. It is important to note that an event may have more than one cause, and that not necessarily each one of these causes is sufficient to produce the effect.

## 14.2 Causal Bayesian Networks

A Causal Bayesian network (CBN) is a directed acyclic graph, $G$, in which each node represents a variable and the arcs in the graph represent *causal* relations; that is, the relation $A \rightarrow B$ represents some physical mechanism such that the value of $B$ is directly affected by the value of $A$. This relation can be interpreted in terms of *interventions* –setting of the value of some variable or variables by an external agent. For example, assume that $A$ stands for a water sprinkler (OFF / ON) and $B$ represents

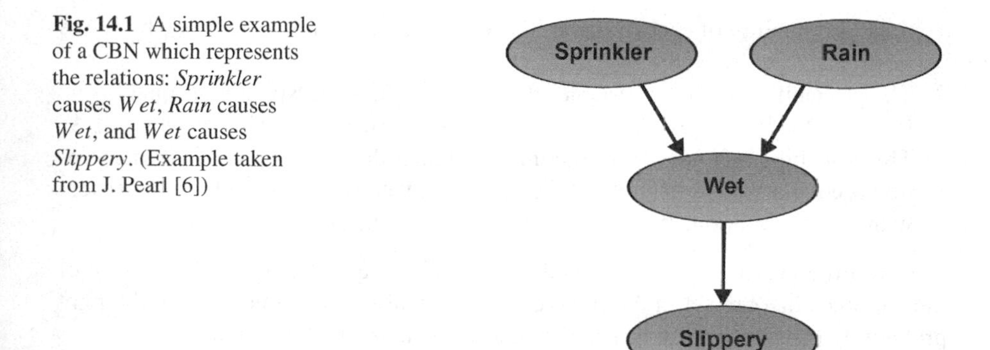


**Fig. 14.1** A simple example of a CBN which represents the relations: *Sprinkler* causes $Wet$, *Rain* causes $Wet$, and $Wet$ causes *Slippery*. (Example taken from J. Pearl [6])

if the grass is wet (FALSE / TRUE). If the grass ($B$) is originally not WET and the sprinkler is set to ON by an intervention, then $B$ will change to TRUE.

As in BNs, if there is an arc from $A$ to $B$ ($A$ is a direct cause of $B$), then $A$ is a parent of $B$, and $B$ is a child of $A$. Given any variable $X$ in a CBN, $Pa(X)$ is the set of all parents of $X$. Also, similarly to BNs, when the direct or immediate causes –parents– of a variable are known, the more remote causes (or ancestors) are irrelevant. For example, once we know that the grass is WET, this makes it SLIPPERY, no matter how the grass became wet (we turned on the sprinkler or it rained).

Causal networks represent stronger assumptions than Bayesian networks, as all the relations implied by the network structure should correspond to causal relations. Thus, all the parent nodes, $Pa(X)$, of a certain variable, $X$, correspond to the direct causes of $X$. This means that if any of the parent variables of $X$, or any combination of them, is set to a certain value via an intervention, this will have an effect on $X$. In a CBN, a variable which is a root node (variable with no parents) is called *exogenous*, and all other variables are *endogenous*.

A simple example of a CBN is depicted in Fig. 14.1, which basically encodes the following causal relations: (i) *Sprinkler* causes $Wet$, (ii) *Rain* causes $Wet$, (iii) $Wet$ causes *Slippery*. In this case, *Sprinkler* and *Rain* are exogenous variables, and $Wet$ and *Slippery* are endogenous variables

If a $P(\mathbf{X})$ is the joint probability distribution of the set of variables $\mathbf{X}$, then we define $P_{\mathbf{y}}(\mathbf{X})$ as the distribution resulting from setting the value for a subset of variables, $\mathbf{Y}$, via an intervention. This can be represented as $do(\mathbf{Y} = \mathbf{y})$, where $\mathbf{y}$ is a set of constants. For instance, given the CBN of Fig. 14.1, if we set the sprinkler to ON, $do(Sprinkler = ON)$, then the resulting distribution will be denoted as $P_{Sprinkler=ON}(Sprinkler, Rain, Wet, Slippery)$.

Formally, a Causal Bayesian Network can be defined as follows [6]:

A CBN $G$ is a directed acyclic graph over a set of variables $\mathbf{X}$ that is compatible with all the distributions resulting from interventions on $\mathbf{Y} \subseteq \mathbf{X}$, in which the following conditions are satisfied:

1. The probability distribution $P_{y}(\mathbf{X})$ resulting from an intervention is Markov compatible with the graph $G$; that is, it is equivalent to the product of the condi-

tional probability of each variable $X \in G$ given its parents: $P_y(\mathbf{X}) = \prod_{X_i} P(X_i \mid Pa(X_i))$.
2. The probability of all the variables that are part of an intervention is equal to one for the value it is set to: $P_y(X_i) = 1$ if $X_i = x_i$ is consistent with $Y = y$, $\forall X_i \in \mathbf{Y}$.
3. The probability of each of the remaining variables that are not part of the intervention is equal to the probability of the variable given its parents and it is consistent with the intervention: $P_y(X_i \mid Pa(X_i)) = P(X_i \mid Pa(X_i))$, $\forall X_i \notin \mathbf{Y}$.

Given the previous definition, and in particular the fact that the probability of the variables that are set in the intervention is equal to one (condition 2), the joint probability distribution can be calculated as a *truncated* factorization:

$$P_y(\mathbf{X}) = \prod_{X_i \notin \mathbf{Y}} P(X_i \mid Pa(X_i)) \tag{14.1}$$

such that all $X_i$ are consistent with the intervention $\mathbf{Y}$.

Another consequence of the previous definition, is that once all the parents of a variable $X_i$ are set by an intervention, setting any other variable does not affect the probability of $X_i$:

$$P_{(Pa(X_i),\mathbf{W})}(X_i) = P_{Pa(X_i)}(X_i) \tag{14.2}$$

such that $\mathbf{W} \cap (X_i, Pa(X_i)) = \emptyset$.

Considering again the example in Fig. 14.1, if we make the grass wet by any mean, $do(Wet = TRUE)$, then the probability of *Slippery* is not affected by *Rain* or *Sprinkler*.

According to the definition of a causal Bayesian network and the intervention ($do$) operator we can give a formal definition to the manipulationist interpretation of cause and effect. Given a Causal Graphical Model, $G$, and $P_G$ its corresponding probability distribution; and let $X$ and $Y$ variables in the model, $X$ causes $Y$ if $P_G(Y \mid do(X = x)) \neq P_G(Y \mid do(X = x')), x \neq x'$.

### 14.2.1 Gaussian Linear Models

*Gaussian Linear Models* (GLM), a common type of graphical causal model, are those ones that consider linear relations between variables and Gaussian noise. This type of models are related to *Structural Equation Models* [8], and are easier to learn from observable data (see Chap. 15).

In a GLM we have a linear equation that relates a variable to its direct causes (parents):

$$Y = \beta_1 X_1 + \beta_2 X_2 + \cdots + \beta_q X_q + N_x \tag{14.3}$$

where $X_1 \ldots X_q$ are the direct causes of $Y$, $\beta_1 \ldots \beta_q$ are constants, and $N_x$ is a Gaussian variable with zero mean. Thus:

$$P(Y \mid X_1, \ldots, X_q) = \mathcal{N}(\beta_1 x_1 + \cdots, +\beta_q x_q; \sigma^2) \tag{14.4}$$

The joint distribution of a GLM can be represented as a multivariate Gaussian distribution over a set of $n$ continuous random variables, $\mathbf{X} = \{X_1, \ldots, X_n\}$, by an $n$-dimensional **mean vector**, $\mu$, and a symmetric $n \times n$ **covariance matrix**, $\Sigma = (\sigma_{ij})$. This parameterization of the multivariate Gaussian distribution is called the **covariance form**, in which the multivariate Gaussian density function is defined as:

$$P(\mathbf{x}) = \frac{1}{(2\pi)^{p/2}|\Sigma|^{1/2}} \exp\Big[-\frac{1}{2}(\mathbf{x}-\mu)^T\Sigma^{-1}(\mathbf{x}-\mu)\Big]. \tag{14.5}$$

For multivariate Gaussians, independence is easy to determine directly from the parameters of the distribution. In concrete, if $\mathbf{X} = \{X_1, \ldots, X_n\}$ have a joint normal distribution $\mathcal{N}(\mu; \Sigma)$. Then $X_i$ and $X_j$ are independent if and only if $\sigma_{i,j} = 0$.

For instance consider a simple model with two variables, $X$ and $Y$, and the causal model depicted in Fig. 14.2a. $X$ could represent the amount of rain and $Y$ the level of a dam. So the equations for this model could be:

$$X = N_x$$

$$Y = 8X + N_y$$

with $N_x, N_y \sim N(0, 1)$, which are independent noise variables. This model induces bivariate normal distribution:

$$(X, Y) = \mathcal{N}\begin{pmatrix}0\\8\end{pmatrix}\begin{pmatrix}1 & 8\\8 & 65\end{pmatrix}$$

Now let's see what happens if we make interventions in this model. First, consider that we make $X = 10$ (let say we make it rain 10 mm), so the equations now change to:

$$X = 10$$

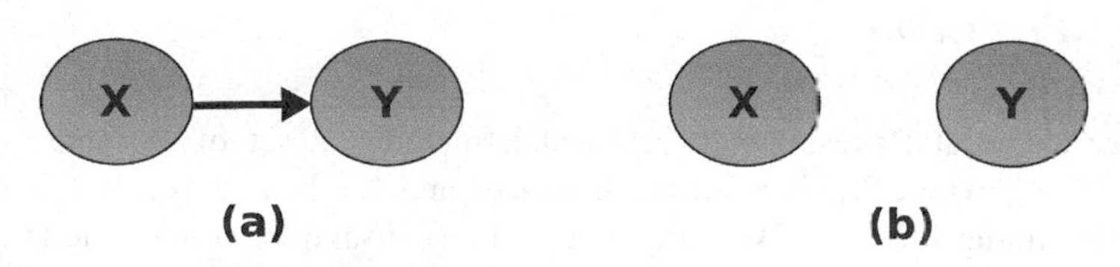


**Fig. 14.2** **a** A simple example of a GLM which represents the relation: $X$ causes $Y$. **b** The model resulting after an intervention on $Y$

$$Y = 80 + N_y$$

And the causal model remains the same as in Fig. 14.2a.

Now, we intervine $Y$ and make it equal to $N(2, 2)$ (assume we can fill the dam via other means). This alters the graphical model, as $Y$ is not influenced by $X$, so we modified the model as shown in Fig. 14.2b. The joint distribution becomes:

$$(X, Y) = \mathcal{N}\begin{pmatrix}0\\2\end{pmatrix}\begin{pmatrix}1 & 0\\0 & 2\end{pmatrix}$$

So $X$ and $Y$ are independent. Thus, if we make an intervention in $Y$ this does not affect the value of $X$; however, as in the previous case, if we intervine $X$ it has a causal effect on $Y$.

From this example, we can see that an intervention can change the joint distribution of the model; something that does not happen when we do an observation. Thus, the effect of an intervention in a variable is different from conditioning based on observing a variable. In this example:

$$P_{do(Y=y)}(X) \neq P(X \mid Y = y)$$

## 14.3 Causal Reasoning

Causal reasoning has to do with answering causal queries from a causal model, known as *causal effect estimation*, and in our case in particular from graphical causal models. There are several types of causal queries we might consider, we will start by analyzing causal predictions, and then we will analyze counterfactuals.

In general, if we want to estimate the causal effect of some variable $X$ on other variable $Y$, the result can be affected if there are *cofactors*, that is, common ancestors of $X$ and $Y$; so this should be taken into account in the inference process. Initially we present a simplified version of the prediction and counterfactual procedures without considering cofactors, and later, in Section 14.4, we describe the criteria to consider these cofactors via what is called an *adjustment set*.

### *14.3.1 Prediction*

Consider a causal Bayesian network which includes a set of variables: $\mathbf{X_G} = \{\mathbf{X_C}, \mathbf{X_E}, \mathbf{X_O}\}$; where $\mathbf{X_C}$ is a subset of *causes* and $\mathbf{X_E}$ is a subset of *effects*; $\mathbf{X_O}$ are the remaining variables. We want to perform a causal query on the model: What will the effect be on $\mathbf{X_E}$ when setting $\mathbf{X_C} = \mathbf{x_C}$? That is, we want to obtain the probability distribution of $\mathbf{X_E}$ that results from the intervention $\mathbf{X_C} = \mathbf{x_C}$:

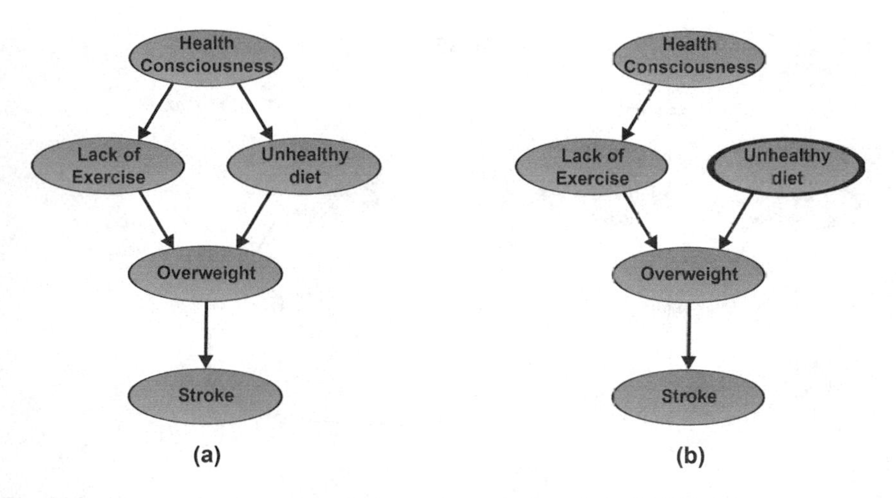


**Fig. 14.3** An example of causal prediction. **a** A simple, hypothetical CBN that represents some causal relations related to a stroke. **b** The resulting graphical model obtained from **a** by the intervention $do(unhealthy - diet = TRUE)$

$$P_C(\mathbf{X_E} \mid do(\mathbf{X_C} = \mathbf{x_C})) \tag{14.6}$$

To perform causal prediction with a CBN, $G$, the following procedure is followed:

1. Eliminate all incoming arrows in the graph to all nodes in $\mathbf{X_C}$, thus obtaining a modified CBN, $G_r$.
2. Fix the values of all variables in $\mathbf{X_C}$. $\mathbf{X_C} = \mathbf{x_C}$.
3. Calculate the resulting distribution in the modified model $G_r$ (via probability propagation as in Bayesian networks).

For example, consider the hypothetical CBN depicted in Fig. 14.3a that represents certain causal knowledge about a stroke. If we want to measure the effect of an "unhealthy diet" in the variable "stroke", then, according to the previous procedure, we eliminate the link from "health consciousness" to "unhealthy diet", resulting in the model in Fig. 14.3b. Then we should set the value of "unhealthy diet" to $TRUE$, and by probability propagation obtain the distribution of "stroke". If the distribution of "stroke" changes depending on the value of "unhealthy diet", we can conclude according to this model that it does have an effect.

An interesting question is: When is the distribution resulting from an intervention equal to the distribution resulting from an observation? In mathematical terms, is $P(X_E \mid X_C) = P_C(X_E \mid do(X_C = x_C))$? Both are equal if $X_C$ includes all the parents of $X_E$ and none of its descendants; given that any variable in a BN is independent of its non-descendants given its parents. In other cases they are not necessarily equal, they will depend on other conditions. Thus, in general the joint probability distribution of a CBN is different from the original model after an intervention.

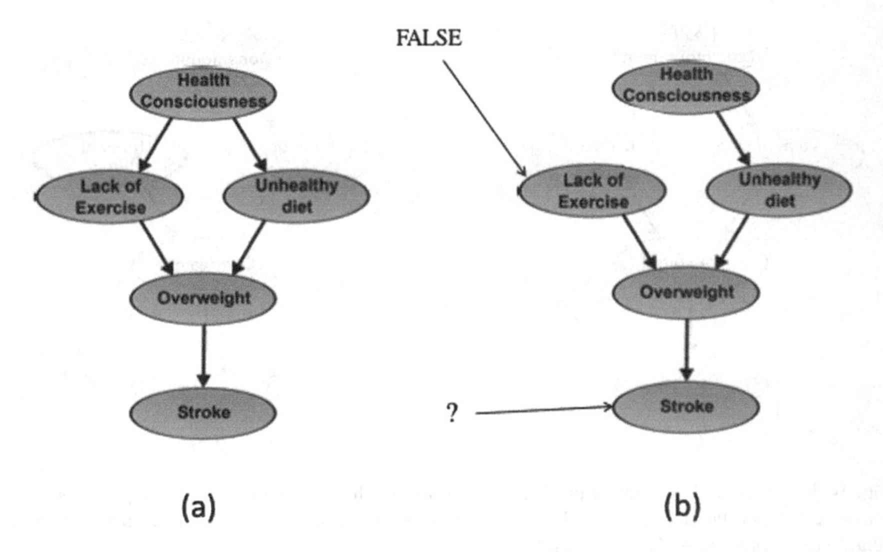


**Fig. 14.4** An example of a conterfactual question. **a** An hypothetical CBN that represents some causal relations related to a stroke. **b** The resulting graphical model obtained from **a** by the counterfactual: If some person suffered a stroke, would he still have suffered the stroke ("stroke" = TRUE) if he had exercised more ("lack of exercise" = FALSE)?

## *14.3.2 Counterfactuals*

Counterfactuals are a way of reasoning that we commonly perform in our life. For instance, consider the causal model of Fig. 14.3a. A typical counterfactual question would be: If some person suffered a stroke, would he still have suffered the stroke ("stroke" = TRUE) if he had exercised more ("lack of exercise" = FALSE)?

Counterfactual inference involves 3 main steps:

1. Abduction: modify the model in terms of the new evidence (in the example, modify the value of "stroke" to unknown).
2. Action: perform the minimal intervention in the model according to the hypothetical evidence (in the example, set "lack of exercise" to FALSE, removing the link from "health consciousness" to "lack of exercise").
3. Prediction: perform probabilistic inference on the modified model and obtain the probability of the variable(s) of interest (in the example, perform probability propagation to estimate $P_{lack-exercise}(stroke \mid do(lack - exercise = FALSE))$.

The model resulting from the previous counterfactual question is depicted in Fig. 14.4

When we estimate the causal effect of some variable $X$ on other variable $Y$, either as a prediction or a counterfactual, based on the joint (observational) probability distribution, the result can be affected if there is another variable, or set of variables, $\mathbf{W}$, that are ancestors of $X$ and $Y$ (cofactors). To control for these covariates it is necessary to condition on $Z \subseteq W$, known as an *adjustment set*:

$$P(Y \mid do(X = x)) = \sum_{z} P(Y \mid X = x, Z = z)P(Z = z) \tag{14.7}$$

A valid adjustment set are the parents of $X$, $Pa(X)$, but there are other alternatives. In general, this issue can be resolved using the criteria described next.

## 14.4 Front Door and Back Door Criterion

Given a CBN, we want to determine the results of an intervention, $P(Y \mid do(X = x))$, and there are some cofactors, $\mathbf{Z}$, can we always identify the causal effect from the joint distribution? The answer is "yes" when the covariates $\mathbf{Z}$ contain all the other relevant variables. If some variables are not observed, then the issue of which causal effects are observationally identifiable is considerably trickier. The basic principle is that we would like to condition on the adequate control variables, which will block paths linking $X$ and $Y$ other than those which would exist in the altered graph where all paths into $X$ have been removed. If other unblocked paths exist, then there could be some confounding of the causal effect of $X$ on $Y$. There are two main criteria we can use to get an adequate control; they are called the *back-door criterion* and the *front-door criterion* [6].

### 14.4.1 Back Door Criterion

If we desire to know the casual effect of $X$ on $Y$ and have a set of variables $\mathbf{Z}$ as the control, then $\mathbf{Z}$ satisfies the back-door criterion if:

1. $\mathbf{Z}$ blocks every path from $X$ to $Y$ that has an arrow into $X$ (blocks the back door), and
2. No node in $\mathbf{Z}$ is a descendant of $X$.

Then:

$$P(Y \mid do(X = x)) = \sum_{z} P(Y \mid X = x, \mathbf{Z} = \mathbf{z})P(\mathbf{Z} = \mathbf{z}) \tag{14.8}$$

where $P(Y \mid X = x, \mathbf{Z} = \mathbf{z})$ and $P(\mathbf{Z} = \mathbf{z})$ are observational conditional probabilities. Some examples of the back-door criteria are illustrated in Fig. 14.5.

### 14.4.2 Front Door Criterion

If we want to know the causal effect of $X$ on $Y$ and have a set of variables $\mathbf{Z}$ as the control, then $\mathbf{Z}$ satisfies the front-door criterion if:

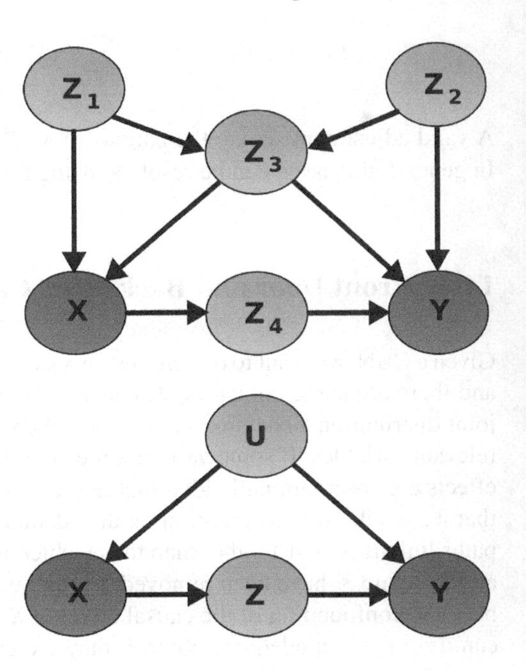


**Fig. 14.5** Given the CBN in the figure, the sets $\{Z_1, Z_3\}$ and $\{Z_2, Z_3\}$ satisfy the back-door criteria for the causal effect $X \rightarrow Y$; but the set $\{Z_3\}$ does not satisfy it as it does not block the trajectory $(X, Z_1, Z_3, Z_2, Y)$

**Fig. 14.6** Given the CBN in the figure, the set $\{\mathbf{Z}\}$ satisfies the front-door criteria for the causal effect $X \rightarrow Y$. $U$ is an unobserved variable that is a cofactor of $X$ and $Y$

1. $\mathbf{Z}$ blocks all directed paths from $X$ to $Y$,
2. there are no unblocked back-door paths from $X$ to $\mathbf{Z}$, and
3. $X$ blocks all back-door paths from $\mathbf{Z}$ to $Y$.

Then:

$$P(Y \mid do(X = x)) = \sum_{\mathbf{z}} P(\mathbf{Z} = \mathbf{z} \mid X = x) \sum_{x'} P(Y \mid X = x', \mathbf{Z} = \mathbf{z}) P(X = x') \tag{14.9}$$

To understand the front-door criteria we can analyze it by parts. By clause (1), $\mathbf{Z}$ blocks all directed paths from $X$ to $Y$, so any causal dependence of $Y$ on $X$ must be mediated by a dependence of $Y$ on $\mathbf{Z}$. Clause (2) says that we can estimate the effect of $X$ on $\mathbf{Z}$ directly; and clause (3) says that $X$ satisfies the back-door criterion for estimating the effect of $\mathbf{Z}$ on $Y$ (the inner sum in Eq. 14.9 is the back-door criterion estimate). An example of of the front-door criteria is illustrated in Fig. 14.6.

Both the back-door and front-door criteria are sufficient for estimating causal effects from probabilistic distributions, but are not necessary. Necessary and sufficient conditions for the identifiability of causal effects are described [6].

## 14.5 Applications

There are many practical applications in which a causal model is useful. Just to mention a few:

- Public policy making.
- Diagnosis of physical systems.
- Generating explanations.
- Understanding causal expressions.
- Online marketing.
- Medical treatments.

Next we will present two applications of causal models: (i) analyze unfairness in machine learning systems, and (ii) accelerate reinforcement learning.

### *14.5.1 Characterizing Patterns of Unfairness*

Data used to train machine learning systems may contain different types of bias (social, cognitive) that can lead to bad decisions. A way to reduce this problem is to analyze the the relations between sensitive attributes and the system's outputs; which is a challenging task. DeepMind [3] proposed the use of causal Bayesian networks as a tool to formalize and measure potential unfairness scenarios underlying a data set.[1]

The main variables and their relationships in a machine learning system, and in particular those that are sensitive to bias, are modeled as a causal Bayesian network. For example, consider a hypothetical college admission system in which applicants are admitted, $A$, based on qualifications, $Q$, choice of department, $D$, and gender, $G$; and in which female applicants apply more often to some departments. Figure 14.7 depicts a CBN of the admission process. Gender has a direct influence on admission through the causal path $G \rightarrow A$ and an indirect influence through the causal path $G \rightarrow D \rightarrow A$. The direct influence represent the fact that applicants with the same qualifications applying to the same department might be treated differently based on their gender. The indirect influence captures differing admission rates between female and male applicants due to their different department choices. The direct influence of the sensitive attribute (gender) on admission is considered unfair for social and legal reasons, the indirect influence could be considered fair or unfair depending on other factors.

CBNs can also be used to quantify unfairness in a dataset and to design techniques for alleviating unfairness in the case of complex relationships in the data. Contrafactual techniques can estimate the influence that a sensitive attribute has on other variables along specific sets of causal paths. This can be used to measure the

[1] This example is based on the information in https://deepmind.com/blog/article/Causal_Bayesian_Networks.

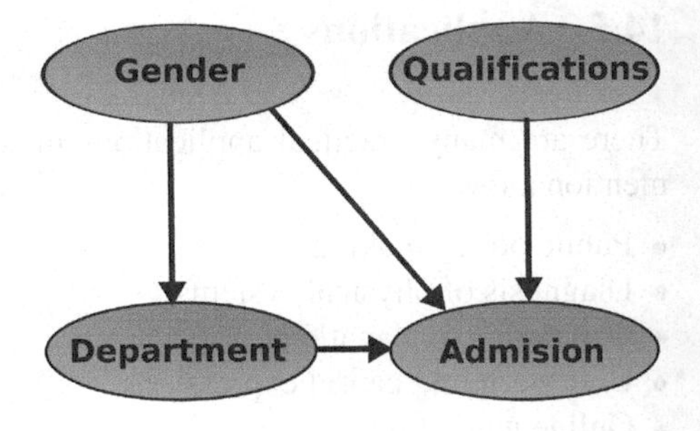


**Fig. 14.7** Causal Bayesian network for the college admission process

degree of unfairness on a given dataset in complex scenarios in which some causal paths are considered unfair while other causal paths are considered fair. In the college admission example in which the path $G \rightarrow D \rightarrow A$ is considered fair, path-specific techniques would enable the measurement of the influence of $G$ on $A$ restricted to the direct path $G \rightarrow A$ over the whole population, in order to obtain an estimate of the degree of unfairness contained in the dataset.

The use of counterfactual inference techniques would also make possible to ask if a specific individual was treated unfairly, for example by asking whether a rejected female applicant would have obtained the same decision in a counterfactual world in which her gender were male.

In addition to answering questions of fairness in a data set, path-specific counterfactual inference could be used to design methods to alleviate the unfairness of machine learning systems [2].

### 14.5.2 Accelerating Reinforcement Learning with Causal Models

Reinforcement learning (RL) is the study of how an agent can learn to choose actions that maximize its future rewards through interactions with its environment [10]. In general, RL takes a lot of time (interactions) to learn, as it has to explore all the potential actions in all the possible states, a very large search space in realistic problems. However, if the agent could use additional knowledge to guide this process, the time to learn can be considerably reduced as it can avoid useless actions and can focus on productive ones. Actions are like interventions, so causal knowledge can be considered, as it can represent relevant causal relations in the domain; and it is *generic*, so it can be applied to different tasks.

In this work [5] it is proposed a method to guide the action selection in a RL algorithm using one or more causal models as oracles. The agent can consult these oracles to not perform actions that can lead to unwanted states or to choose the best option. This helps the agent to learn faster since it will not move blindly. Through interventions in the causal model, it can make queries of the type *What if I do ...?*

(e.g.., If I drop the passenger off here, will my goal be achieved?). This type of interventions can help to reduce the search space.

#### 14.5.2.1 Causal Q-Learning

To illustrate the use of causal knowledge for RL, the classic *Taxi problem* is considered [4]. Jaime is a taxi driver whose main goal is to pick up a passenger at a certain point (passenger position), take him to the desired destination (destination position) and drop him off there. Jaime uses common sense rules to achieve its goal: he does not try to pick a passenger when there is no passenger, drop him off when he has not arrived to the goal position, etc. We can create a causal model from the rules that guide Jaime. The parameters of the causal model can be defined as boolean variables in a set of structural equations:

$$\begin{aligned}
\text{pickup} &= u_1 \\
\text{dropoff} &= u_2 \\
\text{cabPosition} &= u_3 \\
\text{destinationPosition} &= c_4 \\
\text{passengerPosition} &= c_5 \\
\text{onDestinationPosition} &= [(\mathit{destinationPosition} = \mathit{cabPosition}) \vee u_6] \wedge \neg u'_6 \\
\text{onPassengerPosition} &= [(\mathit{passengerPosition} = \mathit{cabPosition}) \vee u_7] \wedge \neg u'_7 \\
\text{inTheCab} &= [(\mathit{pickup} = \mathit{True} \wedge \mathit{onPassengerLocation} = \mathit{True}) \vee u_8)] \wedge \neg u'_8 \\
\text{goal} &= [(\mathit{dropoff} = \mathit{True} \wedge \mathit{inTheCab} = \mathit{True} \wedge \\
&\quad \mathit{onDestinationLocation} = \mathit{True}) \vee u_9] \wedge \neg u'_9
\end{aligned}$$

where $u_1, u_2 \in \{True, False\}$, $u_3, c_4, c_5$ can take some value that characterizes some position in the environment. The rest of $u_i, u'_i \in \{True, False\}$ variables represent unusual behaviors. Suppose the case when $onDestinationLocation = False$, even when the taxi is on the same position as the passenger, maybe the passenger position has been updated without notifying the taxi driver, in this scenario $u'_6 = True$. The counterpart happens when $u_6 = True$, then the taxi is on the passenger position. The corresponding causal structure is shown in Fig. 14.8.

To take advantage of the causal knowledge, the classic RL algorithm, *Q-learning*, is modified. The agent observes its current state and queries the causal model; if the model provides an action according to the state it selects this action. Otherwise, it selects an action based on to the traditional exploration strategy of Q-learning ($\epsilon$-greedy). The general procedure is shown in Algorithm 14.1. It receives the set of states, set of actions, the reward function and the causal model; and returns the $Q(s, a)$ table, that specifies the "best" action, $a$ for each state, $s$.

#### 14.5.2.2 Results

Figure 14.9 depicts an instance of the Taxi problem considering a $5 \times 5$ grid world. There are four locations in this world, marked as R, B, G, and Y. The taxi problem

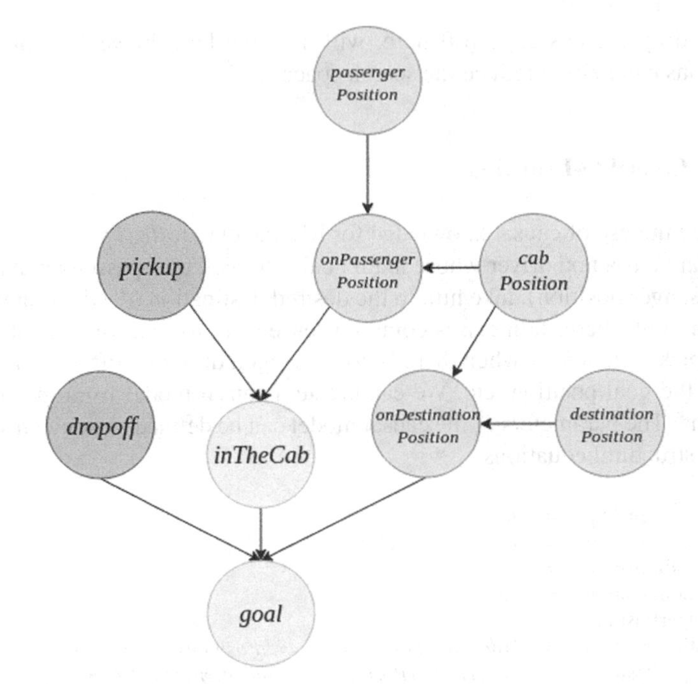


**Fig. 14.8** Causal Bayesian network for the Taxi problem. The color of the nodes indicates to which set of variables it corresponds. Red for actions, yellow for target variables, and blue for state variables

---

**Algorithm 14.1** Causal Q-Learning

---

**Require:** States, Actions, Rewards, Causal Model
1: Initialize $Q(s, a)$ arbitrarily
2: **repeat**
3: Initialize $s$
4: **repeat**
5: $a \leftarrow$ intervention based on Causal Model for $s$
6: **if** $(a = null)$ **then**
7: Choose $a$ for $s$ using policy derived from $Q$ ($\epsilon$-greedy)
8: **end if**
9: Take action $a$, observe $r$, $s'$
10: $Q(s, a) \leftarrow Q(s, a) + \alpha[r + \gamma max_{a'} Q(s', a')) - Q(s, a))]$
11: $s \leftarrow s'$
12: **until** end of episode
13: **until** $s$ is a terminal state
14: **return** Table $Q$ (policy)

---

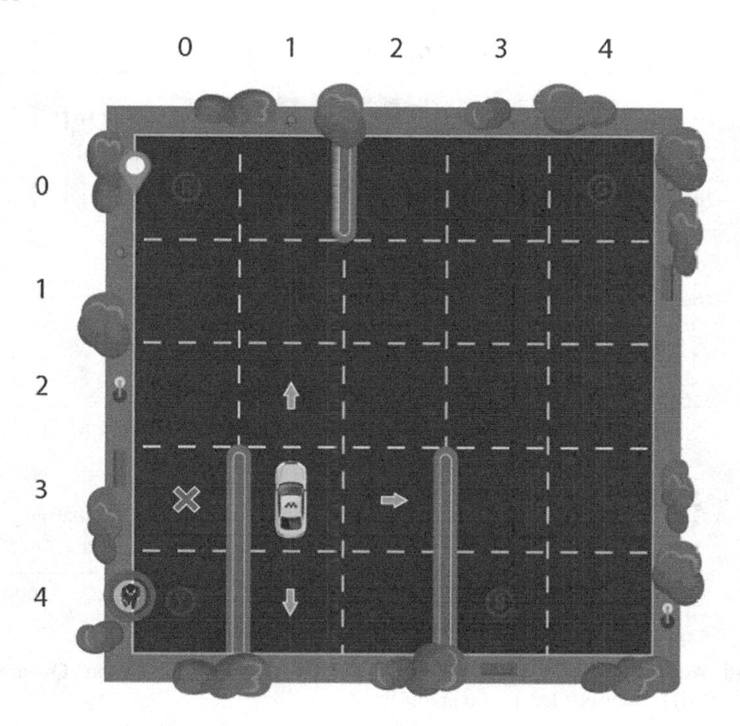


**Fig. 14.9** Sketch of the taxi environment in a $5 \times 5$ grid world. The taxi is shown in one cell (state), with arrows that depict the possible actions for that state

is episodic. In each episode, the taxi starts in a randomly-chosen square. There is a passenger at one of the four locations (chosen randomly), and that passenger wishes to be transported to one of the four locations (also chosen randomly). The taxi must go to the passenger's location, pick up the passenger, go to the destination location, and drop the passenger there. The episode ends when the passenger is deposited at the destination location. There are six primitive actions in this domain: (a) four navigation actions that move the taxi one square North, South, East, or West; (b) a Pickup action; and (c) a Drop off action. The six actions are deterministic. There is a reward of $-1$ for each action and an additional reward of $+20$ for successfully delivering the passenger. There is also a 10 point penalty for illegal pick-up and drop-off actions.

Causal Q-learning was compared to the basic Q-learning algorithm for the Taxi problem. Each version of the algorithm was executed 50 times, and in each execution the average reward per episode was computed. The results are summarized in Fig. 14.10, which shows the average reward per episode of both algorithms. We observe that Causal Q-learning starts with a higher reward and tends to stabilize faster. To quantify the difference, it is considered that an algorithm has reached an *optimal reward* once the average reward is equal to 9. On average, basic Q-learning

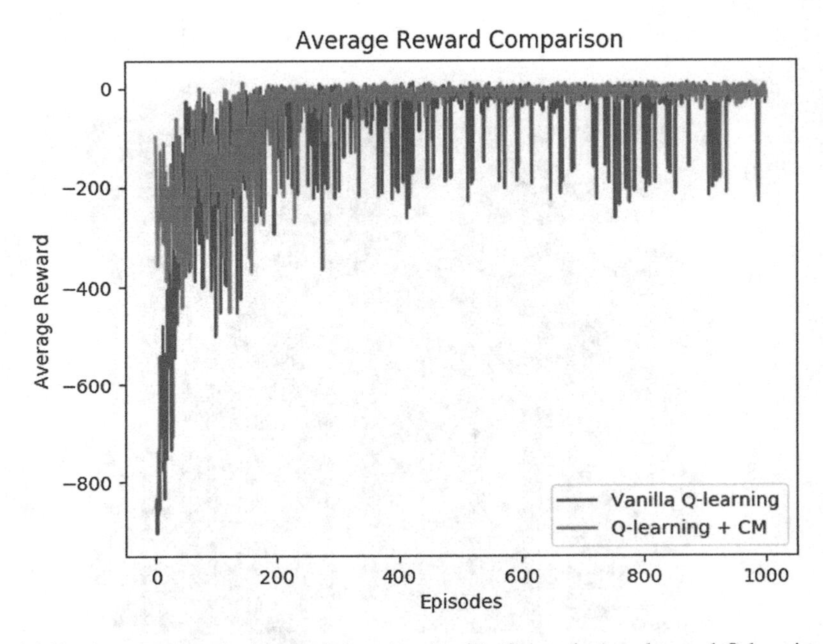


**Fig. 14.10** Average reward of basic Q-learning (vanilla Q-learning) and causal Q-learning (Q-learning + CM) per episode for the Taxi problem

reaches that reward in 95 episodes and causal Q-learning in 65 episodes. So, at least in this case, augmenting Q-learning with causal knowledge results in faster learning.

## 14.6 Additional Reading

Graphical causal modeling was originally introduced in genetics [11]. An accesible introduction to causal models is presented in [7]. Two comprehensive books on graphical causal models are [6, 9]. Peters et al. [8] give a general introduction to causality, with emphasis on its relation with statistics and machine learning.

## 14.7 Exercises

1. What are the differences between a directed graphical model, such as a Bayesian network, and a causal model?
2. Given the CBN in Fig. 14.3a, and the following CPTs (HC-Health Consciousness, LE-Lack of Exercise, UD-Unhealthy diet, O-Overweight, S-Stroke):

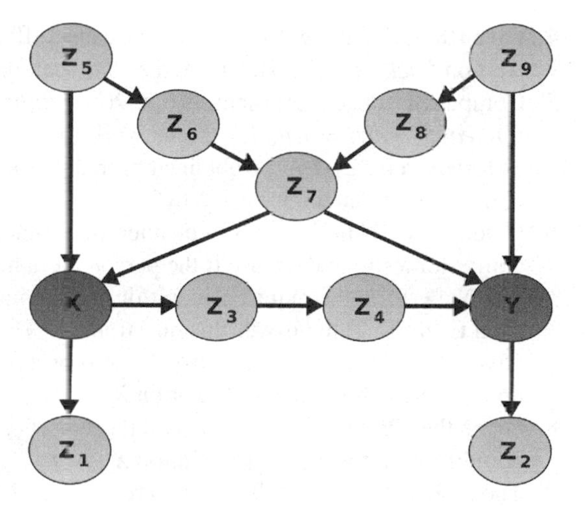


**Fig. 14.11** A CBN

$P(LE \mid HC) =$

| HC | | F | T |
|---|---|---|---|
| LE | F | 0.2 | 0.7 |
| | T | 0.8 | 0.3 |

$P(UD \mid HC) =$

| HC | | F | T |
|---|---|---|---|
| UD | F | 0.4 | 0.9 |
| | T | 0.6 | 0.1 |

$P(O \mid LE, UD) =$

| LE,UD | | F, F | F, T | T, F | T,T |
|---|---|---|---|---|---|
| O | F | 0.8 | 0.6 | 0.7 | 0.2 |
| | T | 0.2 | 0.4 | 0.3 | 0.8 |

$P(S \mid O) =$

| S | | F | T |
|---|---|---|---|
| O | F | 0.8 | 0.6 |
| | T | 0.2 | 0.4 |

(a) Given the intervention $do(Unhealthy - diet = true)$, calculate the posterior probability of *Overweight* and *Stroke*. (b) Repeat for $do(Unhealthy - diet = false)$.

3. Based on the results from the previous problem, what is the causal effect of *Unhealthy−diet* on *Overweight*? And on *Stroke*? (use the same parameters as in Exercise 2).

4. Given the CBN in Fig. 14.3, what is the causal effect of *Stroke* on $Unhealthy{-}diet$? And on *Lack-of-Exercise*? (Use the same parameters as for Exercise 2.)
5. Compute the posterior probability of *Overweight* and *Stroke* given the observation $Unhealthy{-}diet = true$ for the network in Fig. 14.3a (use the same parameters as for Exercise 2). Do the obtained probabilities coincide with the ones from the corresponding intervention? Why?
6. Given the CBN in Fig. 14.3a, consider the counterfactual question: Will the probability for a stroke decrease if the person has a healthy diet ($Unhealthy - diet = false$)? Perform the required operations on the model (using the same parameters as in Exercise 2) to answer this question.
7. Given the CBN in Fig. 14.11, list all the subsets of variables that satisfy the back-door criterion for the causal relation $X \rightarrow Y$.
8. Given the CBN in Fig. 14.11, list all the subsets of variables that satisfy the front-door criterion for the causal relation $X \rightarrow Y$.
9. The table below shows data from a real medical study [1] comparing the success rates of two treatments for kidney stones. The table includes the success rate (Recovery, $R$) of the patients according to the Treatment ($T$) (A or B) and the Size of Stone ($S$). The numbers in parentheses indicate the number of success cases over the total size of the group. In total there are 700 patients, 350 were given treatment A and 350 treatment B.

| Stone Size | Treatment A | Treatment B |
|---|---|---|
| Small stones | 93% (81/87) | 87% (234/270) |
| Large stones | 73% (192/263) | 69% (55/80) |
| Both | 78% (273/350 | 83% (289/350) |

There is an apparent paradox, known as Simpon's Paradox, as the data seems to indicate that treatment A is better for each group, small and large stones, however considering both, treatment B seems better! A way to solve this paradox is via a causal Bayesian network, its structure is shown in Fig. 14.12. (a) Given the data estimate the CPTs for the CBN. (b) Given this model, apply the do-calculus to obtain $P_{do(T=A)}(R = true)$ and $P_{do(T=B)}(R = true)$. (c) Which treatment is better, i.e. has a higher recovery rate?

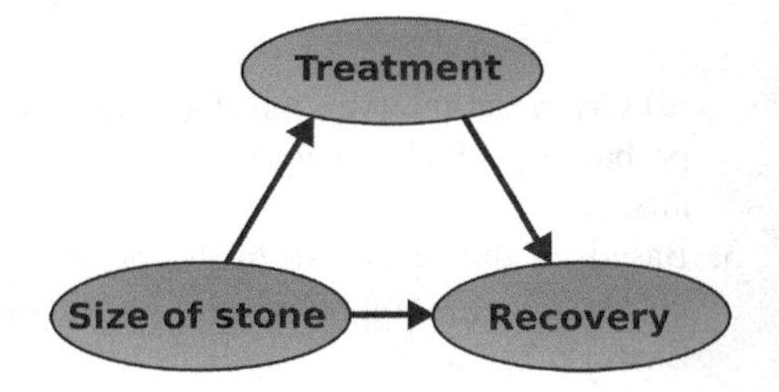


**Fig. 14.12** Causal model for the kidney stones problem

10. *** Develop a program that implements the the do-calculus for discrete causal Bayesian networks.

## References


1. Charig, C. R., Webb, D. R., PayneS. R., Wickham, J. E.: Comparison of treatment of renal calculi by open surgery, percutaneous nephrolithotomy, and extracorporeal shockwave lithotripsy. Brit. Med. J. **292**(6524), 879–882 (1986)
2. Chiappa, S., Gillam, T. P. S.: Path-Specific Counterfactual Fairness, arXiv-stat.ML (2018)
3. Chiappa, S., William, S.I.: A Causal Bayesian Networks Viewpoint on Fairness. IFIP Adv. Inf. Commun. Technol. **3–20** (2019)
4. Dietterich, T.G.: Hierarchical reinforcement learning with the MAXQ value function decomposition. J. Artif. Intell. Res. **13**(1), 227–303 (2000)
5. Méndez-Molina, A., Feliciano-Avelino, I. Morales, E.F, Sucar, L.E.: Causal Based Q-Learning, Causal Reasoning Workshop, Mexican International Conference on Artificial Intelligence (2019)
6. Pearl, J.: Causality: Models, Reasoning and Inference. Cambridge University Press, New York (2009)
7. Pearl, J., Mackienze, D.: The Book of Why. Basic Books, New York (2018)
8. Peters, J., Janzing, D., Scholkpf, B.: Elements of Causal Inference. MIT Press, Cambridge (2017)
9. Spirtes, P., Glymour, C., Scheines, R.: Causation, Prediction, and Search. MIT Press, Cambridge (2000)
10. Sutton, R.S, Barto, A.G.: Reinforcement Learning: An Introduction. The MIT Press, Cambridge (1998)
11. Wright, S.: Correlation and causation. J. Agric. Res. **20**, 557–585 (1921)

# Chapter 15
# Causal Discovery

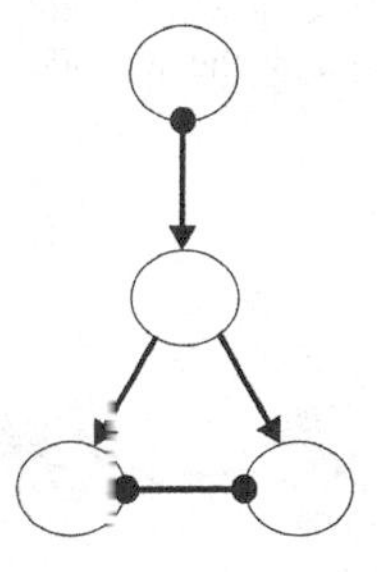

## 15.1 Introduction

Learning causal models from observational data (without direct interventions) poses many challenges. If we discover that there is a certain dependency between two variables, $X$ and $Y$, we can not determine, without additional information, if $X$ *causes* $Y$ or vice versa. Additionally, there could be some other variable (cofactor) that produces the dependency between these two variables. For instance, consider that based on data we discover that people that drink wine tend to have less heart attacks. Then we might be inclined to conclude that drinking wine tends to decrease the probability of a heart attack (Fig. 15.1a). However, there might be another variable that produces this apparent causal relation, known as a *latent common cause*. It could be that both, wine drinking and heart attacks, have to do with the income level of the person, as persons with high income tend to drink more wine and at the same time have a lower probability of heart attack because of better medical services (see Fig. 15.1b). Thus, a difficulty of learning causal relations lies in how to include in the model all the relevant factors.

Reichenbach [9] established a relation between graphical causal models and dependence as the *common cause principle*: If two random variables $X$ and $Y$ are statistically dependent, then there exists a third variable $Z$ that causally influences both. As a special case, $Z$ may coincide with either $X$ or $Y$. This principle is illustrated in Fig. 15.2



L. E. Sucar, *Probabilistic Graphical Models*, Advances in Computer Vision and Pattern Recognition, https://doi.org/10.1007/978-3-030-61943-5_15

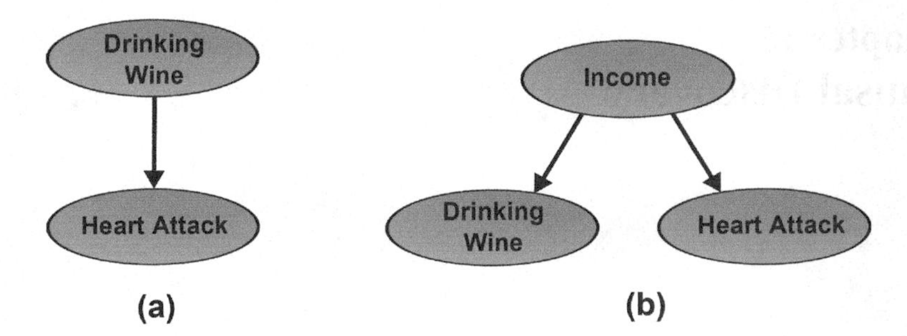


**Fig. 15.1** An example of the difficulty of learning causal models. **a** The initial network that shows an apparent causal relation between "drinking wine" and "heart attack". **b** An alternative network with a common cause, "income", that explains the dependency between "drinking wine" and "heart attack"

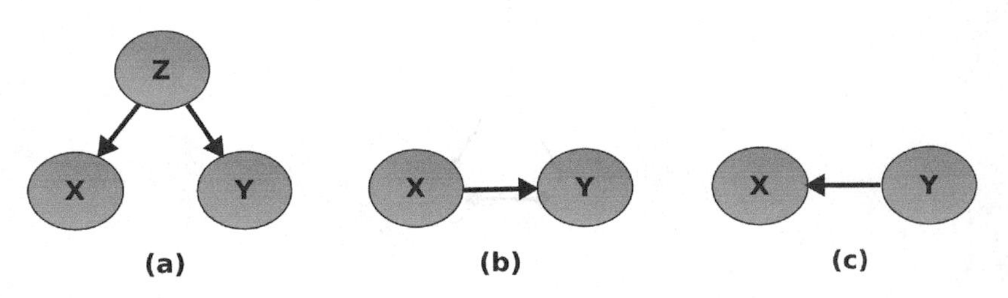


**Fig. 15.2** Common cause principle. A statistical dependence between $X$ and $Y$ indicates that both are caused by a third variable $Z$ (**a**) or $X$ causes $Y$ (**b**) or $Y$ causes $X$. It is also possible that there is a direct causal relation between $X$ and $Y$ and also a cofactor $Z$

In general, it is not possible to obtain a unique structure for a causal Bayesian network given only observational data; we obtain what is known as a *Markov equivalence class* (MEC). For instance, if we consider three variables, and the following *skeleton*: $X - Y - Z$, there are four possible directed graphs, as shown in Fig. 15.3. Three of these graphs represent the same conditional independence relation, $X$ and $Z$ are independent given $Y$, so they correspond to a Markov equivalence class. The other one, is a different MEC, in which $X$ and $Z$ are not independent given $Y$.

A Markov equivalence class includes all the graphs that entail the same set of D-separations—same conditional independences under the *faithfulness* assumption (described below). All the graphs in a MEC satisfy the following two properties:

1. They have the same skeleton, that is the same underlying undirected graph (if all directed edges are made undirected).
2. They have the same $V$ structures, that is subgraphs of the form $X \rightarrow Y \leftarrow Z$, such that there is no arc between $X$ and $Z$.

If causal sufficiency is not assumed, then what we can learn are *maximal ancestral graphs* (MAGs), and the set of equivalent MAGs known as *partial ancestral graphs*. In the following section, we discuss in detail the different types of graphs.

To obtain the structure of a causal model from observational data, additional assumptions are required. In general, the following assumptions are used when learning the structure of causal networks:

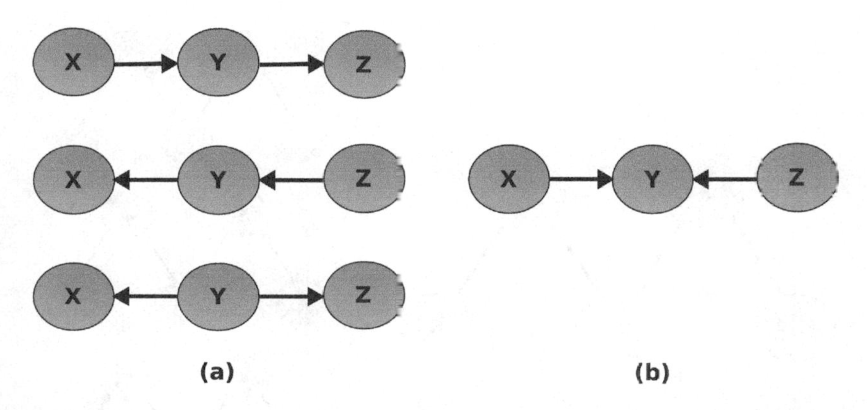


**Fig. 15.3** Dependence graphs for three variables with the skeleton $X - Y - Z$. **a** These graphs represent that $X$ and $Z$ are conditionally independent given $Y$. **b** This graph corresponds to the case in which $X$ and $Z$ are not independent given $Y$

Causal Markov Condition: a variable is independent of its non-descendants given its direct causes (parents in the graph).
Faithfulness: there are no additional independencies between the variables in the model that are not implied by the causal Markov condition.
Causal Sufficiency: there are no common unobserved confounders (latent or hidden variables) of the observed variables in the model.

Other assumptions can be made, such as to consider a particular type of relations between a variable and it causes (e.g.., linear models) and/or particular distributions about the uncertainty/noise (e.g., Gaussian noise).

Next we present the different types of graphs used when learning causal models; and later the main types of causal discovery algorithms.

## 15.2 Types of Graphs

A causal Bayesian network (CBN) is represented as a directed acyclic graph or DAG (see Chap. 14). CBNs include only directed arcs in the graph that represent *causal* relations; such as $A \rightarrow B$. As for Bayesian networks, the conditional independence relations can be read directly from the graph via the criterion known as *D-separation*.

Causal discovery algorithms can not always obtain a unique causal model, that is a DAG, so other types of graphs are used to represent the (partial) structure obtained. Next we introduce the representations used when assuming, or not, causal sufficiency.

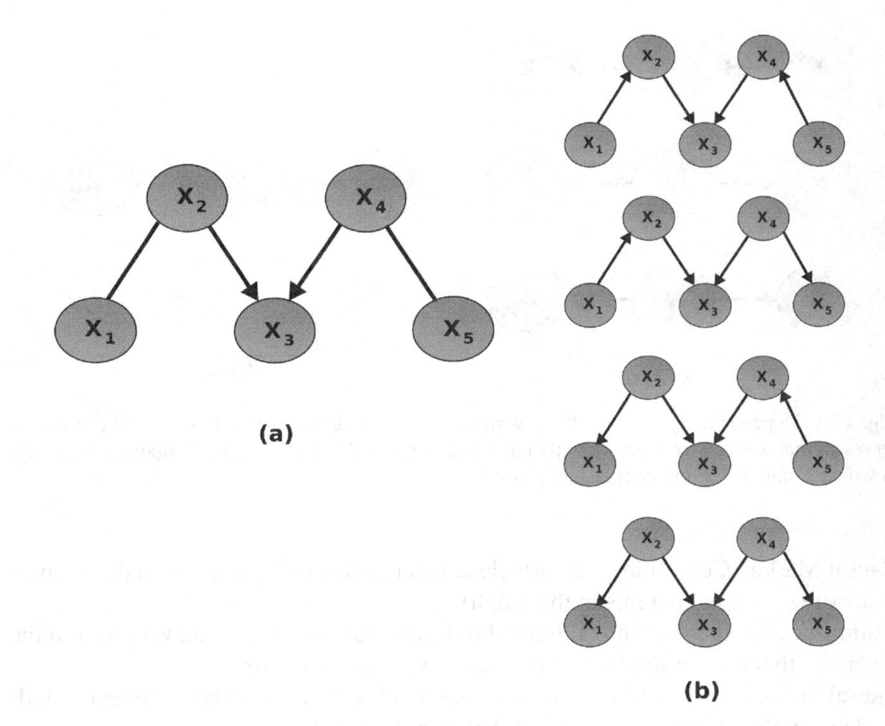


**Fig. 15.4** An example of a CPDAG (**a**) and the set of DAGs it encodes (**b**). Note that in all DAGs, $I(X_1, X_2, X_3)$ and $I(X_3, X_4, X_5)$

## *15.2.1 Markov Equivalence Classes Under Causal Sufficiency*

Under causal sufficiency, in general what we can learn from observational data is a Markov equivalence class.

Several DAGs can encode the same conditional independencies via $d$-separation. Such DAGs form a **Markov Equivalence Class (MEC)** which can be described uniquely by a **Completed Partially Directed Acyclic Graph (CPDAG)**. A CPDAG $\mathcal{C}$ has the same adjacencies as any DAG in the MEC encoded by $\mathcal{C}$. A directed edge $X \to Y$ in a CPDAG $\mathcal{C}$ corresponds to a directed edge $X \to Y$ in every DAG in the MEC described by $\mathcal{C}$. For any non-directed edge $X - Y$ in a CPDAG $\mathcal{C}$, the Markov equivalence class described by $\mathcal{C}$ contains a DAG with $X \to Y$ and a DAG with $X \leftarrow Y$. Thus, CPDAGs contain directed and non-directed edges. An example of a MEC and the corresponding DAGs is depicted in Fig. 15.4.[1] Note that all DAGs in the same Markov equivalence class have the same skeleton (undirected graph obtained if we make all directed edges undirected) and $V$ structures (that is, subgraphs of the form $X \to Y \leftarrow Z$).

[1]Recall the $I(X_1, X_2, X_3)$ means that $X_1$ is independent of $X_3$ given $X_2$.

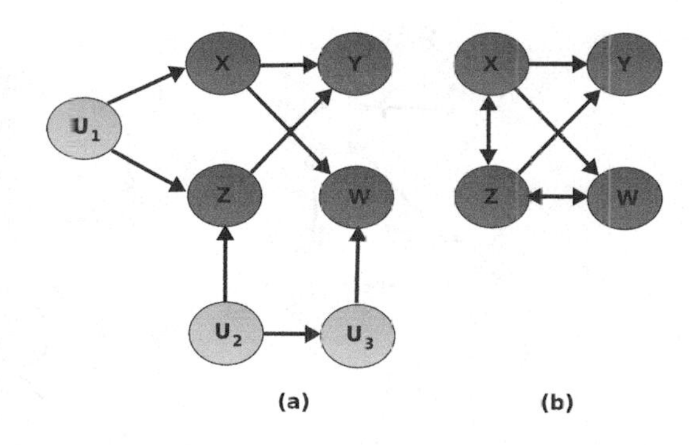


**Fig. 15.5** An example of a DAG (**a**) where $U_1, U_2, U_3$ are latent variables and the corresponding MAG (**b**)

## *15.2.2 Markov Equivalence Classes with Unmeasured Variables*

If some variables are unmeasured and can be common causes of the measured variables (cofactors), an alternative representation is required that takes into account these cofactors. We assume a set of variables $\mathbf{V}$, in which $\mathbf{O}$ are the subset of measured or observed variables and $\mathbf{L}$ the unmeasured or latent variables, such that $\mathbf{V} = \mathbf{O} \cup \mathbf{L}$.

*Maximal Ancestral Graphs* (MAGs) can represent conditional independence information and causal relationships in DAGs that include unmeasured (hidden or latent) variables. A MAG represents a DAG after all latent variables have been marginalized out, and it preserves all entailed conditional independence relations among the measured variables. This is, given a DAG $D$ over $\mathbf{V} = \mathbf{O} \cup \mathbf{L}$ there is a MAG $M$ over $\mathbf{O}$ alone, such that for any disjoint sets $\mathbf{X}, \mathbf{Y}, \mathbf{Z} \subseteq \mathbf{O}$, $\mathbf{X}$ and $\mathbf{Y}$ are d-separated by $\mathbf{Z}$ in $D$ if and only if they are *m-separated*[2] by $\mathbf{Z}$ in the MAG $M$. The following construction gives us such a MAG: (i) For each pair of variables $X, Y \in \mathbf{O}$, $X$ and $Y$ are adjacent in $M$ if and only if there is an *inducing path*[3] between them relative to $\mathbf{L}$ in $D$. (ii) For each pair of adjacent variables $X, Y$ in $M$: (a) orient the edges $X \rightarrow Y$ in $M$ if $X$ is an ancestor of $Y$ in $D$; (b) orient it as $X \leftarrow Y$ in $M$ if $Y$ is an ancestor of $X$ in $D$; (c) orient it as $X \leftrightarrow Y$ in $M$, otherwise.

An example of a DAG with latent variables and the corresponding MAG is shown in Fig. 15.5.

Several MAGs can also encode the same conditional independencies via m-separation. Such MAGs form a Markov Equivalence Class (MEC) which can be described uniquely by a *partial ancestral graph* (PAG). A PAG, $P_M$, si a partial mixed graph with three kinds of marks in its edges: arrowhead ($>$), tail ($-$) or circle ($\circ$), such that: (i) $P_M$ has the same adjacencies as all the MAGs in the MEC; (ii) Every non-circle mark in $P_M$ is an invariant mark in in all MAGs in the MAC; i.e., a mark of arrowhead is in $P_M$ if and only if it is shared by all MAGs in the MEC, and a mark of tail is in $P_M$ if and only if it is shared by all MAGs in the MEC; (iii) A mark of circle otherwise (when the same mark is not shared by all MAGs).

[2]m-separation is as d-separation for MAGs, see [12].

[3]See Section 15.3.2.3 for a formal definition of inducing path.

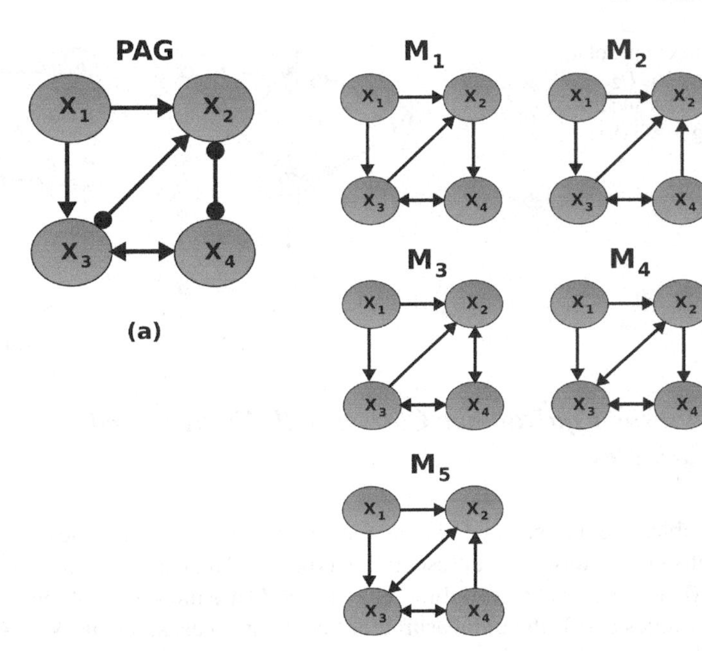


**Fig. 15.6** **a** An example of a PAG; **b** the corresponding MAGs ($M_1$, ...$M_5$) in the Markov equivalence class

Fig 15.6 depicts a PAG and the set of MAGs in the corresponding MEC.

## 15.3 Causal Discovery Algorithms

The type of datasets used in the learning of the causal structure may be: (i) **Observational**, data corresponding to measurements made under natural conditions of a causal system; (ii) **Experimental**, data correspond to measurements made under external interventions; (iii) **Hybird**, it combines observational and experimental data. Ideally, experimental data should be used in the structure learning of causal BNs. Nevertheless, experimental data are not always available or it can be unethical, infeasible, time consuming, or expensive to obtain. On the other hand, observational data, i.e., data associated with processes that cannot be reproduced, are often abundant.

Based on observational data, causal structure learning approaches can be classified according to three dimensions:

- They assume causal sufficiency or consider unmeasured cofounders.
- They consider continuous variables (usually Gaussian linear models) or discrete variables.

- They are based on dependency tests (constraint-based) or in search over the set of possible structures and a score (score-based).

In general these methods recover a Markov equivalence class unless extra assumptions are made or they incorporate background knowledge.

Next we will review some score-based and constraint based techniques for learning CBNs from observational data, initially assuming causal sufficiency, and then relaxing this assumption. Then we will describe an algorithm for learning linear models.

### *15.3.1 Score-Based Causal Discovery*

#### 15.3.1.1 Greedy Equivalence Search

Greedy Equivalence Search (GES) [1, 2] is a score-based (global) algorithm for causal structure learning, that assumes causal sufficiency and discrete variables. GES heuristically searches the structure in the space of Markov equivalence classes (MECs) in two stages. In each step of the algorithm, every candidate MEC is evaluated, and it is selected the MEC with the highest score that improves the score function. In the first stage, starting with an empty graph, GES adds edges to candidate MECs until a local maximum is reached. In the second stage, it removes edges of the MEC found in the first stage; assuming that some *extra* edges might have been added in the first stage given the greedy nature of the search process. The algorithm stops when a local maximum is reached and returns the CPDAG that represents the found MEC. In the large sample limit, GES returns the Markov equivalence class containing the true causal graph.

GES utilizes the $BDeU$ score, which satisfies two important properties of a score function for causal structure learning:

Decomposable: a scoring function, $S$, is decomposable if it can be expressed as the product of local functions that only depend of a node $X_i$ and its parents $Pa(X_i)$: $S = \prod_i f(X_i, Pa(X_i), \mathbf{D})$. This property is critical for an efficient algorithm.

Score Equivalent: a scoring function $S$, is score equivalent if for any pair of equivalent DAGs $G$ and $G'$, it assigns the same score, $S(G) = S(G')$. Since the scoring function is score equivalent, any DAG contained in the candidate MEC could be used for evaluating that MEC.

The *Bayesian Dirichlet equivalent and Uniform* (BDeU) is a decomposable and score equivalent function that evaluates MECs defined over discrete variables with complete data sets $\mathbf{D}$ (without missing values) [6]:

$$BDeU(\mathcal{G}, \mathbf{D}) = \prod_{i=1}^{n} \{score(X_i, \mathbf{Pa}(X_i), \mathbf{D})\} \tag{15.1}$$

$$score(X_i, \mathbf{Pa}(X_i), \mathbf{D}) = \prod_{j=1}^{q_i} \frac{\Gamma(\alpha_{ij})}{\Gamma(\alpha_{ij} + C_{ij})} \prod_{k=1}^{r_i} \frac{\Gamma(\alpha_{ijk} + C_{ijk})}{\Gamma(\alpha_{ijk})} \tag{15.2}$$

where $n$ is the number of nodes in $G$, $q_i$ is the number of values of $\mathbf{Pa}(X_i)$, $r_i$ is the number of values of $X_i$, $C_{ijk}$ is the number of cases in which $X_i = k$ and its parents $\mathbf{pa}(X_i = k) = j$, $N_{ij} = \sum_k C_{ijk}$, and $\alpha_{ijk} = \frac{1}{r_i q_i}$ is a Dirichlet prior parameter with $\alpha_{ij} = \sum_k \alpha_{ijk}$.

## 15.3.2 Constraint-Based Causal Discovery

### 15.3.2.1 PC Algorithm

The PC algorithm is probably the most well-known constraint-based structure learning algorithm. The PC-algorithm proceeds in two stages, the first searches for adjacencies between variables to obtain the skeleton, while in the second stage it finds the orientations of the edges, where possible. See Chap. 8 for a detailed description.

The PC algorithm assumes causal sufficiency, and an independence oracle that supplies the independence constraints true in the distribution that generated the data; if so, it recovers as much information about the true causal structure as is possible with the available independence tests. However it can not provide guarantees for a limited sample size.

Next we will cover two extensions of the PC algorithm that relax the causal sufficiency assumption.

### 15.3.2.2 Bayesian Constraint-Based Causal Discovery

The Bayesian Constraint-Based Causal Discovery (BCCD) [3] algorithm is an extension of the PC algorithm that consists of two main phases:

1. Start with a completely connected graph and estimate the reliability of each causal link, $X - Y$, by measuring the conditional independence between $X$ and $Y$. If a pair of variables are conditionally independent with a reliability above a certain threshold, it then deletes the edge between these variables.
2. The remaining causal relations (undirected edges in the graph) are ordered according to their reliability. Then the edges in the graph are oriented starting from the most reliable relations, based on the conditional independence test for variable triples.

To estimate the reliability of a causal relation, $R = X \rightarrow Y$, the algorithm uses a Bayesian score:

$$P(R \mid D) = \frac{P(D \mid M_R)P(M_R)}{P(D \mid M)P(M)} \tag{15.3}$$

where $D$ is the data, $M$ are all the possible structures, and $M_R$ are all the structures that contain the relation $R$. Thus, $P(M)$ denote the prior probability of a structure $M$ and $P(D \mid M)$ the probability of the data given the structure. Calculating 15.3 is very costly, so it is approximated by the marginal likelihood of the data given the structure, and usually restricted to a maximum number of variables in the network.

Depending on the reliability threshold, the resulting network can have undirected edges, $-$, which means that there is not enough information to obtain the direction of the link, and bi-directed edges, $\leftrightarrow$, indicating that there is a common cofounder. So in general, BCCD returns a PAG.

In the Sect. 15.4 we will illustrate the BCCD algorithm in a real-world application.

#### 15.3.2.3 FCI Algorithm

*Fast Casual Inference* (FCI) is an extension of the PC algorithm for causal structure learning, considering variables that are causally insufficient. The Fast Causal Inference Algorithm [13] tolerates, and sometimes discovers, unknown confounding variables, and it has been shown to be asymptotically correct even in the presence of confounders.

Before describing the FCI algorithm, we need to define some graphical concepts. Given a DAG over the set of variables $\mathbf{V}$, where $\mathbf{O} \subset \mathbf{V}$ are the observed variables, an undirected path between $A, B, \in \mathbf{O}$ is an *inducing path* relative to $\mathbf{O}$ if and only if every member of $\mathbf{O}$ in the trajectory is a collider except the endpoints, and every collider is an ancestor of either $A$ or $B$. It can be shown that if there is an inducing path between $A$ and $B$, they can not be d-separated by any subset of variables of $\mathbf{O}$; and if there si not such inducing path, there is at least one subset of $\mathbf{O}$ that d-separates them [13].

$G'$ is an *inducing path graph* over $\mathbf{O}$ for a DAG $G$ if and only if $\mathbf{O}$ is a subset of the vertices in $G$, there is an edge between variables $A$ and $B$ with an arrowhead at $A$ if and only if $A$ and $B$ are in $\mathbf{O}$, and there is an inducing path in $G$ between $A$ and $B$ relative to $\mathbf{O}$ that is into $A$. In an inducing path graph, there are two kinds of edges: $A \rightarrow B$ implies that every inducing path over $\mathbf{O}$ between $A$ and $B$ is out of $A$ and into $B$, and $A \leftrightarrow B$ entails that there is an inducing path over $\mathbf{O}$ that is into $A$ and into $B$. This last kind of edge can only occur when there is a latent common cause of $A$ and $B$ [13].

An example of an inducing path and inducing path graph is depicted in Fig. 15.7. Some additional definitions are required:

Collider: $B$ is a collider along the path $< A, B, C >$ if and only if $A* \rightarrow B \leftarrow *C$.

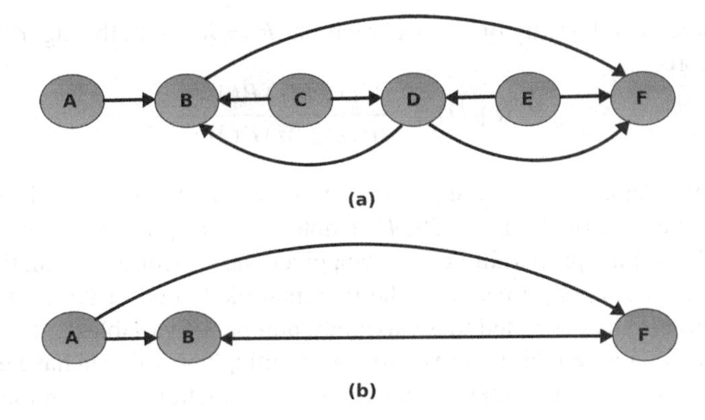


**Fig. 15.7** Given the DAG in **a**, the path $< A, B, C, D, E, F >$ is an inducing path over $O = \{A, B.F\}$; in **b** it is shown the corresponding induced path graph over $O$ (Figure based on [13])

Definite non-collider: $B$ is a definite non-collider on undirected path $U$ if and only if either $B$ is an endpoint of $U$, or there exist vertices $A$ and $C$ such that $U$ contains one of the subpaths $A \leftarrow B * - * C$, $A * - * B \rightarrow C$, or $A * - * \underline{B} * - * C$. ($A * - * \underline{B} * - * C$ means that edges $A, B$ and $A, C$ do not collide in $B$.)

Edge into $A$: An edge between $B$ and $A$ is into $A$ if and only if $A \leftarrow *B$.

Edge out of $A$: An edge between $B$ and $A$ is out of $A$ if and only if $A \rightarrow B$.

Definite discriminating path: In a partially oriented inducing path graph, $U$ is a definite discriminating path for $B$ if and only if $U$ is an undirected path between $X \neq B$ and $Y \neq B$ containing $B$, every vertex on $U$ except for $B$ and the endpoints is a collider or a definite non-collider on $U$, and (i) if $V$ and $V'$ are adjacent on $U$, and $V$ is between $V$ and $B$ on $U$, then $V* \rightarrow V$ on $U$; (ii) if $V$ is between $X$ and $B$ on $U$ and $V$ is a collider on $U$ then $V \rightarrow Y$, else $V \leftarrow *Y$; (iii) if $V$ is between $Y$ and $B$ on $U$ and $V$ is a collider on $U$ then $V \rightarrow X$, else $V \leftarrow *X$; and (iv) $X$ and $Y$ are not adjacent.

Triangle: A triangle in a graph $G$ is a complete subgraph of $G$ with three vertices; i.e., vertices $X, Y, Z$ form a triangle if and only if $X$ and $Y$ are adjacent, $Y$ and $Z$ are adjacent and $X$ and $Z$ are adjacent.

In what follow we consider the same types of marks as in PAGs for the edges: arrowhead (>), tail (−) or circle (∘); plus an (∗) that indicates any of the previous marks. Sepset($X, Y$) contains the subsets of variables that d-separate $X, Y$, and Adjacencies($G, X$) the variables that are adjacent to $X$ in the graph $G$. To estimate d-separation from the probability distributions, the FCI algorithm performs conditional independence tests for discrete variables or partial correlations for linear, continuos variables. Next we present the Causal Inference (CI) algorithm (antecedent of FCI):

1. Build a complete undirected graph $G$ over the set of variables $V$.
2. If $A, B \in V$ are d-separated given any subset $S$ of $V$, remove the edge between $A$ and $B$, and record $S$ in Sepset($A, B$) and Sepset($B, A$).

3. Let $G'$ be the graph resulting from step (2). Orient each edge as $\circ - \circ$. For each triple of vertices $A, B, C$ such that the pair $A, B$ and the pair $B, C$ are each adjacent in $G'$ but the pair $A, C$ are not adjacent in $G'$, orient $A * - * B * - * C$ as $A* \rightarrow B \leftarrow *C$ if and only if $B$ is not in Sepset($A, C$), and orient it as $A * - * \underline{B} * - * C$ if and only if $B$ is in Sepset($A, C$). ($A * - * \underline{B} * - * C$ means that edges $A, B$ and $A, C$ do not collide in $B$.)
4. Repeat:
   - If there is a directed path from $A$ to $B$, and an edge $A * - * B$, orient $A * - * B$ as $A* \rightarrow B$,
   - Else if $B$ is a collider along the path $< A, B, C >$ in $G'$, $B$ is adjacent to $D$, and $D$ is in Sepset($A, C$), then orient $B * - * D$ as $B \leftarrow *D$,
   - Else if $U$ is a definite discriminating path between $A$ and $B$ for $M$ in $G'$, and $P$ and $R$ are adjacent to $M$ on $U$, and $P - M - R$ is a triangle, then:
     - If $M$ is in Sepset($A, B$) then $M$ is marked as a noncollider on subpath $P * - * M * - * R$,
     - Else $P * - * M * - * R$ is oriented as $P* \rightarrow M \leftarrow *R$.
   - Else if $P* \rightarrow M * - * R$ then orient as $P* \rightarrow M \rightarrow R$.

   until no more edges can be oriented.

The CI algorithm is not practical for large numbers of variables because of the way the adjacencies are constructed. While it is correct to remove an edge between $A$ and $B$ from the complete graph if and only if $A$ and $B$ are d-separated given some subset of $V$, this is impractical for two reasons: (i) there are too many subsets of $V$ on which to test the conditional independence of $A$ and $B$, (ii) for discrete distributions, unless the sample sizes are enormous there are no reliable tests of independence of two variables conditional on a large set of other variables.

The FCI algorithms follows a strategy similar to PC to avoid testing independence on all subsets of $V$. Given the initial complete undirected graph, it removes an edge between $X$ and $Y$ if they are d-separated given subsets of vertices adjacent to $X$ or $Y$ in $G$. This will eliminate many, but perhaps not all of the edges that are not in the inducing path graph. Second, it orients edges by determining whether they collide or not, just as in the PC algorithm. In a third step, FCI finds possible additional d-separations not found in the first step.

The FCI algorithm:

1. Build a complete undirected graph $G$ over the set of variables $V$.
2. $n = 0$

   Repeat:
   - Repeat:
     - Select an ordered pair of variables $X$ and $Y$ that are adjacent in $G$ such that Adjacencies($G, X$) has cardinality greater than or equal to $n$, and a subset $S$ of Adjacencies($G, Y$) of cardinality $n$, and if $X$ and $Y$ are d-separated given

$S$ delete the edge between $X$ and $Y$ from $G$, and record $S$ in Sepset($X, Y$) and Sepset($Y, X$).

until all ordered variable pairs of adjacent variables $X$ and $Y$ such that Adjacencies($G, X$) has cardinality greater than or equal to $n$ and all subsets $S$ of Adjacencies($G, X$) of cardinality $n$ have been tested for d-separation;

- $n = n + 1$

until for each ordered pair of adjacent vertices $X, Y$, Adjacencies($G, X$) is of cardinality less than $n$.

3. Let $G'$ be the undirected graph resulting from step (2), orient each edge as $\circ - \circ$. For each triple of vertices $A, B, C$ such that the pair $A, B$ and the pair $B, C$ are each adjacent in $G'$, but the pair $A, C$ are not adjacent in $G'$, orient $A * - * B * - * C$ as $A* \rightarrow B \leftarrow *C$ if and only if $B$ is not in Sepset($A, C$).
4. For each pair of variables $A$ and $B$ adjacent in $G'$, if $A$ and $B$ are d-separated given any subset $S$ of Possible-D-SEP($A, B$) or any subset $S$ of Possible-D-SEP($B, A$) in $G$ remove the edge between $A$ and $B$, and record S in Sepset($A, B$) and Sepset($B, A$).

FCI then reorients an edge between any pair of variables $X$ and $Y$ as $X \circ - \circ Y$, and proceeds to reorient the edges in the same way as steps (3) and (4) of the CI algorithm.

An example of a model (a PAG) obtained with FCI algorithm is shown in Fig. 15.8, considering that the independence tests are correct. For more details and the theoretical bases of the CI and FCI algorithms see [13].

## 15.3.3 *Casual Discovery with Linear Models*

### 15.3.3.1 LiNGAM

LiNGAM (Linear, Non-Gaussian, Acyclic Model) [10] proposes a different approach for causal discovery considering linear non-Gaussian models. It is based on an statistical tool to identify linear models: *Independent Component Analysis* (ICA) [4]. In a linear model the variables $\mathbf{X} = \{X_1, ..., X_N\}$ are linear functions of their parents and the error terms, here represented by $e$, which are assumed to be continuous random variables with non-Gaussian distributions and independent of each other.

LiNGAM assumes: (i) causal sufficiency, (ii) linear relationships, (iii) non-Gaussian noise with non-zero variance. It also considers a causal ordering of the variables, $X_1, X_2, ... X_N$, such that a variable $X_i$ can not be a cause of a variable $X_j$, $j < i$ according to the causal order; which is not required to be known *a priori* and can be estimated. The structural equation of each variable is of the following form:

$$X_i = \sum_j b_{ij} X_j + e_i + c_i \tag{15.4}$$

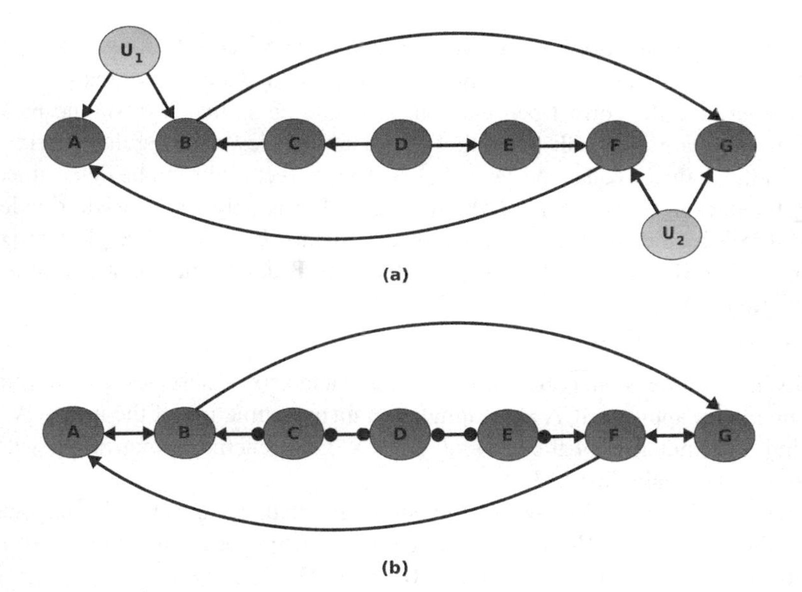


**Fig. 15.8** **a** Original causal Bayesian network, where $A, B, C, D, E, F, G$ are the observed variables and $U_1, U_2$ are unobserved or hidden variable. **b** The PAG that will be obtained by the FCI algorithm assuming correct independence tests (Figure based on [13])

where $b_{ij}$ are constants, $e_i$ is a non-Gaussian noise term and $c_i$ is an optional constant term. The noise terms, $e_i$, are independent with non-zero variance.

In the LiNGAM-algorithm the causal model is represented by a matrix equation:

$$\mathbf{X} = \mathbf{BX} + e \tag{15.5}$$

where the matrix $\mathbf{B}$ is an $N \times N$ matrix representing the edge coefficients on the corresponding graph. If the columns of $\mathbf{B}$ correspond to the hierarchical order of the variables in the graph, then $\mathbf{B}$ is lower triangular. The basic idea of LiNGAM is to recover the $\mathbf{B}$-matrix from a data matrix. Solving the above equation for $\mathbf{X}$ we get:

$$\mathbf{X} - \mathbf{BX} = e$$

So:

$$\mathbf{X} = (\mathbf{I} - \mathbf{B})^{-1} e \tag{15.6}$$

Which can be written as:

$$\mathbf{X} = \mathbf{A} e \tag{15.7}$$

where $A = (\mathbf{I} - \mathbf{B})^{-1}$.

Equation 15.7 together with the above assumptions fits the ICA framework, and it follows that $\mathbf{A}$ is identifiable. ICA is able to estimate $\mathbf{A}$, however with indetermination regarding the permutations of the variables and a scale factor. ICA returns $W_{ICA} =$ **PDW**, where **P** is an unknown permutation matrix and **D** a diagonal matrix.

To determine the correct permutation, **P** should be so that **DW** should have no zeros in the main diagonal; given that **B** should be a lower triangular matrix with zeros in the main diagonal and $\mathbf{W} = \mathbf{I} - \mathbf{B}$. The correct scale can be determined by using the ones in the diagonal of **W**. To obtain **W** it is only necessary to divide the rows of **DW** by the corresponding diagonal elements. Finally the weight matrix, **B**, is obtained as $\mathbf{B} = \mathbf{I} - \mathbf{W}$. The non-zero terms in **B** determine the structure of the causal graphical model.

LiNGAM algorithm:

1. Given a $d$-dimensional random vector $X$ and a matrix of observed data of dimension $d \times n$, apply an ICA algorithm to obtain an estimation of the matrix **A**.
2. Find a permutation of the rows of $\mathbf{W} = \mathbf{A}^{-1}$ to generate a matrix $\mathbf{W}'$ without zeros in the main diagonal.
3. Divide each row of $\mathbf{W}'$ by the corresponding element in the main diagonal to obtain a new matrix $\mathbf{W}''$ in which all the elements in the main diagonal are 1.
4. Obtain an estimate $\mathbf{B}'$ of **B** such that $\mathbf{B}' = \mathbf{I} - \mathbf{W}''$.
5. Finally, to estimate the causal order, find the permutation matrix $\mathbf{P}'$ of $\mathbf{B}'$, such that the matrix $\mathbf{B}'' = \mathbf{P}'\mathbf{B}'\mathbf{P}'^{\mathbf{T}}$ is as close as possible to a lower triangular matrix.

Thus, the assumptions of linearity and non-Gaussian independent error terms, enables the discovery of the exact true causal structure, as opposed to only the Markov equivalence class.

## 15.4 Applications

There are several domains in which causal discovery has been applied, including health, brain effective connectivity, finance, economy, among others. Next we describe two applications of learning causal graphical models, one related to attention deficit hyperactivity disorder, and other for decoding brain effective connectivity.

### *15.4.1 Learning a Causal Model for ADHD*

In [11], the authors extended the BCCD algorithm for a mixture of discrete and continuous variables, and apply it to a data set that contains phenotypic information about children with Attention Deficit Hyperactivity Disorder (ADHD). They used a reduced data set that contains data from 223 subjects, with the following nine variables per subject:

1. Gender (male/female)
2. Attention deficit score (continuous)

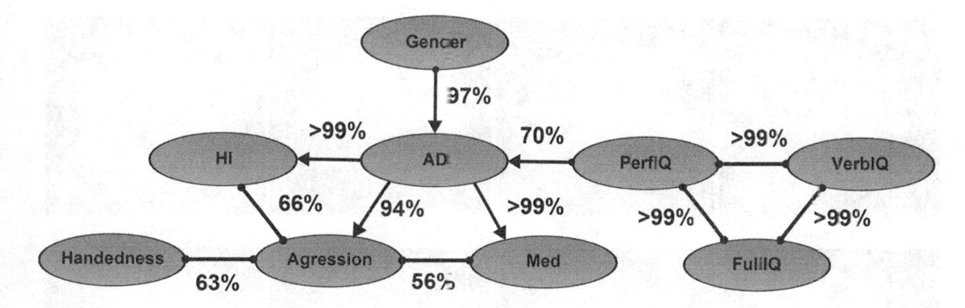


**Fig. 15.9** The causal model obtained from the ADHD data set. The graph represents the resulting PAG in which edges are marked as → or − for invariant relations and as "o" for non-invariant relations. The reliability of each edge is indicated. Notation: AD—attention deficit score, HI—hyperactivity/impulsivity score, PerfIQ—performance IQ, VerbIQ—verbal IQ, Med—medication status (Figure based on [11])

3. Hyperactivity/impulsivity score (continuous)
4. Verbal IQ (continuous)
5. Performance IQ (continuous)
6. Full IQ (continuous)
7. Aggressive behavior (yes/no)
8. Medication status (naïve/not naïve)
9. Handedness (right/left)

Given the small data set, they included some background information, in particular that no variable in the network can cause "gender". Using a reliability threshold of 50%, the network depicted in Fig. 15.9 was obtained; which is represented as a PAG.

Several interesting causal relations are suggested by the resulting network, some of which were already known from previous medical studies [11]:

- There is a strong effect from gender on the level of attention deficit.
- The level of attention deficit affects hyperactivity/impulsivity and aggressiveness.
- Handedness (left) is associated with aggressive behavior.
- The association between performance IQ, verbal IQ and full IQ is explained by a latent common cause.
- Only the performance IQ has a direct causal link with attention deficit.

### *15.4.2 Decoding Brain Effective Connectivity Based on fNIRS*

Montero et al. [7] explore the use of causal graphical models for decoding the effective connectivity of the brain from functional optical neuro-imaging (fNIRS). The basic idea is that directions of arcs of the connectivity network left undecided by existing learning algorithms can be resolved by exploiting prior structural knowledge from

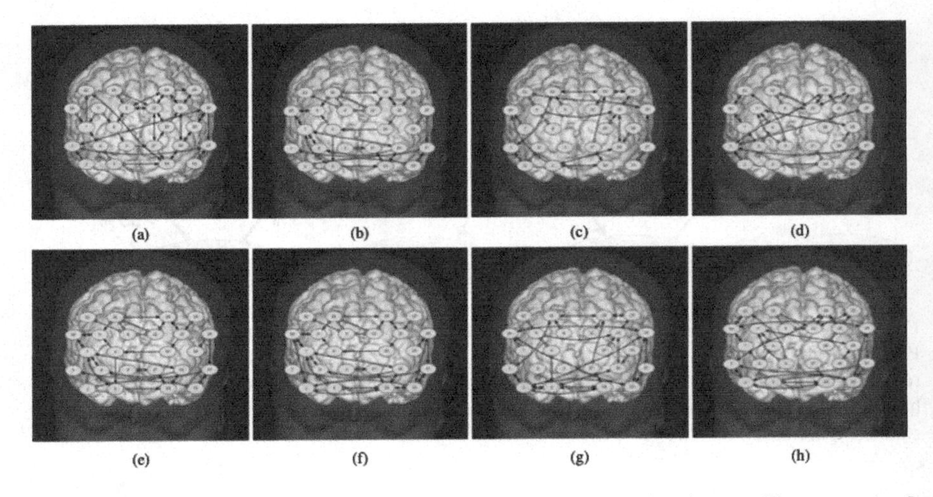


**Fig. 15.10** Effective connectivity networks obtained for the knot-tying data set. The top row **a–d** depicts the results from the original FCI algorithm, while the bottom row **e–h** shows the results with sFCI that incorporated the prior knowledge from the connectome. Columns represent the different expertise groups: novices (**a**, **e**), trainees (**b**, **f**), consultants (**c**, **g**) and all (**d**, **h**) (Figure based on [7])

the *human connectome*. A variant of the fast causal inference algorithm, seeded FCI, is therefore proposed to handle prior information.

To accomplish complex tasks, the functionally specialized brain regions have to collaborate among them either by temporally coordinating their actions (functional connectivity), or causally regulating their activity (effective connectivity). Modeling the effective connectivity is important to understand the brain behavior and the expression of neural circuits. Effective connectivity is concerned with decoding cooperating brain regions, and most importantly, determining the direction of the flow of information; thus it can be considered as a causal model, in which the activation of certain region in the brain, *causes* the activation of other(s) region(s). The brain regions can be modeled as variables within a system; and their activity can be estimated based on fNIRS neuro-imaging, which is able to take a snapshot of the cortical activity across brain regions while a person is performing certain task.

In fNIRS, infrared light is irradiated at the scalp that travels through extra cerebral tissues to reach the cortex, and a fraction of it returns to surface due to backscattering. Changes in light attenuation are attributed to changes in concentration of hemoglobin, in turn, assumed to be representative of underlying neural activity.

The structural connections of the human brain establishes a set of anatomical constraints with respect to the possible paths in its connectivity. This set of constraints can be obtained from the so called human connectome, which establishes the expected physical links in the human brain. So the human connectome provides prior knowledge for learning the causal structure related to the effective connectivity in the brain, by limiting the potential connections (links) in the causal network.

**Table 15.1** Number of undefined links in the learned PAGs for each group, using the FCI and sFCI algorithms

| Algorithm | Novices | Trainees | Experts | All |
|---|---|---|---|---|
| FCI | 11 | 19 | 18 | 26 |
| sFCI | 8 | 16 | 14 | 21 |

To incorporate prior knowledge, a modified version of the FCI algorithm was developed, named *seeded FCI* (sFCI). The addition of prior knowledge may resolve some of the undecided directions present in the output of the FCI algorithm, i.e. a partial ancestral graph. This prior knowledge can be added in form of restrictions by defining necessary or impossible relations. The basic idea in sFCI is to start with a complete undirected graph and a set of invariant links $L$ (prior information). This additional information is incorporated in the original FCI algorithm and used as restrictions when removing and directing edges in the different phases.

The sFCI algorithm was applied to uncover effective connectivity in a fNIRS neuro-imaging dataset collected at Imperial College to question about experience-dependent differences on pre-frontal activity for a group of surgeons while knot-tying. 62 surgeons (19 consultants, 21 trainees and 22 medical students) participated in the study, and brain activity was monitored with a 24-channel fNIRS system. A total of eight causal networks, shown in Fig. 15.10, were learned considering four groups (novices, trainees, experts and all subjects) and two variants of the FCI algorithm, with and without prior knowledge. The results are summarized in Table 15.1. As expected, the number of undefined links decreases with the utilization of prior information.

## 15.5 Additional Reading

A comprehensive discussion of causal graphical models, and in particular of the PC and FCI algorithms can be found in [13]. Reference [8] focuses on the analysis of cause-effect relations for two variables, as well as its relation to machine learning. Causal strcuture learning for linear non-Gaussian models is described in [10]. The BCCD algorithm for learning causal graphs is introduced in [3]. A review of alternative approaches for learning causal networks can be found in [5].

## 15.6 Exercises

1. Given the following conditional independence relations: $I(Y, X.Z)$, $I(X, YZ.W)$ and $\neg I(Y, W, Z)$: (a) draw the CPDAG that represents these relations, (b) draw all the DAGs in the Markov equivalence class.
2. Given the PAG in Fig. 15.9, show all the MAGs included in this PAG.

3. Different DAGs can be associated to a MAG. A particular one is called a *canonical DAG, D*, and is obtained from a MAG by the following transformations: (i) $X \rightarrow Y$ in the MAG, then $X \rightarrow Y$ in $D$; (ii) $X \leftrightarrow Y$ in the MAG, then $X \leftarrow U_{XY} \rightarrow Y$ in $D$. Show the canonical DAG for the MAG in Fig. 15.5b.
4. Consider the BN structure of the golf example given in Fig. 8.1 and the associated data (in Chap. 8), calculate the BDeU score for this example.
5. Given the same data as in the previous problem, and an alternative structure for the golf example that is in the same Markov equivalence class, calculate the BDeU score. Is it the same as in the previous problem?
6. Given the data on the golf example (in Chap. 8), apply the standard PC algorithm to learn the structure of a BN. Then apply the BCCD algorithm, using an estimate of the reliability of each causal link to order the causal relation, and setting a threshold to determine if a link has certain direction or remains undirected. Compare the structures obtained with PC and BCCD.
7. Given the DAG in Fig. 15.7a, obtain the inducing paths over $O = \{A, B, D, E, F\}$ and $O = \{A, B, D, F\}$.
8. Consider a linear system with 4 variables $X_1, ...X_4$, and non-Gaussian noise $e_1, ...e_4$. Given the following equations: $X_1 = X_4 + e_1, X_2 = 0.2X_4 + e_2, X_3 = 2X_2 + 5X_1 + e_3, X_4 = e_4$. (a) Write the equations in matrix form, $\mathbf{X} = \mathbf{BX} + e$, reordering the variables so the matrix $\mathbf{B}$ is in lower triangular form. (b) Draw the corresponding graphical model.
9. *** Develop a program that implements the BCCD algorithm for discrete variables.
10. *** Obtain data from some real-world domain and use the program from the previous exercise to obtain a causal model. Vary the reliability threshold and compare the models obtained.

## References


1. Alonso-Barba, J.I., Gámez, J.A., Puerta, J.M., et al.: Scaling up the greedy equivalence search algorithm by constraining the search space of equivalence classes. Int. J. Approx. Reason. **54**(4), 429–451 (2013)
2. Chickering, D.M.: Optimal structure identification with greedy search. J. Mach. Learn. Res. **3**, 507–554 (2002)
3. Claassen, T., Heskes, T: A Bayesian Approach to Constraint Based Causal Inference. In: Proceedings of Uncertainty in Artificial Intelligence (UAI), AUAI Press, pp. 207–216 (2012)
4. Comon, P.: Independent component analysis - a new concept? Signal Process. **36**, 287–314 (1994)
5. Aitken, J.S.: Learning Bayesian networks: approaches and issues. Knowl. Eng. Rev. **26**(2), 99–157 (2011)
6. Heckerman, D., Geiger, D., Chickering, D.M.: Learning Bayesian networks: the combination of knowledge and statistical data. Mach. Learn. **20**(3), 197–243 (1995)
7. Montero-Hernandez, S.A., Orihuela-Espina, F., Herrera-Vega, J., Sucar, L.E.: Causal probabilistic graphical models for decoding effective connectivity in functional near infrared spectroscopy. In: The Twenty-Ninth International Flairs Conference, AAAI Press (2016)

8. Peters, J., Janzing, D., Scholkpf, B.: Elements of Causal Inference. MIT Press, Cambridge (2017)
9. Reichenbach, H.: The Direction of Time. University of California Press, Berkeley (1956)
10. Shimizu, S., Hyvärinen, A., Kano, Y., Hoyer, P. O.: Discovery of non-gaussian linear causal models using ICA. In: Proceedings of the 21st Conference on Uncertainty in Artificial Intelligence (UAI-05), pp. 525–533. AUAI Press (2005)
11. Sokolova, E., Groot, P., Classen, T., Heskes, T.: Causal Discovery from Databases with Discrete and Continuous Variables. In: Probabilistic Graphical Models (PGM). Springer, pp. 442–457 (2014)
12. Spirtes, P., Glymour, C., Scheines, R.: Causation, Prediction, and Search. MIT Press, Cambridge (2000)
13. Spirtes, P., Glymour, C., Scheines, R.: Causation, Prediction, and Search, 2nd edn. MIT Press, Cambridge (2001)

# Chapter 16
# Deep Learning and Graphical Models

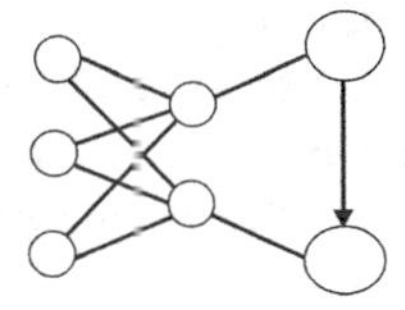

## 16.1 Introduction

In recent years, *Deep learning* has had a great impact in several areas of artificial intelligence and computing in general, such as computer vision, speech recognition, natural language processing, robotics, bio-informatics, among others [10]. Deep models, in particular deep neural networks (DNNs), consist of several layers of simple non-linear processing units that can be trained to learn complex input-output functions. These techniques basically learn a *representation* of the input data, or more precisely, several representations at different levels of abstraction, so they are also known as *representation learning*. Thus, in contrast to traditional techniques for classification and pattern recognition which rely on hand-crafted features, DNNs learn these automatically from data.

As probabilistic graphical models (PGMs), deep neural networks are based on distributed representations that can solve complex problems based on local operations, and which tend to be robust and flexible. DNNs can learn any non-linear function and in this respect are more powerful than graphical models such as Bayesian networks that are limited in the type of functions that can be represented (see Sect. 16.3). However, PGMs have certain advantages: (i) it is easier to incorporate background knowledge, including conditional independence relations, (ii) in general they provide more *transparent models* which can be interpreted and explained. Thus, it could be useful to explore potential combinations of these two paradigmas to take advantage of their complementary advantages.



L. E. Sucar, *Probabilistic Graphical Models*, Advances in Computer Vision and Pattern Recognition, https://doi.org/10.1007/978-3-030-61943-5_16

After a brief introduction to neural networks and deep learning, we present an analysis of the similarities and differences between graphical models and neural networks. Then we explore some different ways to combine DNNs and PGMs, and present some applications.

## 16.2 Review of Neural Networks and Deep Learning

### *16.2.1 A Brief History*

Artificial Neural Networks (ANNs) were developed at the start of the computer era and were one of the first techniques proposed for a computer to learn from data. They were inspired by simple model of the biological neural networks. ANNs are composed basically of simple processing units, named *neurons*, that usually have real-valued inputs and calculate a weighted sum of their inputs:

$$z_j = \sum_i w_{ij} x_i \tag{16.1}$$

where $w_{ij}$ are the weights and $x_i$ the inputs. The output, $y_j$, is a non-linear function of the weighted sum, $z_j$:

$$y_j = f(z_j) \tag{16.2}$$

The first and simplest neural network (NN) is the *Perceptron* [12] which consists of a single neuron and the output is a threshold function:

$$y = f(z) = \begin{cases} 1, & z + b > 0 \\ 0, & otherwise \end{cases} \tag{16.3}$$

where $b$ is the bias that shifts the decision threshold. A Perceptron is a binary classifier, which classifies the input $\mathbf{X}$ as positive (1) or negative (0).

The Perceptron learns from a set of examples, $D$, by adjusting the weights so that the error—difference between the actual and expected output—is minimized. For each example, the output is calculated and the weights are adjusted by adding to its previous value a quantity that is proportional to the error according to a predefined constant known as the *learning rate* ($r$):

$$w_{ij}(t+1) = w_{ij}(t) + r(d_j - y_j)x_i \tag{16.4}$$

where $d_j$ is the desired output and $y_j$ the actual output; this is done for all inputs $x_i$, and repeated until the error is less than a pre-specified threshold.

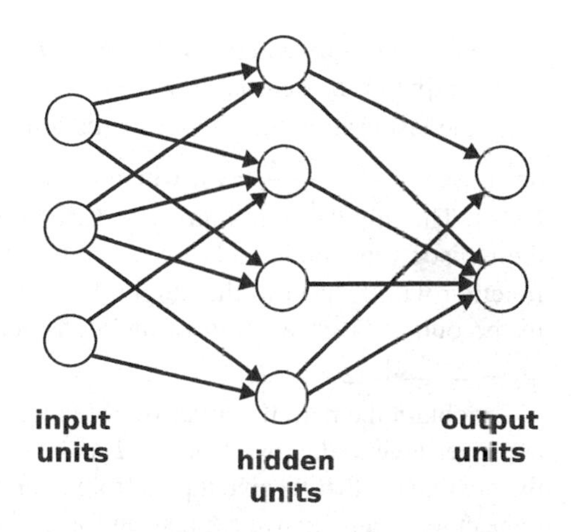


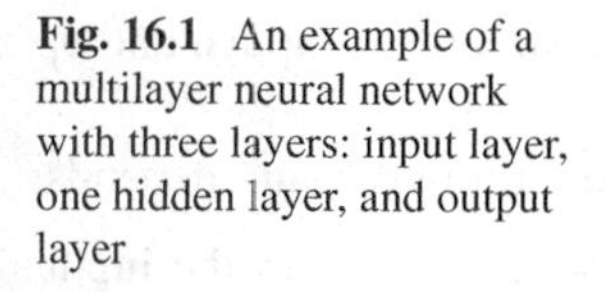
**Fig. 16.1** An example of a multilayer neural network with three layers: input layer, one hidden layer, and output layer

The Perceptron is *linear classifier*, that is, it can distinguish between two classes if these can be separated by an hyperplane in the feature (input) space. If the training set is linearly separable, then the perceptron is guaranteed to converge. As it is a linear classifier, the Perceptron can not learn non-linear functions such as an *XOR*, as shown by Minsky and Papert [11]. However, this limitation is overcomed if several Perceptrons or neurons are stacked in several layers (at least two), so the outputs of the previous layer are the inputs to the next layer. These models, that combine several units or neurons in layers, are known as *multi-layer neural networks*, and are the bases of the current deep learning models, see Fig. 16.1.

A critical step for learning multilayer NNs was the development of the back-propagation algorithm [13] to *propagate* the error from to output layer to the previous layers and in this way update the weights. For this is necessary that the output non-linear functions are differentiable, so instead of using a step threshold, other *activation functions* are used such as the logistic function:

$$f(z) = 1/(1 + e^{-z})$$

Learning a multilayer NN consists on estimating all the weights in the network, assuming a fixed *architecture* (number of layers, neurons in each layer and connections between neurons). This is an optimization problem that minimizes a loss function: the difference between the target output and the computed output for all the training examples. For this, a commonly used algorithm is *gradient descent*, which involves calculating the derivative of the loss function (error) with respect to the weights. To calculate the partial derivative of the weight with respect to the error we apply the *chain rule* of derivatives:

$$\frac{\partial E}{\partial w_{ij}} = \frac{\partial E}{\partial o_j}\frac{\partial o_j}{\partial w_{ij}} = \frac{\partial E}{\partial o_j}\frac{\partial o_j}{\partial net_j}\frac{\partial net_j}{\partial w_{ij}} \tag{16.5}$$

where $E$ is the squared error function, $net_j$ is the weighted sum of its inputs, and $o_j$ is the output of the neuron.

In the last term in the previous equation, $\frac{\partial net_j}{\partial w_{ij}}$, only one term in the sum depends on $w_{ij}$, so $\frac{\partial net_j}{\partial w_{ij}} = \frac{\partial w_{ij} o_i}{\partial w_{ij}} = o_i$, where $o_i$ is the input $i$ to the neuron ($x_i$ in the input layer). The middle term is just the partial derivative of the activation function; for the logistic function is $o_j(1 - o_j)$. The first term is the partial derivative of the loss function with respect to the output, this can be calculated directly for the neurons in the output layer, as $o_j = y_j$, and considering a square error loss function, then $\frac{\partial E}{\partial y} = \frac{\partial \frac{1}{2}(d-y)^2}{\partial y} = d - y$.

To obtain the partial derivative of the loss function with respect to the output for the inner layers, the loss function $E$ can be written as a function of all neurons (in the next layer) that receive inputs from neuron $j$, and taking the total derivative with respect to $o_j$, a recursive expression for $\frac{\partial E}{\partial o_j}$ is obtained. This expression depends on the derivatives with respect to the outputs of the next layer; thus the gradient can be propagated from the output layer backwards to the inner layers, until the input layer is reached. See [13] for more details. In general, back-propagation cannot guarantee to find the global minimum of the error function, but only a local minimum.

In the preceding we have assumed a particular type of NN architecture known as feed-forward; there are other types of architectures, see the additional reading section.

### 16.2.2 Deep Neural Networks

Deep neural networks refer to multilayer NNs with *many* layers, usually between 5 and 20, or even more. Although these deep architectures have been proposed before, there was limited success in training a deep neural network with many layers, until the 2000s. One of the breakthroughs that make possible training deep models was a semi-supervised learning procedure, in which unsupervised learning is used to learn the initial weights one layer at a time; which are then fine-tuned with labeled data.

A technique for learning a NN in an unsupervised way are autoencoders [8]. An autoencoder learns a *representation* of a data set, typically of lower dimensions. An autoencoder is a NN that learns to copy its input to its output. It has an internal (hidden) layer that describes a code used to represent the input, and it is constituted by two main parts: an encoder that maps the input into the code, and a decoder that maps the code to a reconstruction of the original input.

Several improvements in the learning algorithms and architectures make it possible to learn deep NNs without the need of unsupervised pre-training. One of these developments is the use of the ReLU activation function: $f(z) = max(z, 0)$, which typically learns more faster in NNs with many layers. Other improvement is the use of *stochastic gradient descent*. It consists on training by groups of few examples, for which the average gradient is calculated, and the weights area adjusted accord-

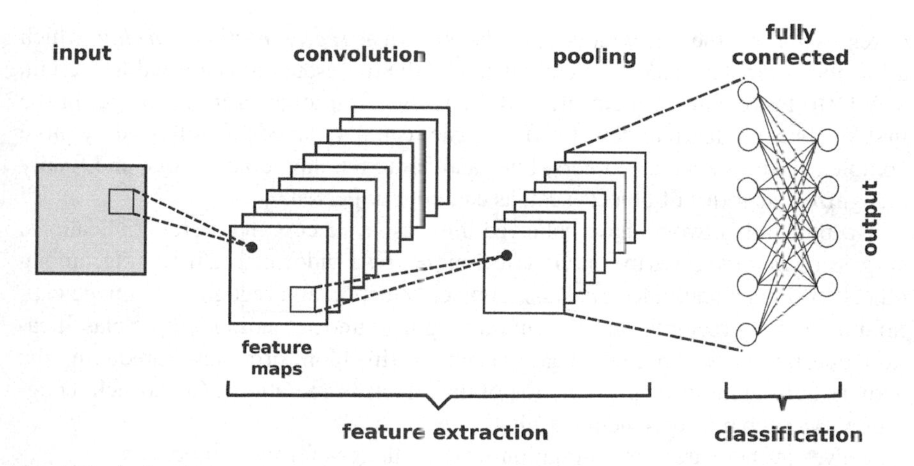


**Fig. 16.2** A simple example of a convolutional neural network with one convolution layer, one pooling layer and a fully connected layer

ingly. A third important element, in particular for processing data that comes in form of arrays, such as signals and sequences (1D), images (2D) and videos (3D), are *convolutional neural networks* (ConvNets).

A typical ConvNet is composed of a series of stages. The first stages alternate convolutional and pooling layers. The convolutional layers are organized in feature maps, in which each unit is connected to local patches form the previous later, forming a *filter bank*. All the units in a feature map share the same filters; in a layer there are several feature maps with different filters. The operation performed by these filters is a discrete convolution, similarly to filters in traditional signal processing, but with the difference that in this case the weights of the filters are learned. The pooling layers merge semantically similar features and at the same time reduce the dimensionality of the input. Typically pooling units compute the maximum of a local set of units in a feature map. The last stages in a ConvNet are usually fully connected as in more traditional feed-forward NNs. A simple example of a ConvNet is depicted in Fig. 16.2.

Besides algorithmic developments, two other factors have contributed to the success of deep learning: (i) the availability of huge amounts of data for training the models, in particular in Internet (i.e., images, text documents, videos, etc.), and (ii) the increase in computer power that makes possible to train very large models (millions of weights) in relatively short times; in particular Graphic Processing Units (GPUs).

The capability of training deep models with many layers has made possible to use as input the raw data instead of using manually engineering features as in the case of the initial shallow NNs or more traditional pattern recognition and classification techniques. Thus we can think that the network is learning representations of the data at different levels of abstraction, and these representations are then used by the final layers of the network to perform the desired task, such as detection, classification or regression. So these techniques are also known as *representation learning*, which allow the computer to automatically discover the representations needed for certain task [10]. For example, a NN trained for recognizing objects in an image, in the first layers could learn to detect certain basic features like edges, followed by more complex features such as corners, lines and contours, then object parts, and finally recognize categories of objects such as cat, dog, or person.

Deep neural networks have achieved impressive success in several applications, surpassing in some cases the previous techniques by a wide margin, including, among others: (i) image and video analysis—object detection and recognition, image captioning, activity recognition; (ii) natural language understanding—topic classification, question answering, language translation; (iii) bioinformatics—predicting the activity of drug molecules, prediction of the effects of mutations; (iv) speech recognition, (v) deep reinforcement learning.

However, there are some limitations and challenges for these models:

- Explainability. Deep NNs tend to be *black boxes*, so it is difficult to analyze or explain how they are achieving certain results; this aspect could be important for some applications such as medical diagnosis.
- Robustness. Related to the previous point, as it is difficult to know which aspect of the data they are using to take decisions, they could be deceived by certain minor changes in the input or could they be learning certain features which are not the most relevant for the task (e.g., the background of an image instead of the object of interest).
- Quantification of uncertainty. In general, DNNs can not provide a measure of the uncertainty or confidence of their outputs.
- Incorporation of prior knowledge. Although the models could be initialized from some previously learned task to learn other similar tasks (transfer learning); other types of prior knowledge are difficult to incorporate, such as causal relations, prior probabilities, etc.

Probabilistic graphical models can help to alleviate these shortcomings, thus there is an opportunity to integrate these approaches to take advantage of their complementary strengths.

## 16.3 Graphical Models and Neural Networks

### *16.3.1 Naives Bayes Classifiers Versus Perceptrons*

There is an interesting analogy between the basic NN, the Perceptron, and the naive Bayes classifier (NBC).

As we saw in Chap. 4, the NBC estimates the posterior probability of the class, $C$, based on Bayes rule, assuming that the attributes, $\mathbf{A}$, are conditionally independent given the class:

$$P(C \mid A_1, A_2, \ldots, A_n) = P(C)P(A_1 \mid C)P(A_2 \mid C) \ldots P(A_n \mid C)/P(\mathbf{A}) \quad (16.6)$$

where $P(\mathbf{A})$ can be considered a normalization constant, so:

$$P(C \mid A_1, A_2, \ldots, A_n) \sim P(C)P(A_1 \mid C)P(A_2 \mid C) \ldots P(A_n \mid C) \quad (16.7)$$

If we are interested in finding the class of maximum probability, $Arg_C[MaxP(C \mid A_1, A_2, \ldots, A_n)]$, we can express Eq. 16.7 in terms of any function that varies monotonically with respect to $P(C \mid \mathbf{A})$, such as:

$$Arg_C[Max[(logP(C) + logP(A_1 \mid C) + logP(A_2 \mid C) \ldots + logP(A_n \mid C)] \quad (16.8)$$

The term to maximize is:

$$Y(C) = logP(C) + \sum_i logP(A_i \mid C) \quad (16.9)$$

If we consider the binary classification case, $C = \{c_o, c_1\}$, then:

$$Y(c_0) = logP(c_0) + \sum_i logP(A_i \mid c_0)$$

$$Y(c_1) = logP(c_1) + \sum_i logP(A_i \mid c_1)$$

Considering equal misclassification costs, we will decide $c_1$ if $Y(c_1) > Y(c_0)$, and $C_0$ otherwise.

In the case of the Perceptron, as we saw before:

$$y = \begin{cases} 1, & \sum_i w_{ij}x_i + b > 0 \\ 0, & otherwise \end{cases} \quad (16.10)$$

So basically both are computing a weighted linear combination of the inputs or attributes, and we can consider that the bias represents the prior probability. It can be shown that linear separability is related to conditional independence, that is, hyperplanes can optimally classify data when the inputs are conditionally independent [14].

So a Perceptron with a sigmoid output function: $f(z) = 1/(1 + e^{-z})$, can calculate the posterior probability of the input when the neuron's inputs are proportional to the probability of the respective input [14], with weights:

$$w_i = log \frac{P(x_i \mid c_0)}{P(x_i \mid c_1)} \quad (16.11)$$

An bias:

$$b = log \frac{P(c_0)}{P(c_1)} \tag{16.12}$$

Thus, to deal with more complex problems in which the data distributions are not conditionally independent/linearly separable, we need more complex graphical models such as Bayesian networks and, in the case of NNs, multi-layer architectures. Next we analyze the relation between these more powerful models.

### 16.3.2 Bayesian Networks Versus Multi-layer Neural Networks

Regarding the similarities and differences between probabilistic graphical models such as Bayesian networks, and function-based approaches such as neural networks, [3] highlights the following observations:

- A query posed to a BN can be viewed as evaluating a function, which can be represented as an *arithmetic circuit.*
- BNs can be trained discriminately using labeled data, similarly to NNs, except that a NN represents only one function, while a BN represents many functions (one for query).
- Neural networks appear to outperform BNs even when trained discriminately.

To explain the last point we need to analyze the class of functions that can be represented by neural networks and Bayesian networks.

Assuming a single output, a neural network represents a function $Y = f(X_1, \ldots, X_n)$. A NN with a single hidden layer and sigmoid activation functions can approximate any continuous function to within an arbitrary error. Such neural networks are called *universal approximators*. A network with a single hidden layer is sufficient, but may require exponentially many neurons. A deep neural network may be more succinct [3]. Other activation functions also make a universal approximator, such as the ReLU.

We now consider a BN with some evidence variables, $\mathbf{E}$, and a query variable $Y$. The probability $P(y \mid \mathbf{e})$ can be seen as a function $f(\mathbf{E})$ that maps the evidence to a number in the interval [0, 1]. This function $f(\mathbf{E})$ can be represented by an arithmetic circuit only containing multipliers and adders [5]. Thus, each Bayesian network query is the result of evaluating a function. Such functions are polynomials; in particular, multi-linear functions; so the class of functions induced by BNs are less expressive than the ones represented by NNs. This explains why a BN can not outperform a NN for the same query.

To construct a universal approximator, a non-polynomial activation function is required; this is achieved by the non-linear activation functions of the neurons in ANNs, such as the sigmoid or the ReLU. For instance, the ReLU implements a test, $f = 0$ if $x < 0$. So an alternative to make BNs more expressive is to incorporate *testing units*, as we will see in the next section.

## 16.4 Hybrid Models

There are different ways in which we can combine graphical models and deep NNs. Two basic type of approaches can be distinguished:

*Fusion*: they are combined in a tightly-coupled form, such that one empowers the other. That is, both approaches are combined in a single model or architecture, that could be an extension of a graphical model or a deep neural network.

*Integration*: they are combined in a loosely-coupled way, in which different parts of an architecture are based on one or the other, and they complement each other. Generally there are several parts or modules based on one paradigm, that interchange data to solve a complex problem.

Next we describe an approach that makes Bayesian networks more expressive by incorporating some features of neural networks; and then some strategies to integrate deep models within graphical models. An example of fusion is presented in Sect. 16.5.2, in which a type of graphical model (factor graphs) and graph neural networks are combined.

### 16.4.1 Testing Bayesian Networks

The main idea of testing Bayesian networks (TBNs) [5] is that some variables in the model have a dynamic conditional probability table (CPT), that is, a variable has two CPTs such that one is selected at inference time depending on the values of some other variables; instead of having a fixed conditional probability table as in *standard* Bayesian networks. For instance, the CPT of a node $X$ will be determined dynamically based on a test of the form $P(\mathbf{u} \mid \mathbf{e_X}) > T_{X|\mathbf{u}}$, where $\mathbf{u}$ are the current states of the parents of $X$, $\mathbf{e_X}$ is the evidence from its ancestors, and $T_{X|\mathbf{u}}$ is the threshold for $X$.

A TBN has two types of nodes, *regular* and *testing*. A regular node is the same as in standard BNs with a static CPT. A testing node has a dynamic CPT, with one set of parameters if $P(\mathbf{u} \mid \mathbf{e_X}) > T_{X|\mathbf{u}}$, and other set otherwise; so it has twice as many parameters as a regular node. A testing CPT corresponds to a set of standard CPTs, one of which is selected depending on the evidence. All root nodes are regular.

While a BN represents a single joint distribution for the variables in the model, a TBN represents a set of distributions, one of which is selected depending on the evidence. Once the CPTs for each testing node are selected according to the evidence, the TBN is transformed to a regular BN and the same inference techniques can be applied. Thus, to compute the posterior for some query variable $Q$ given some evidence $\mathbf{e}$, a distribution (of the several ones represented in the TBN) is selected and used to compute $P(Q \mid \mathbf{e})$.

Figure 16.3 shows a simple example of a TBN with three nodes, $X$ and $Z$ are regular nodes, and $Y$ is a testing node. The model represents the output of a sensor

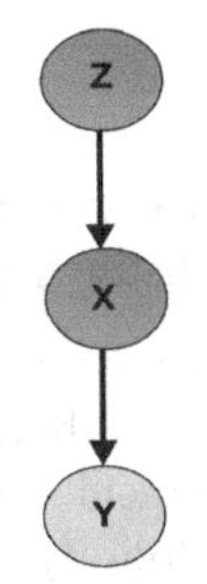


**Fig. 16.3** An example of a TBN; $Z$ and $X$ are regular nodes, and $Y$ is a testing node (in yellow) (Color online)

that saturates when the control variable ($Z$) is above certain threshold, $T$. If the control is below this threshold, the output ($Y$) is the same as the input ($X$) with some noise, otherwise the output takes the maximum value before saturation. Assuming $X$ and $Y$ can take the following values: {0, 1, 2, 3, 4}, then $Y$ will have the following testing CPT:

$P(Y \mid X), Z \leq 1$

| Y / X | 0 | 1 | 2 | 3 | 4 |
|---|---|---|---|---|---|
| 0 | 0.8 | 0.05 | 0.05 | 0.05 | 0.05 |
| 1 | 0.05 | 0.8 | 0.05 | 0.05 | 0.05 |
| 2 | 0.05 | 0.05 | 0.8 | 0.05 | 0.05 |
| 3 | 0.05 | 0.05 | 0.05 | 0.8 | 0.05 |
| 4 | 0.05 | 0.05 | 0.05 | 0.05 | 0.8 |

$P(Y \mid X), Z > 1$

| Y / X | 0 | 1 | 2 | 3 | 4 |
|---|---|---|---|---|---|
| 0 | 0.8 | 0.05 | 0.05 | 0.05 | 0.05 |
| 1 | 0.05 | 0.8 | 0.05 | 0.05 | 0.05 |
| 2 | 0.05 | 0.05 | 0.8 | 0.8 | 0.8 |
| 3 | 0.05 | 0.05 | 0.05 | 0.05 | 0.05 |
| 4 | 0.05 | 0.05 | 0.05 | 0.05 | 0.05 |

A TBN can simulate a Perceptron, that is a neuron with a step activation function, with the structure depicted in Fig. 16.4a. In this TBN, $X_i$ (the inputs) and $H$ are regular nodes, and $Y$ is a testing binary node. $H$ is a noisy-OR and $Y$ has the following testing CPT:

$$P(Y \mid H) = \begin{cases} 1, & P(h \mid x_1, x_2, x_3, x_4) \geq T \\ 0, & otherwise \end{cases} \tag{16.13}$$

where $T$ is a threshold (bias in the Perceptron). The parameters of the TBN can be determined from the weights of the neuron. Several of this TBNs can be combined to simulate a layer of a NN, as shown in Fig. 16.4b; and several of these structures can be stacked to simulate a multi-layer NN.

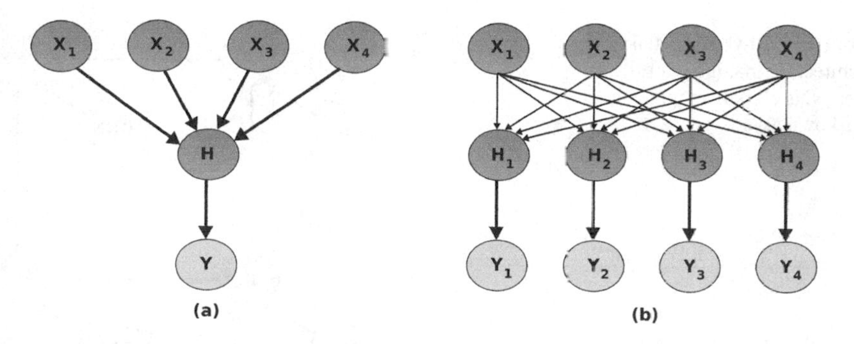


**Fig. 16.4** **a** A TBN that simulates a Perceptron, $Y$ is a testing node. **b** Several structures as the one in (**a**) are combined to simulate a layer of a NN

A TBN based on the previous structure was built to recognize digits. For this a NN with two hidden layers was trained with the MNIST digit classification data set,[1] which has binary images ($28 \times 28$ pixels) of handwritten digits, 60,000 examples for training, and 10,000 for testing. The weights of the NN after training were used to parameterized the TBN, and the both models were evaluated on the testing data set. They achieved practically the same performance with an accuracy above 95% [5].

It can be shown that TBNs are as powerful as NN for approximating functions [5]. A remaining challenge is how to learn the parameters of a TBN directly from data; and eventually also the structure.

### *16.4.2 Integrating Graphical and Deep Models*

Probabilistic graphical models provide a well-founded and practical way to represent dependence relations between variables, including spatial and temporal relations. Deep neural networks are very good for classification and regression tasks. Thus, one way to integrate deep NNs and PGMs is by representing the structure of a complex problem through a PGM, and then using deep NNs as local classifiers for the different elements of the problem. The deep models, trained on labeled data on particular elements or variables, provide an initial estimation; then this initial estimates are combined and improved through belief propagation in the graphical model. This approach can also make more efficient the training of the deep models, as each one only considers a particular data set.

Spatial analysis is a class of problems in which this type of hybrid system could be useful. For example, human pose estimation, in which there several body parts have certain spatial structure; this structure provides geometric constraints which can be encoded by a graphical model. Other example is in mapping a physical environment, in which there are spatial relations between the different landmarks in the environ-

[1] http://yann.lecun.com/exdb/mnist.

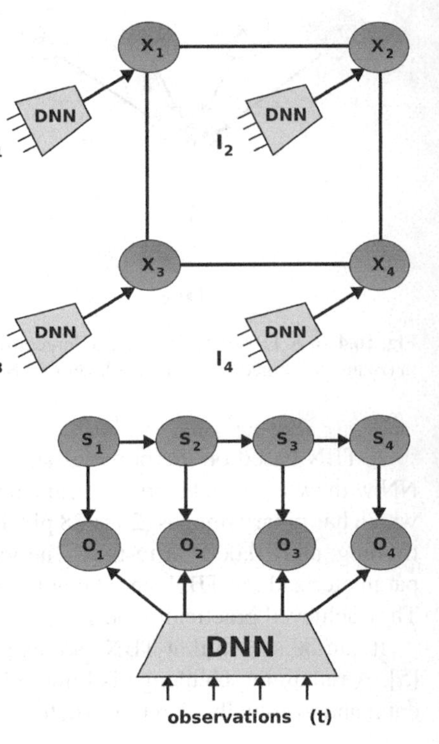


**Fig. 16.5** A hybrid spatial architecture that integrates deep NNs classifiers within a Markov network

**Fig. 16.6** A hybrid temporal architecture that combines NN classifiers and a hidden Markov model

ment. The spatial constraints between the different elements in the model can be represented as a Markov network, expressing the constraints as the local joint probabilities of neighboring elements. The parts or elements are detected or classified using a deep NN. An hypothetical example of this hybrid architecture is depicted in Fig. 16.5.

Other class of problems is temporal modeling, in which some state variable evolves over time, usually depending on the previous state. Markov chains and hidden Markov models are appropriate for representing this class of problems, In a hybrid architecture, a deep NN can be used to classify the state based on the observations, and a Markov model can code the temporal relations. Potential applications include speech recognition, human activity recognition, among others. Figure 16.6 shows an example of a temporal hybrid model.

In the next section we describe an application of a hybrid architecture for human body pose estimation.

## 16.5 Applications

Next we describe two applications that combine graphical models and deep learning. First, an application that combines CNNs and HMMs for tracking the human body pose in videos. Second, a system that integrates factor graphs and graph neural networks for error correcting codes.

### *16.5.1 Human Body Pose Tracking*

Human pose estimation consists on the localization of the body joints from images or videos and how they are connected between each other. This is a complex problem due to the flexibility of the human body structure that makes the space of possible movements high dimensional, in addition to other external factors such as clothing, illumination changes, occlusions, etc. The problem is particularly challenging if it is realized over single-image videos without markers in realistic conditions. Figure 16.7

**Fig. 16.7** Human body pose estimation. The image shows several examples of the joints and corresponding skeleton for different body postures. It illustrates some of the difficulties as occlusion (**A**), incomplete body (**B**) and a complex position (**D**)

depicts the body joints and a graph that connects them (virtual skeleton) for several persons in the image, showing some of the challenges.

In this work [1] an architecture for tracking the human pose in video sequences was developed. The proposed model is composed by a Convolutional Neural Network based part-detector and a Coupled Hidden Markov Model (CoHMM). The combination of both models allows learning spatial and temporal dependencies. The part-detector recognizes the different joints based on a CNN, and exploits the spatial correlations between neighboring regions through a Conditional Random Field (CRF) [4]. On the other hand, the CoHMMs generates the best movement sequence between interacting processes.

#### 16.5.1.1 Body Joints Detection

To estimate the location of the body joints a CNN is combined with a CRF (CNN-CRF) to model the correlations among neighboring body joints. The CRF is approximated with Gaussian pair-wise potentials as a recurrent neural network, which is plugged as part of the CNN. The proposed architecture starts with a multi-resolution attention module, generating feature maps at different resolutions which are merged for the detection of each body part. These are fed to the CRF that refines the estimates by combining information from neighboring regions.

The input to the system are images of dimension $256 \times 256$, and the output a *heat map* of dimension $64 \times 64$ for each body joint. Each heat map gives a probability distribution for the location of each joint in the image. See [4] for more details.

The skeleton of the body is represented as a graph $G = (V, E)$, where $V$ are the joints and $E$ the links between neighboring joints, so the graph has a tree structure. In this work 16 body joints are considered. This first phase gives an estimate of the position of each joint per image; the next phase integrates these positions and improves the estimate of each body joint trajectory based on a temporal graphical model.

#### 16.5.1.2 Tracking

For tracking the position of each joint, its position is represented relative to the previous joint in the skeleton. $X_i$ is a random variable associated to the $x$ coordinate and $Y_i$ a random variable associated to the $y$ coordinates of joint $i$. Therefor, the coordinates of a joint $i$ are defined as $X_i = x_i - x_j$ and $Y_i = y_i - y_j$, where $j$ is the parent joint.

The sequence of positions of each joint across a video is determined using a Coupled hidden Markov model. A CoHMM is an extension of the basic hidden Markov model (HMM) that includes two HMMs, that are coupled by adding conditional probabilities between the hidden state variables of the different HMMs (see Fig. 5.5, Chap. 5). In this work a CoHMM is used to model the temporal evolution of the position of each joint. Each one is represented by two state variables, $X_i$ and $Y_i$, which

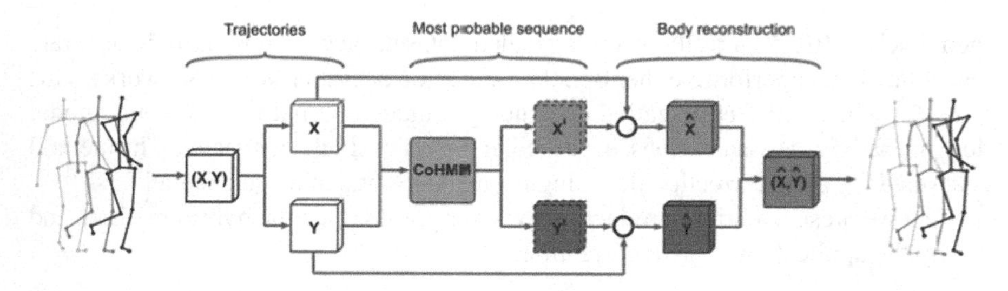


**Fig. 16.8** Tracking the human body joints. The estimates of the position of each joint from the CNN $(X, Y)$ are fed the the CoHMM, that estimates the most probable sequence in relative coordinates $(X', Y')$. These are transformed to absolute coordinates $(\hat{X}, \hat{Y})$ to obtain the body reconstruction [1]

represent the relative position in the corresponding cartesian coordinate. The most probable state sequence for the coupled variables is estimated through an extension of the Viterbi algorithm [2]. Each joint is modeled independently using a CoHMM. Each model receives as observations the position estimates obtained from the CNN, and produces as output the most probable sequence (trajectory of the joint in the video) for each joint. Finally the relative coordinates are transformed to absolute coordinates, to obtain the reconstruction of the body skeleton for each frame in the video. The process is illustrated in Fig. 16.8.

#### 16.5.1.3 Results

The method was tested with the *PoseTrack*[2] data set that includes short videos of persons realizing different activities in the wild. The CNN-CRF joint detector already achieved good results in terms of the Euclidean distance between the estimated position of each joint and the ground truth. However, in some frames it could loose completely some joints or obtain a wrong estimate due to uncommon body configurations. The inclusion of the CoHMM for tracking the joints reduces these errors, providing in general smoother trajectories.

### *16.5.2 Neural Enhanced Belief Propagation for Error Correction*

In [7] they present a hybrid model that combines messages from belief propagation (BP) on a factor graph and from a graph neural network (GNN). The GNN messages are learned from data and complement the BP messages. The GNN receives as input the messages from BP at every iteration and delivers as output a refined version of

[2]https://posetrack.net.

them back to BP. As a result, given a labeled dataset, they propose a more accurate algorithm that outperforms either Belief Propagation or Graph Neural Networks. This is useful when belief propagation does not guarantee optimal results as the model does not satisfy the assumptions, in particular if it is not singly connected. The method is applied for error correction decoding for a non-Gaussian noise channel.

First we present a brief introduction to factor graphs, then the hybrid method, and finally the application to error correction.

#### 16.5.2.1 Factor Graphs

A factor graph (FG) [9] is a bipartite graph used to represent the factorization of a probability distribution function, enabling efficient computations, such as the computation of the marginal distributions through belief propagation. Factor graphs provide a unifying representation for both, Markov random fields (see Chap. 6) and Bayesian networks (see Chap. 7), by explicitly expressing the factorization structure of the corresponding probability distribution.

Factor graphs are a class of probabilistic graphical model that allow an efficient representation of a function. A FG is a bipartite undirected graph with two types of nodes: *factor nodes* (represented as rectangles) and *variable nodes* (represented as ovals). The edges in the graph connect the variables to the factors, representing the dependency between factors and variables. For instance, consider the following factorization of a probability distribution:

$$P(V, W, X.Y, Z) = (1/z)f_1(V, W)f_2(X, W)f_3(X, Y)f_4(X, Z)$$

where $f_1, f_2, f_3, f_4$ are the factors and $z$ a normalizing constant. The corresponding factor graph is shown in Fig. 16.9.

In general the joint probability distribution can be written as the product of $M$ factors:

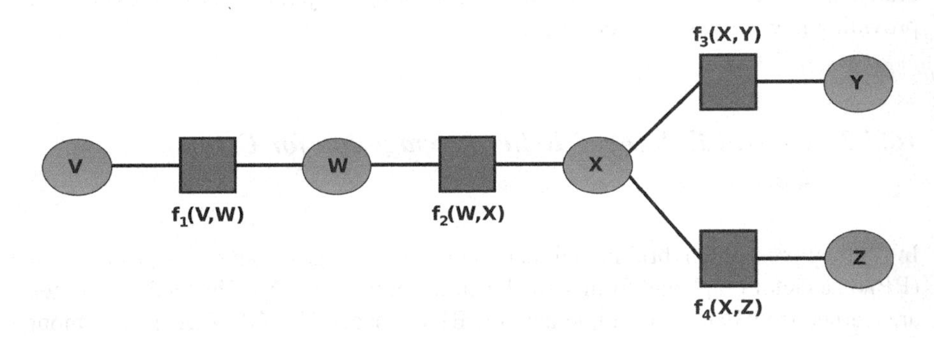


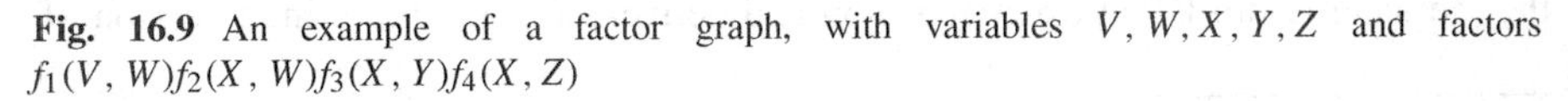

**Fig. 16.9** An example of a factor graph, with variables $V, W, X, Y, Z$ and factors $f_1(V, W)f_2(X, W)f_3(X, Y)f_4(X, Z)$

$$P(X_1, X_2, \ldots X_N) = (1/z) \prod_i f_i(X_j, X_k, \ldots X_l) \tag{16.14}$$

Given a tree-structured factor graph the marginal probabilities of each variable can be calculated efficiently using *belief propagation*, a message passing scheme similar to probability propagation in singly-connected Bayesian networks. In the case of multi-connected graphs, an iterative approach analogous to loopy belief propagation (see Chap. 7) can be applied, which provides an estimate of the marginal probabilities when it converges; however there are no guaranties of convergence.

#### 16.5.2.2 Graph Neural Networks and Factor Graphs

A hybrid method to improve belief propagation by combining it with graph neural networks is proposed. Both methods can be seen as message passing on a graph. However, BP sends messages that follow directly from the definition of the graphical model, while messages sent by GNNs must be learned from data. To achieve seamless integration of the two message passing algorithms, GNNs are modified to be compatible with factor graphs.

Graph Neural Networks [6] operate on graph-structured data by modeling interactions between pairs of nodes. The parameters of a GNN are learned from data, and messages are sent between the nodes in the graph, similarly to BP. A mapping between a factor graph and the graph of a GNN is illustrated in Fig. 16.10. All factors from the factor graph are assumed to be factor-nodes in the GNN, with exception for factors connected to only one variable which are considered as attributes of that variable-node.

The inference procedure is a hybrid method that runs co-jointly with belief propagation on a factor graph, denoted as *Neural Enhanced Belief Propagation* (NEBP). The procedure is as follows: after every belief propagation iteration, the BP messages are input into the FG-GNN. The FG-GNN then runs for two iterations and updates

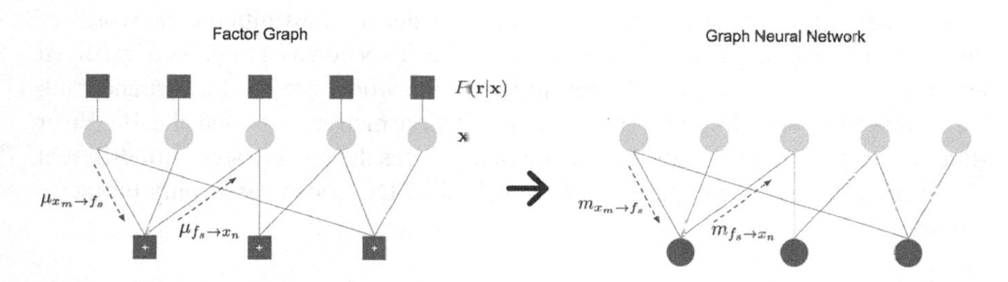


**Fig. 16.10** An example of a factor graph (left) and its equivalent representation as a graph neural network (right). The dark circle nodes in the GNN correspond to the factors in the FG, and light circle nodes in the GNN to the variable nodes in the FG. The factors that are connected to the one variable are not included in the GNN (Figure taken from [7])

the BP messages. This step is repeated recursively for $N$ iterations. After that, the marginals of the variables in the factor graph are refined.

The training of the GNN is based on minimizing a loss function (binary cross entropy loss) computed from the estimated marginals and the ground truth marginals. During training it back-propagates through the whole multi-layer estimation model (with each layer an iteration of the hybrid model), updating the FG-GNN weights. The number of training iterations is chosen by cross-validating.

#### 16.5.2.3 Error Correction Decoding

The method is applied to error correction decoding. Low Density Parity Check (LDPC) codes are linear codes used to correct errors in data transmitted through noisy communication channels. The sender encodes the data with redundant bits while the receiver has to decode the original message. In LPDC a parity check matrix $H$ is designed. $H$ can be interpreted as an adjacency matrix that connects $n$ variables (the transmitted bits) with ($k$) factors, i.e. the parity checks that must sum to 0. The entries of $H$ are 1 if there is an edge between a factor and a variable.

At the receiver it gets a noisy version of the code-word, $r$. The posterior distribution of the transmitted code can be expressed product of factors, where some factors are connected to multiple variables expressing a constraint among them, while other factors are connected to a single variable expressing a prior distribution for that variable. In order to infer the transmitted code-word $x$ given $r$, usually belief propagation is applied to the factor graph. So error correction with LDPC codes can be interpreted as an instance of BP applied to its associated factor graph.

LDPC codes have shown excellent results for Gaussian channels. However, In real world scenarios, the channel may differ from Gaussian, leading to sub-optimal estimates. An advantage of neural networks is that, in such cases, they can learn a decoding algorithm from data. The proposed hybrid method was applied to a *bursty* node channel, a common situation in practice. For example, radars may cause bursty interference in wireless communications.

The method was evaluated in simulation considering transmitted code words of 96 bits and 48 factors. The training dataset for the GNN consists of pairs of received and transmitted code-words. The transmitted code-words are used as ground truth for training the decoding algorithm. As evaluation metric it is used the Bit Error Rate. The results show that the hybrid method reduces the bite error rate for different signal to noise ratios, compared to the standard LPDC codes based only on factor graphs.

## 16.6 Additional Reading

A recent book on deep learning that presents an overview of the field is [8]. Reference [10] gives a concise introduction to deep learning. The Perceptron was introduced in [12], and the back propagation learning algorithm in [13]. An analysis of the expressiveness of Bayesian networks and neural networks is described in [3]. Some works that combine deep neural networks and graphical models include: conditional random fields as recurrent neural networks [16], Bayesian deep learning [15], convolutional neural networks and hidden Markov models [1], neural enhanced belief propagation in factor graphs [7], among others.

## 16.7 Exercises

1. Design a Perceptron that implements a logical AND function with 3 inputs and one output. Specify the weights and bias.
2. Repeat the previous problem for an OR function.
3. Design a multilayer NN to implement the XOR function with two inputs, specify the architecture, weights of each neuron, and activation functions.
4. Given the data for the golf example in Chap. 4, Table 4.3, train a Perceptron which has as output *Play*, and as an inputs the other four variables.
5. Compare the performance of the Perceptron of the previous problem with a Gaussian Naive Bayes classifier trained on the same data set. For the comparison do leave-one-out cross validation. That is, trained with all the data except one register, and test on this register; repeat for all the registers and report the average accuracy.
6. Given the testing Bayesian network in Fig. 16.3, and the associated testing CPT: (a) What will be the posterior distribution of $Y$ if (a1) $Z = 1, X = 3$, (a2) $Z = 2$, $X = 3$? (b) Repeat (a) considering that $X$ is unknown, and $Z$ is a binary variable with values 1, 2. Define the missing parameters of the TBN if required.
7. Design a testing BN that is equivalent to the multilayer NN that implements the XOR function.
8. *** Prove that data can separated by hyperplanes when the inputs are conditionally independent given the class. Specify under which conditions the proof is valid.
9. *** Propose a model that combines NNs and BNs for hierarchical classification.
10. *** Investigate other applications for which the combination of deep learning and graphical models can provide certain advantages.

## References


1. Barrita, R.: Human Body Pose Tracking Based on Spatio-Temporal Joints Dependency Learning. Master Thesis, INAOE (2018)
2. Brand, M., Oliver, N., Pentland, A.: Coupled hidden Markov models for complex action recognition. In: Proceedings of IEEE Computer Society Conference on Computer Vision and Pattern Recognition, pp. 994–999 (2017)

3. Choi, A., Darwiche, A.: On the relative expressiveness of Bayesian and neural networks. In: Proceedings of Machine Learning Research, PGM 2018, vol. 72, pp. 157–168 (2018)
4. Chu, X., Yang, W., Ouyang, W., Ma, C., Yuille, A.L., Wang, X.: Multi-context attention for human pose estimation. In: IEEE Conference on Computer Vision and Pattern Recognition (CVPR), pp. 5669–5678 (2017)
5. Darwiche, A.: A logical approach to factoring belief networks. In: Knowledge Representation, pp. 409–420 (2002)
6. Defferrard, M., Bresson, X., Vandergheynst, P.: Convolutional neural networks on graphs with fast localized spectral filtering. In: Advances in Neural Information Processing Systems, pp. 3844–3852 (2016)
7. Garcia, V., Welling, M.: Neural Enhance Belief Propagation on Factor Graphs (2020), arXiv:2003.01998v4
8. Goodfellow, I., Bengio, Y., Courville, A.: Deep Learning. MIT Press, Cambridge (2016)
9. Kschischang, F.R., Frey, B.J., Loeliger, H.A.: Factor graphs and the sum-product algorithm. IEEE Trans. Inf. Theory **47**(2), 498–519 (2001)
10. LeCun, Y., Bengio, Y., Hinton, G.E.: Deep learning. Nature **521**, 436–444 (2015)
11. Minsky, M.L., Papert, S.A.: Perceptrons. MIT Press, Cambridge (1969)
12. Rosenblatt, F.: The Perceptron–a perceiving and recognizing automaton. Report 85-460-1. Cornell Aeronautical Laboratory (1957)
13. Rumelhart, D.E., Hinton, G.E., Williams, R.J.: Learning representations by back-propagating errors. Nature **323**, 533–536 (1986)
14. Tishby, N., Zaslavsky, N.: Deep learning and the information bottleneck principle. In: IEEE Information Theory Workshop, pp. 1–5 (2015)
15. Wang, H., Yeung, D.: Towards Bayesian deep learning: a survey (2016). arXiv:1604.01662v2
16. Zheng, S. et al.: Conditional random fields as recurrent neural networks. IEEE Int. Conf. Comput. Vis. 1529–1537 (2015)

# Appendix A
# A Python Library for Inference and Learning

## A.1 Introduction

*PGM_PyLib* is a toolkit that contains a wide range of Probabilistic Graphical Models algorithms implemented in Python, and serves as a companion of this book.

Although there are many implementations of PGMs, most of them are focused on particular types of models. The objective of developing this new toolkit is to gather different types of PGMs algorithms in one place and in the same language.

The algorithms included, so far, in the toolkit are the following:
Bayesian classifiers:

- Naive Bayes
- Gaussian Naive Bayes
- BAN
- Semi-Naive Bayes
- Multidimensional Classifiers
- Hierarchical Classifiers

Hidden Markov models:

- Evaluation
- State estimation (Viterbi)
- Parameter Learning (Baum–Welch)

Markov random fields:

- Iterative conditional modes (ICM)
- Metropolis
- Simulated annealing

Bayesian networks:

- Tree learning (Chow and Liu algorithm) and
- Tree learning based on conditional mutual information



L. E. Sucar, *Probabilistic Graphical Models*, Advances in Computer Vision and Pattern Recognition, https://doi.org/10.1007/978-3-030-61943-5

- Structure learning (PC algorithm)

Markov Decision Processes:

- Value iteration
- Policiy iteration

Other general functions:

- Mutual information
- Conditional mutual information
- Estimation of probabilities

In the near future it will include algorithms for:

- Inference and Learning of Bayesian Networks
- Dynamic and Temporal Bayesian Networks
- Influence diagrams

These will be incorporated in the library and the User Manuel updated.

## A.2 Requirements

The library was implement to work correctly in Python 3.[1] Our test were run on Python 3.5.2, so the library should work correctly in newer versions of Python. The library requires the Numpy[2] package. Our test were run on Numpy 1.14.5, however, the library should work correctly in newer versions.

## A.3 Installation

First, you have to download the package in the following link:

https://github.com/jona2510/PGM_PyLib.

After downloading the zip file, you have to decompress it. Inside the zip file you will find different examples which were written in Python. Also you will find a folder called PGM_PyLib which is the library and contains the different algorithms implemented. The easiest way to use the library is to copy the full folder (PGM_PyLib) in your working directory, then you can use the different algorithms

The library includes a User Manuel which describes in detail all the classes implemented and examples on how to use them.

[1] https://www.python.org/download/releases/3.0/.

[2] https://numpy.org/.

# Glossary

**Abduction** A form of reasoning in which a plausible explanation is generated given a set of observation or facts. In the context of probabilistic reasoning, it corresponds to finding the maximum posterior probability of some variables given some evidence.

**Bayesian Classifier** A classifier that assigns probabilities to the different object labels based on Bayes rule.

**Bayesian network** A directed cycling graph that represents the joint distribution of a set of random variables such that each variable is conditionally independent of its non-descendants given its parents in the graph.

**Canonical model** Represent the CPT of a variable in a Bayesian network using fewer parameters, when the probabilities conform to certain canonical relations with respect to the configurations of its parents.

**Causal Bayesian Network** A directed acyclic graph in which the nodes represent random variables and the arcs correspond to direct causal relations.

**Causal discovery** The process of learning a causal model from observational and/or interventional data.

**Causal reasoning** A procedure for answering causal queries from a causal model.

**Chain Classifier** A method for multidimensional classification that incorporates class dependencies by linking the classes in a chain, such that each classifier incorporates the predicted classes by the previous classifiers as additional attributes.

**Classifier** A method or algorithm that assigns labels to objects.

**Clique** A completely connected subset of nodes in a graph that is maximal.

**Completed Partially Directed Acyclic Graph** A hybrid graph that represents a Markov equivalence class of DAGs.

**Conditional independence** Two variables are conditionally independent given a third variable if they become independent when the third variable is known.

**Conditional Probability** Probability of certain event given that another event or events have occurred.

**Conditional Random Field** A random field in which all the variables are globally conditioned on the observations.



L. E. Sucar, *Probabilistic Graphical Models*, Advances in Computer Vision and Pattern Recognition, https://doi.org/10.1007/978-3-030-61943-5

**Counterfactuals** Conditionals which discuss what would have been true under different circumstances (What would have happened if?).

**Decision Tree** A tree that represents a decision problem and has three types of nodes: decisions, uncertain events and results.

**Decision Theory** Provides a normative framework for decision making under uncertainty.

**Directed Acyclic Graph** A directed graph that has no directed circuits (a directed circuit is a circuit in which all edges in the sequence follow the directions of the arrows).

**D-separation** A graphical criteria for determining if two subsets of variables are conditionally independent given a third subset in a Bayesian network.

**Dynamic Bayesian Network** An extension of Bayesian networks to model dynamic processes; it consists of series of time slices, each time slice represents the state of all variables at certain time.

**Expectation-Maximization** An statistical technique used for parameter estimation when there are non-observable variables.

**Expected Utility** Average gain of all the possible results of a decision, weighted by their probability.

**Factor graph** A bipartite graph used to represent the factorization of a probability distribution function, enabling efficient computations.

**Factored MDP** A compact representation of an MDP in which the states are described via a set of random variables.

**Gaussian Bayesian Classifier** A classifier that assigns probabilities to the different object labels based on Bayes rule considering continuous attributes with Gaussian distributions.

**Gaussian Hidden Markov Model** A type of HMM where the observation probabilities given the state are modeled as Gaussian distributions.

**Gibbs Sampling** An algorithm for obtaining a sequence of samples to approximate a multivariate probability distribution.

**Graph** A graphical representation of binary relations between a set of objects.

**Graph Isomorphism** Two graphs are isomorphic if there is a one to one correspondence between their vertices and edges, so that the incidences are maintained.

**Hidden Markov Model** A Markov chain in which the states are not directly observable.

**Hierarchical Classifier** A multidimensional classifier in which the classes are arranged in a predefined structure, typically a tree, or in general a directed acyclic graph (hierarchy).

**Independence Axioms** In the context of Bayesian networks, a set of rules to derive new conditional independence relations from other conditional independence relations.

**Independent variables** Two random variables are independent if knowing the value of one of them does not affect the probability distribution of the other one.

**Influence Diagram** A graphical model for solving decision problems. It is an extension of Bayesian networks that incorporates decision and utility nodes.

**Information** In abstract terms it can be though as the resolution of uncertainty. In Information Theory it refers to a measure of what a message communicates, and it is inversely proportional to its probability.

**Interpretation of Probability** Meaning of the primitive symbols of the formal theory of probability.

**Joint Probability** The probability of a conjunction of $N$ propositions.

**Junction Tree** A tree in which each node corresponds to a subset (group or cluster) of variables.

**Limited Memory Influence Diagram** An influence diagram in which the variables known when making a decision are not necessarily remembered for future decisions.

**Maximal Ancestral Graphs** A hybrid graph that represents conditional independence and causal relationships in DAGs that include unmeasured (hidden or latent) variables.

**Marginal Probability** The probability of an event independent of information on other events.

**Markov Blanket** A set of variables that make a variable independent of all other variables in a probabilistic graphical model.

**Markov Chain** A state machine in which the transition between states are non-deterministic and satisfy the Markov property.

**Markov Decision Process** A model for sequential decision making composed of a finite set of states and actions, in which the states follow the Markov property.

**Markov Equivalence** Graphical models that codify the same conditional independence relations.

**Markov Logic Network** A generalization of Markov networks which consists of a set formulas in first-order logic and a weight associated to each formula.

**Markov Network** A random field represent as an undirected graph that satisfies the locality property –each variable in the field is independent of all other variables given its neighbors in the graph.

**Markov Property** The probability of the next (future) state is independent of the previous states (past) given the current (present) state.

**Markov Random Field** Markov network.

**Most Probable Explanation** Abduction.

**Multidimensional Classifier** A classifier that can assign more than one label to each object.

**Naive Bayes Classifier** A Bayesian classifier that assumes that all attributes are independent given the class variable.

**Partial Abduction** In the context of probabilistic reasoning, it corresponds to finding the maximum posterior probability of some (not all) of the variables in a model given some evidence.

**Partially Observable Markov Decision Process** A Markov decision process in which the states are not directly observable.

**Particle Filters** A sampling technique for Bayesian filtering, that is used to predict the next state based on past observations and controls.

**Policy** A function that maps states to actions.

**Polytree** A singly-connected directed acyclic graph.

**Probabilistic Graphical Model** A compact representation of a joint probability distribution of a set of random variables composed by a graph and a set of local probability distributions.

**Probabilistic Inference** A procedure for calculating the posterior probability of the unknown variables in a probabilistic graphical model given certain evidence (a subset of known or instantiated variables).

**Probabilistic Relational Models** An extension of Bayesian networks that provide a more expressive, object-oriented representation.

**Probability** A function that assigns a real number to each event (subset of a sample space) and satisfies certain axioms known as the probability axioms.

**Random Field** A collection of random variables indexed by sites.

**Random Variable** A mapping from a sample space to real numbers.

**Rational Agent** An agent that selects its decisions to maximize its expected utility according to its preferences.

**Relational Probabilistic Graphical Models** An extension of probabilistic graphical models that are more expressive by incorporating some type of relational representation.

**Sample Space** The set of possible outcomes of an experiment.

**Semi-Naive Bayes Classifier** A Bayesian classifier that, based on the Naive Bayes classifier, eliminates or joins attributes to improve classification accuracy.

**Temporal Event Network** A Bayesian network for modeling dynamic processes in which some nodes represent the time of occurrence of an event or state change of certain variable.

**Testing Bayesian Networks** An extension of Bayesian networks such that some variables in the model have a dynamic conditional probability table, that is, a variable has two CPTs such that one is selected at inference time depending on the values of some other variables.

**Total Abduction** In the context of probabilistic reasoning, it corresponds to finding the maximum posterior probability all of the variables in a model given some evidence.

**Transfer Learning** To transfer knowledge and/or data from related domains or tasks to learn another task.

**Tree** A connected graph that does not have simple circuits.

# Index




L. E. Sucar, *Probabilistic Graphical Models*, Advances in Computer Vision and Pattern Recognition, https://doi.org/10.1007/978-3-030-61943-5

Printed in the United States
By Bookmasters